Chapter 11 **SOTB Implementation of a Field Programmable Gate Array with Fine-Grained Vt Programmability** 193
Masakazu Hioki, Chao Ma, Takashi Kawanami, Yasuhiro Ogasahara, Tadashi Nakagawa, Toshihiro Sekigawa, Toshiyuki Tsutsumi and Hanpei Koike

Chapter 12 **Embedded Memory Hierarchy Exploration Based on Magnetic Random Access Memory** 206
Luís Vitório Cargnini, Lionel Torres, Raphael Martins Brum, Sophiane Senni and Gilles Sassatelli

Chapter 13 **A Survey of Neural Front End Amplifiers and Their Requirements toward Practical Neural Interfaces** 223
Eric Bharucha, Hassan Sepehrian and Benoit Gosselin

Chapter 14 **An Efficient Radio Access Control Mechanism for Wireless Network-On-Chip Architectures** 247
Maurizio Palesi, Mario Collotta, Andrea Mineo and Vincenzo Catania

Permissions

List of Contributors

Preface

The world is advancing at a fast pace like never before. Therefore, the need is to keep up with the latest developments. This book was an idea that came to fruition when the specialists in the area realized the need to coordinate together and document essential themes in the subject. That's when I was requested to be the editor. Editing this book has been an honour as it brings together diverse authors researching on different streams of the field. The book collates essential materials contributed by veterans in the area which can be utilized by students and researchers alike.

Electronic circuits and systems are integrated into all modern equipments and devices. This book provides comprehensive knowledge of the framework and design of various electronic circuits and systems. The topics covered in this extensive book deal with the significant concepts and applications of electronic circuits such as analog-to-digital converters, power processors, analog and digital signals, etc. This text attempts to bring forth emerging topics for discussion and research like energy storage, low power technologies, etc. Scientists and students actively engaged in this field will find this book full of crucial information.

Each chapter is a sole-standing publication that reflects each author's interpretation. Thus, the book displays a multi-facetted picture of our current understanding of applications and diverse aspects of the field. I would like to thank the contributors of this book and my family for their endless support.

Editor

1

A Digital Auto-Zeroing Circuit to Reduce Offset in Sub-Threshold Sense Amplifiers

Peter Beshay [1,*], Joseph F. Ryan [2] and Benton H. Calhoun [1]

[1] The Charles L. Brown Department of Electrical and Computer Engineering, University of Virginia, Charlottesville, VA 22904, USA; E-Mail: bcalhoun@virginia.edu

[2] Intel Corporation, Hillsboro, OR 97124, USA; E-Mail: joseph.f.ryan@intel.com

* Author to whom correspondence should be addressed; E-Mail: plb3qt@virginia.edu

---

Abstract: Device variability in modern processes has become a major concern in SRAM design leading to degradation of both performance and yield. Variation induced offset in the sense amplifiers requires a larger bitline differential, which slows down SRAM access times and causes increased power consumption. The effect aggravated in the sub-threshold region. In this paper, we propose a circuit that reduces the sense amp offset using an auto-zeroing scheme with automatic temperature, voltage, and aging tracking. The circuit enables flexible tuning of the offset voltage. Measurements taken from a 45 nm test chip show the circuit is able to limit the offset to 20 mV. A 16kB SRAM is designed using the auto-zeroing circuit for the sense amps. The reduction in the total read energy and delay is reported for various configurations of the memory.

Keywords: offset compensation; SRAM; sense amplifier; auto-zeroing

---

1. Introduction

Variation induced offset in the sense amplifiers requires a larger bit-line differential, which slows down SRAM access times and causes increased power consumption. In the sub-threshold (sub-V_T) region of operation, in particular, the effect is more dominant because the threshold voltage (V_T)

variation has an exponential effect on the drive current. It is shown in [1] that the offset gets worse in sub-threshold relative to strong inversion as technology scales. Furthermore, increasing the sizes of devices in the sense amplifier does not yield the reduction of input referred offset according to $1/(WL)^{0.5}$ that is achieved for strong inversion operation. In addition it causes increased cell instability and a severely degraded read-current [1,2]. Several attempts have been made before to tackle the problem of offset voltage in sense amplifiers (SAs) including redundancy [3], transistor upsizing [4], digitally controlled compensation [5] and dynamic compensation [6]. Our approach to eliminating offset is a digital auto-zeroing (DAZ) scheme inspired by analog amplifier offset correction [7]. The main advantages of the approach are the near-zero offset after cancellation, offset tuning, and the automatic temperature, voltage, and aging tracking achievable using a repeated offset calibration phase, which makes the design useful in the sub-threshold region due to the high offset voltage sensitivity to supply voltage and temperature variations. In Verma and Chandrakasan [3], SA redundancy is used. It requires the SRAM bit-line from each column be connected to N different SAs. One SA will be selected whose offset is bound by the high and low logic levels of bit-line. This scheme statistically reduces the reliance on SAs with high offset, but it adds area and test time calibration to select the SA to use from each set. In Pileggi [4], transistor upsizing is used. A linear response surface model was developed to relate the SA offset voltage to the threshold voltages variations. The model is used to determine the statistical optimized transistors sizing. The optimized design resulted in a 25% decrease in the standard deviation of the offset voltage at a cost of 10% increase in active area. In Ryan and Calhoun [1], a methodology is proposed for sizing sub-V_T SAs to minimize offset using SAs for ultra-low power operation. In Bhargava [5] a digitally controlled compensation is used. The scheme is applied to a latch-style and Strong-ARM SA topologies. Measured results from a 45nm test-chip show that the standard deviation of the offset is reduced by 5X. In Sachdev [6], dynamic compensation is used. A group of transistors are selectively coupled to high and low voltage levels via multi-phase timing. This results in a voltage level on nodes of interest that is a function of transistor mismatch. The voltage levels act to compensate for the transistor mismatch. This scheme is similar to the auto-zeroing scheme presented in this work. However the presented scheme uses a compensation capacitor, charge pump and feed-back circuit. Hence, the calibration phase is not necessarily needed prior to every sensing cycle. This improves the SA power consumption as will be illustrated in Section 5.

Section 2 describes the implementation of the auto-zeroing offset compensation scheme. Section 3 illustrates the voltage, temperature and aging tracking. Section 4 illustrates the offset tuning. Section 5 provides details of the power consumption and methods proposed to minimize it. Section 6 illustrates the offset sensitivity. Section 7 illustrates the offset compensation across technologies. Section 8 shows the improvements in read energy and delay gained by utilizing the DAZ SA in a 16kB SRAM. Section 9 provides a comparison to other offset compensation schemes. Section 10 provides measurements of the DAZ SA offset voltage from a 45 nm bulk CMOS test chip.

2. Mismatch Compensation using Auto- Zeroing Circuitry

Our auto-zeroing scheme uses a split-phase clock and charge pump feedback circuit. Figure 1a shows a conventional latch-based sense amp with PMOS inputs (e.g., to support near-V_{SS} sensing on a low swing bus). Figure 1b shows the auto-zeroing circuit attached to the sense amp. The same scheme can apply to a SA with NMOS inputs in an SRAM. The charge pump circuit is shown in Figure 2. ENI and ENO are the input voltage differential and offset tuning phases respectively. ENR1 and ENR2 are reset phases. During ENR1, a zero differential input is applied to the sense amp. The ENO phase then occurs, and the SA resolves based on its intrinsic offset.

Figure 1. (**a**) Latch-based sense amp for near-V_{SS} inputs; (**b**) Auto-zeroing circuit attached to the sense amp [7].

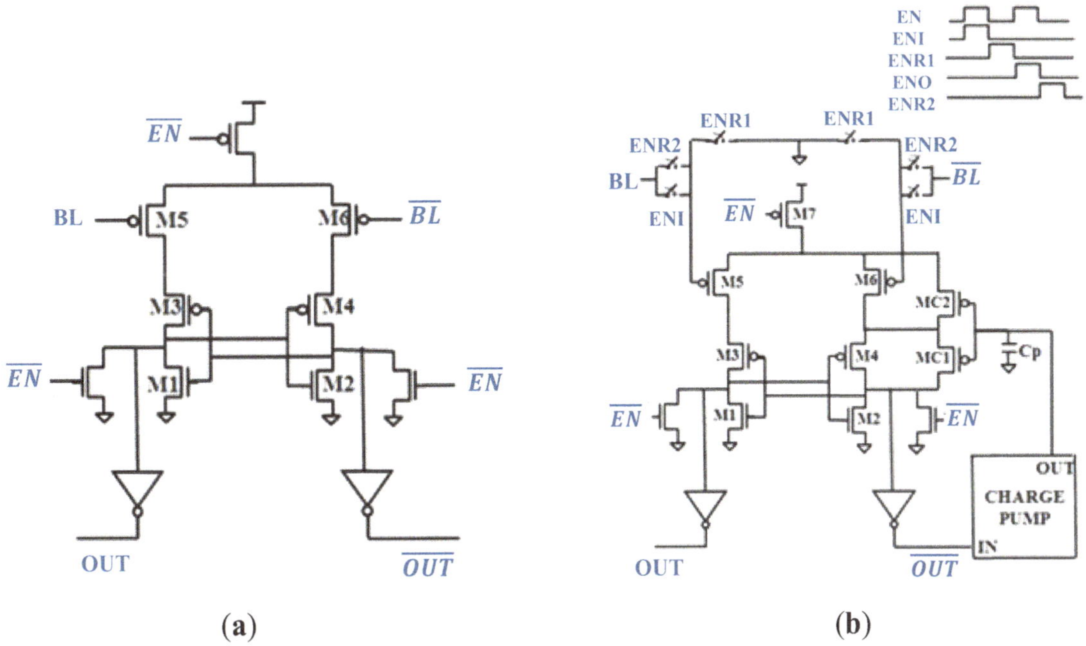

Figure 2. Charge pump circuit for adjusting the voltage on Cp. [7].

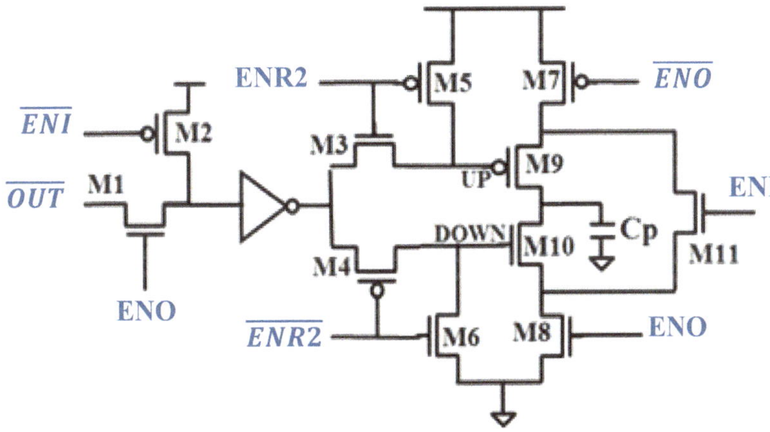

The sense amp output is fed to the charge pump circuit that charges the capacitor, Cp, up or down. During ENR2, the differential input is applied to the sense amp. ENI then occurs, and the SA resolves based on the differential input. Note that phases ENR1 and ENO can be omitted or included based on how often re-calibration is needed. Transistors MC1 and MC2 control the drive strength of the right side of the sense amp to compensate for the offset.

The charge pump controls the drive current in both transistors to equalize the strength of the SA right and left sides to reduce the offset. The offset is compensated with minimal capacitive loading at the output and is independent of input DC bias (V_{INDC}). A supply voltage and clock frequency of 0.5 V and 1MHz are used in the simulations. The output voltage of the sense amp and the voltage on Cp are illustrated in Figure 3 for an input differential of −10 mV. The initial voltage on Cp is zero. This causes an intrinsic positive offset voltage that set the SA output voltage to 1. Simulations indicate that the voltage on Cp required for a zero offset is 142 mV. For a 10 mV offset, the voltage on Cp can vary within ±12 mV.

Figure 3. Simulated output voltage of the sense amp and Cp voltage for a −10 mV differential input voltage at 0.5 V and 1MHz in 45 nm CMOS.

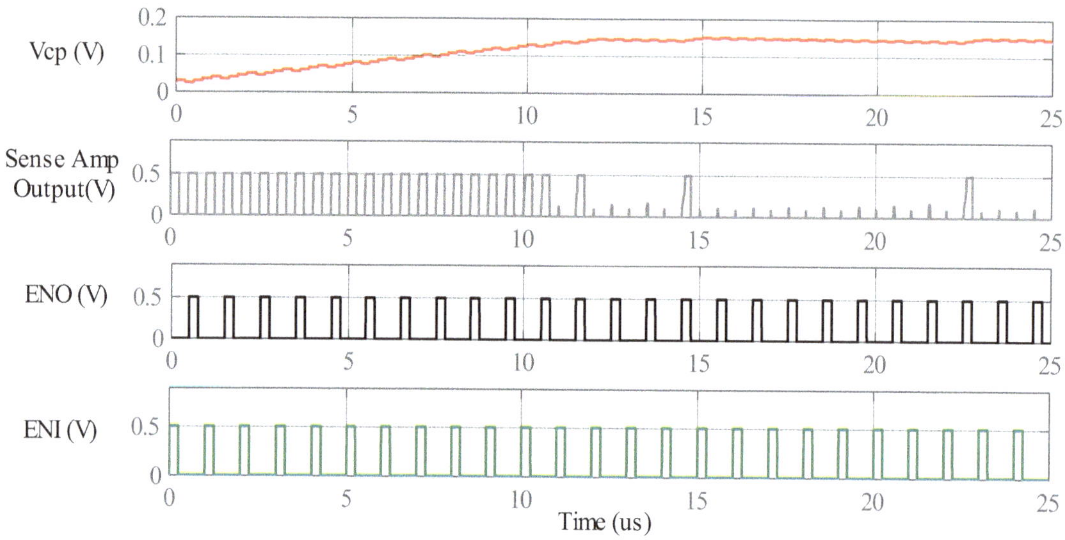

This imposes a minimum and maximum limit on Cp voltage to 130 mV and 154 mV in order to maintain an offset less than 10 mV. The deviation of Cp voltage from to the value corresponding to zero offset (142 mV in this case) is plotted in Figure 4 for desired final offset voltages of 5 mV, 10 mV, 15 mV, and 20 mV. Low offset voltages are usually realized using a higher value of Cp. In Figure 3, the offset compensation is completed when the voltage on Cp settles to its final value within the 130 mV to 154 mV range. The sense amp then resolves its output correctly to 0 during the input phase. A zero differential voltage is applied to the SA input during the offset phase. This sets the SA output to "1" when Cp voltage drops below 142mV and "0" otherwise. In this design, rate at which Cp charges up is higher than its charge down rate. This helps to minimize the power consumption as will be discussed in Section 5.

Figure 4. Variation on Cp voltage *vs.* Minimum Achieved Offset.

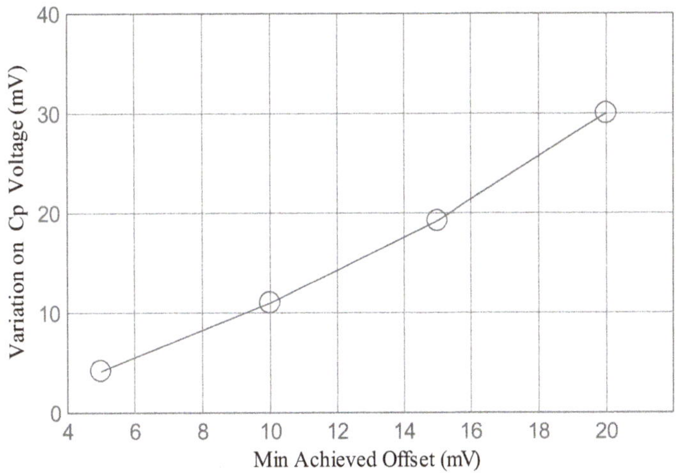

3. Voltage, Temperature, and Aging Tracking

To demonstrate temperature, voltage, and aging tracking, the offset voltage that remains after compensation is calculated for various voltages and temperatures as shown in Figure 5a. Simulations in a commercial 45 nm process show that the circuit maintains a constant offset across temperature. The accuracy of voltage tracking depends on the supply voltage. Higher supply voltage causes more charge to be pumped to Cp during each offset calibration cycle, and this larger change in charge leads to a coarser resolution, as Figure 5a illustrates.

Figure 5. (**a**) Voltage and temperature tracking; (**b**) Aging tracking [7].

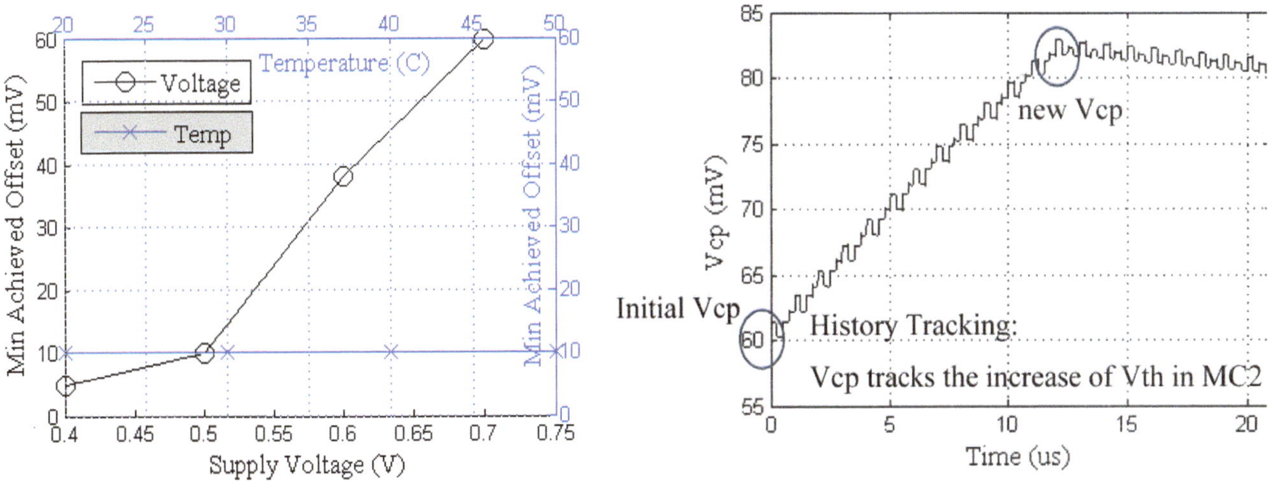

(**a**) (**b**)

The auto-zeroing scheme also has the ability to compensate for any changes in device characteristics after circuit deployment. One common cause for such changes is effective threshold voltage shifting due to Bias Temperature Instability (BTI), hot carrier injection, or other aging effects. To demonstrate how this circuit can compensate for such changes, Figure 5b shows the capacitor voltage after an abrupt increase in the threshold voltage of MC2, to emulate an aging effect. The charge pump boosts the voltage on Cp to decrease the drive strength of MC2 in response and rapidly restores the compensated offset voltage.

4. Offset Tuning

We define settling time as the difference between the time when the zero differential-input is first applied and the time when the voltage of the output capacitor settles as shown in Figure 6. Changing the size of the output capacitor (Cp) affects the amount of charge added during the offset compensation phase (ENO) and so controls both the offset and the settling time. Figure 7 demonstrates the trade-off between accuracy (min achieved offset) and settling time using different values of output capacitors.

Figure 6. Settling time simulation [7]

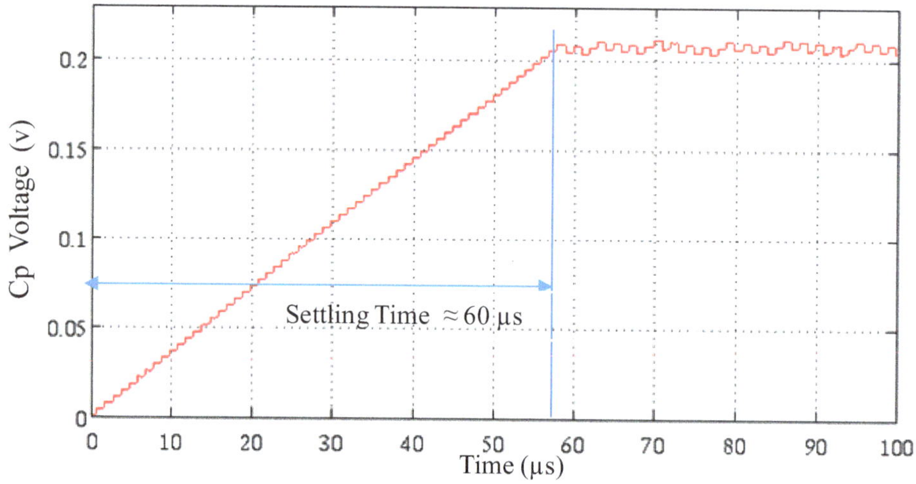

Figure 7. Min achieved offset *vs.* settling time for different values of output capacitor. [7]

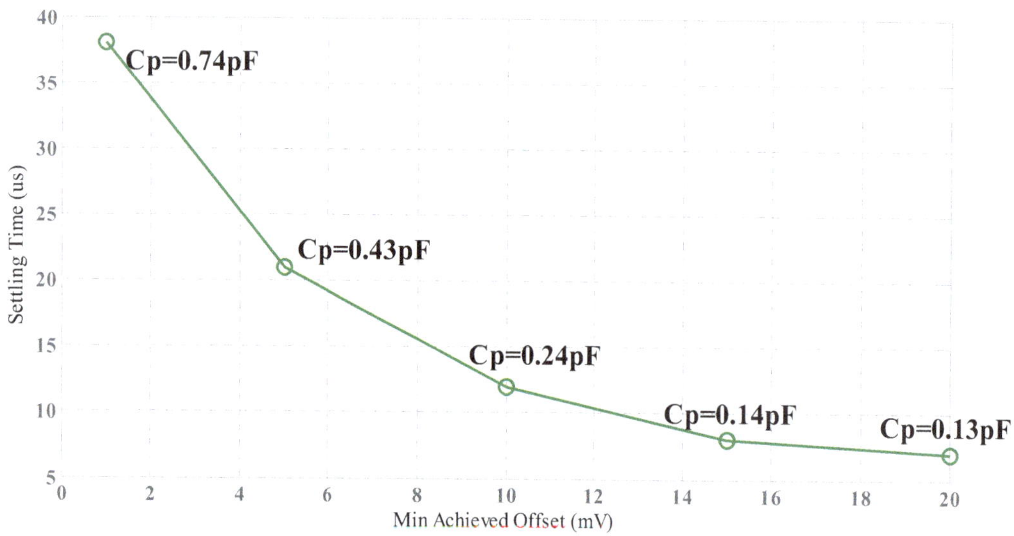

5. Power Consumption

The main contribution to the power consumed by the DAZ SA comes from the continuous calibration. Decreasing the number of cycles of calibration phase (ENI and ENO) relative to the input phase decreases the switching power of the feedback circuit and the power consumed in charging and discharging (Cp) but is limited by the leakage at the output capacitor (Cp). The overhead area of the scheme includes the area of the timing circuit, the charge pump circuit, and the output capacitance (Cp). For an offset voltage of 1mV, a 0.74pF output capacitance is needed. In this case, the area of Cp can dominate the total area overhead. In Figure 8, the offset calibration phase occurs once every 15 clock cycles. The maximum calibration period or the minimum number of offset calibration cycles needed is limited by the leakage on Cp. Simulation results indicated a maximum calibration period of 200 μs. This high period makes the difference in power consumption between the DAZ SA and the Latch SA insignificant. The total power consumption of the DAZ SA and the Latch SA is 2.02 nW and 2 nW respectively. The minimum number of offset calibrations is independent of the required offset or the value of Cp, but it depends on the charge pump current. In this design the charging current is 0.5 μA. High current allows fewer number of calibration cycles. Increasing the charge pump current however increases the dynamic power consumption.

Figure 8. Offset compensation clock period for a 10 mV offset voltage.

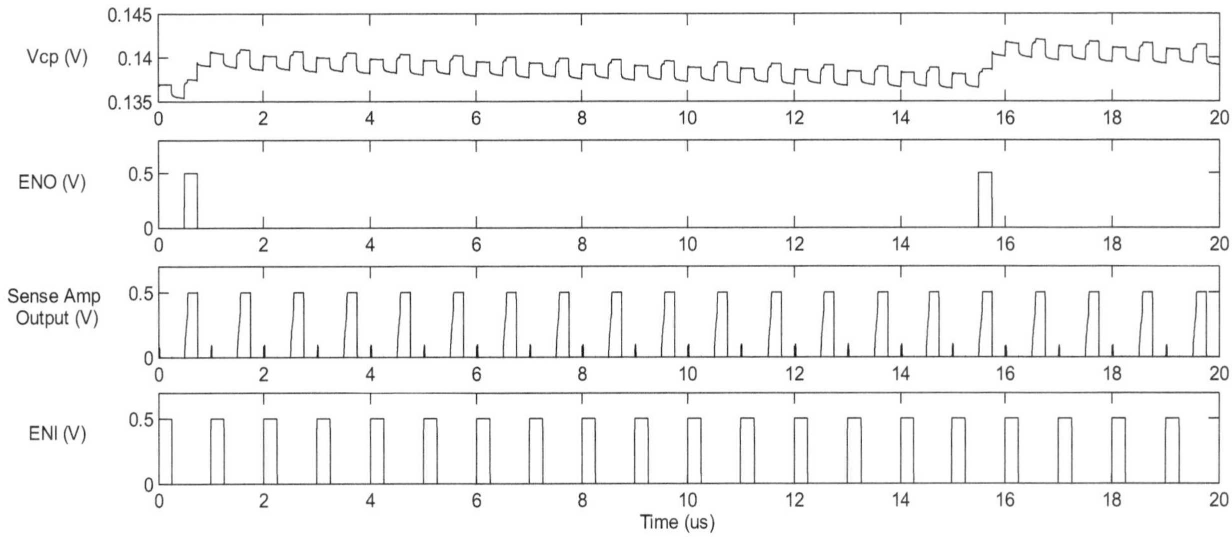

Shorting the output virtual nodes of the charge pump through M11 can decrease the leakage by reducing V_{DS} of the switches and improve settling time as shown in Figure 9. The switching power can also be decreased by strengthening M9 in the charge pump circuit relative to M10 to avoid the continuous toggling of the sense amp output during offset compensation phase (ENO) after settling as shown in Figure 10. Strengthening M9 has the downside effect of increasing the settling time when Cp is moving to lower voltages; the time Cp takes to discharge will increase. However, the compensation usually starts with zero-initial voltage on the capacitor Cp that makes the settling time mostly dependent on the charging rate.

Figure 9. Voltage on Cp with and without shorting the virtual supply nodes of the charge pump [7].

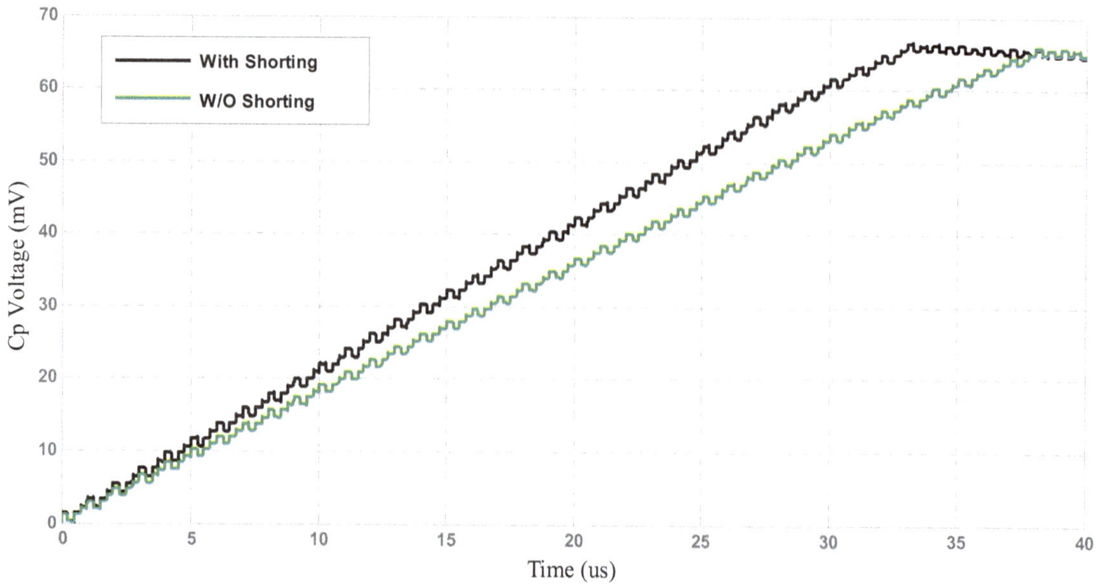

Figure 10. Strengthening the pull up transistor of the charge pump (left column of sims) reduces the rate at which the SA output switches high relative to equal strength devices (right column). This reduces power consumption [7].

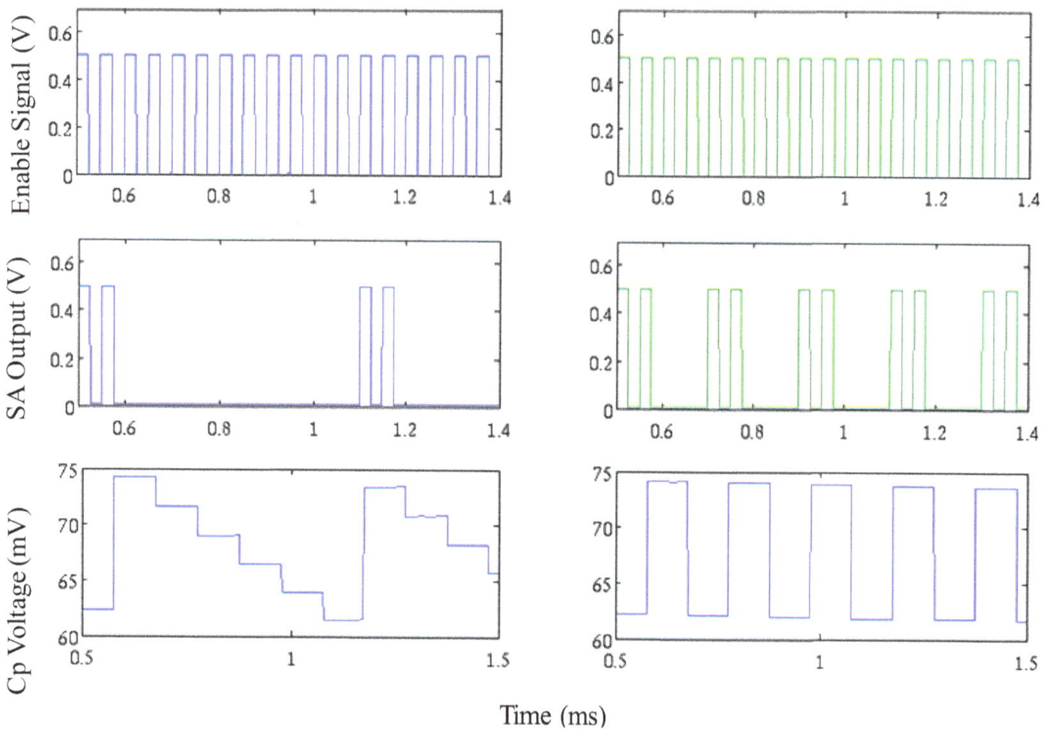

6. Offset Sensitivity

The sensitivity of the offset compensation depends on the split phases, charge pump circuit, and the output capacitance. The accuracy of the split phases has the dominant influence on the resolution. A small overlap between ENO and ENR2 phases can dramatically degrade the accuracy by connecting M1 and M2 to the supply rails during charging. That leads to a significant increase in the charge pump rate degrading the accuracy as shown in Figure 11b, where the min achieved offset is plotted against the error in split phase timing, measured as the percentage of time overlap between ENO and ENR2. The scheme is also sensitive to variations in the M9 and M10 transistors in the charge pump circuit. They are responsible for charging/discharging Cp, and so the one with more drive strength determines the final offset value. Figure 11a shows the sensitivity of the offset voltage to the output capacitance Cp.

Figure 11. (**a**) Offset voltage sensitivity to output capacitance; (**b**) sensitivity to split phases [7].

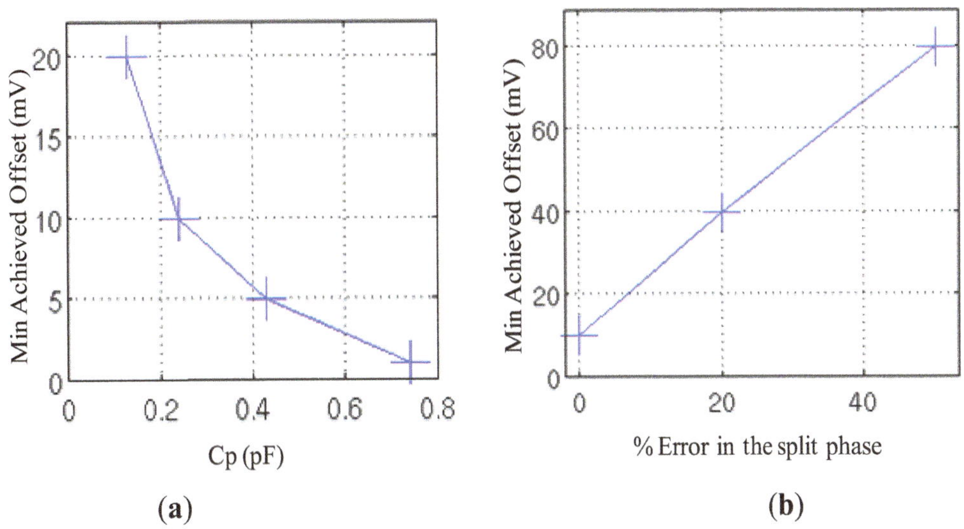

The offset voltage is also sensitive to the frequency of the split phase. The increase in the split phase frequency increases the enable signal switching and degrades the compensated offset voltage.

7. Offset Compensation across Technology Nodes

The intrinsic offset voltage of the sense amplifier relies on the technology of the design. The effect of process variability on the final offset of the auto-zeroing circuit is marginal as explained in Section 1 due to the repeated compensation. To illustrate the variation of the SAs offset across technologies, 1000 Monte-Carlo simulations were performed to evaluate the offset voltage of a latch-based SA using 45, 65, 90 and 130 nm commercial technology models. Figure 12 illustrates the 3σ value of the SA offset voltage across technologies. The results indicate that the largest offset voltage belongs to 45 nm technology, followed by 32 nm, 65 nm, 90 nm and 130 nm respectively. Although the offset behavior is not monotonically increasing with technology scaling, the plot indicates a trend of increased offset in emerging technologies, *i.e.*, 32, 45 nm. Since offset voltage is increasing with newer technologies, compensation becomes more essential as technology scales. The next section demonstrates the benefits gained from using the DAZ SA in a 16kB SRAM. The total energy and delay of the SRAM is calculated and compared to the uncompensated SA case.

Figure 12. 3σ of the intrinsic voltage of the SA across technology nodes.

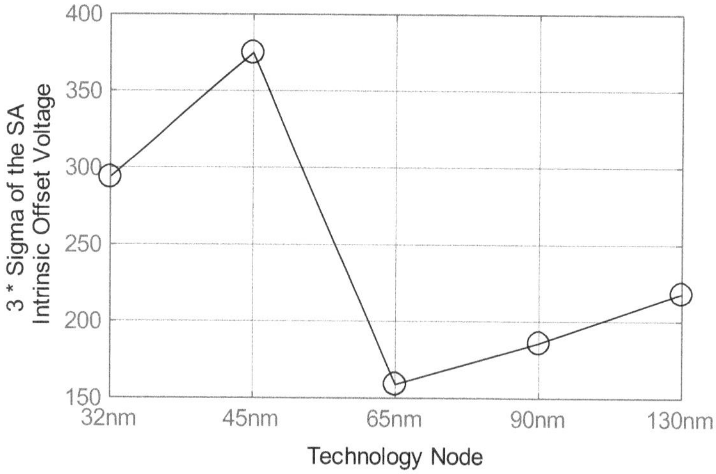

8. 16 kB SRAM Design

In this section, we investigate the effect of utilizing the DAZ SA in a 16kB SRAM memory. The power consumption of the DAZ SA is higher than the uncompensated one due to the clock generator, charge pump and the buffer stages needed for the non-overlapping clock. The sense amplifier delay is also higher due to the high capacitive loading. Reducing the sense amp offset reduces the necessary bit-line swing, which decreases both the precharge and bitcell energy and delay during the read operation. The reduction in the read energy and delay depends on the number of banks, rows and words per row of the memory. The energy and delay of the 16 kB SRAM is calculated using a 20 mV DAZ SA for all possible configurations and plotted in Figure 13. Each point is annotated with (B, R, W) where B is the number of banks, R is the number of rows, and W is the number of words per row. The results indicate a significant improvement in both the energy and delay for cases with large numbers of rows and small improvement or degradation for cases with small number of rows. The design point of 1 bank, 512 rows and 2 words per row has the biggest improvement of 10% in energy and 24% in delay. The design point of 4 banks, 32 rows and 8 words per row has the biggest degradation of 6% in energy and 5% in delay. The DAZ SA created 3 new optimal design points (1, 256, 4), (1, 128, 8) and (8, 128, 1). The improvement in energy and delay for (1, 256, 4) is 12% and 13% respectively. For (1, 128, 8) it improves the energy by 13% and degrades the delay by 5%. The energy of (1, 128, 8) with DAZ SA is the lowest. This could not be achieved using a Latch SA. The system level parameters of the SRAM that satisfy the energy/delay requirements changed (*i.e.*, design point (16, 64, 1) is the minimum energy/delay point with uncompensated SA. Using DAZ SA, the min delay/energy design changed to (1, 128, 8).

Figure 13. Design space of 16kB SRAM Memory with uncompensated and 20 mV digital auto-zeroing (DAZ) SA.

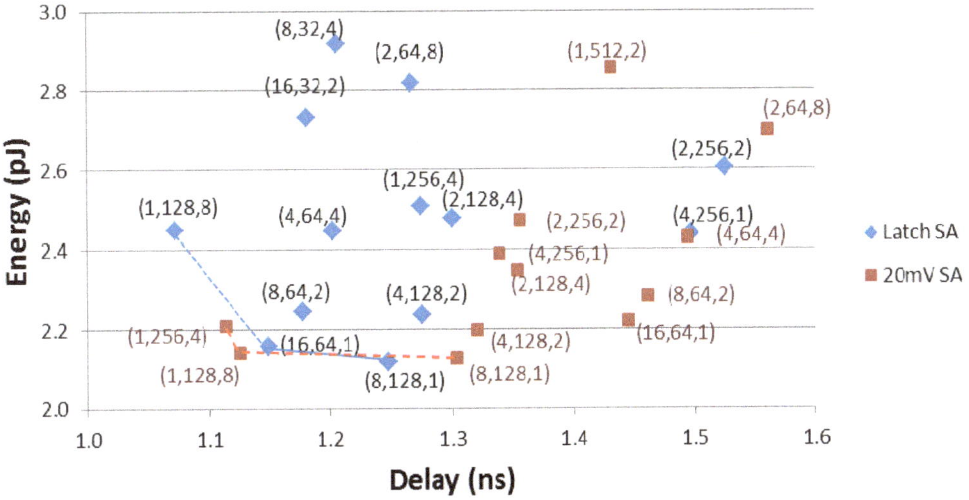

9. Comparison to other Offset Compensation Schemes

The main advantages of the scheme are the continuous calibration that makes it specifically useful for sub-threshold operation and the flexibility to tune the offset voltage. The latter provides different design options that can be utilized in the SRAM design process. Approaches like redundancy [3], transistor upsizing [4], and digitally controlled compensation [5] do not support continuous calibration and hence would not be tolerant to voltage and temperature variation. The approach in [6] provides continuous calibration. The power consumed in this approach is only for compensation clock phase generation. There is no charge pump or additional circuitry. However this approach requires the calibration phase to essentially occur before every sensing cycle. As explained in Section 5, the DAZ SA can perform compensation every N number of cycles. High charge pump current can be used to increase N at a cost of higher dynamic and leakage power of the charge pump circuit. The power consumption of [6] is compared to that of DAZ SA with the offset calibration phase occurring every cycle (cycle period = 1 μs) and every 200μs with a controllable offset phase. The controllable offset phase logic is employed to force calibration every cycle at the beginning. The logic then enables calibration every N cycles once the voltage on Cp settled to its final value. The results are shown in Table 1. The settling time of both schemes is compared. The DAZ SA with controllable offset phase consumes the lowest power with a 12 μs settling time.

Table 1. Power consumption of dynamic offset compensation and auto-zeroing circuit.

Offset Compensation Scheme	Power Consumption	Settling time
DAZ SA	6 nW	12 μs
Dynamic Compensation [6]	4 nW	0.5 μs
DAZ SA with controllable offset phase	2.5 nW	12 μs

10. 45 nm Test Chip Measurements

A test chip fabricated in 45 nm technology is used to verify the scheme. The chip contains one regular SA array for benchmarking and another array that uses SAs with the auto-zeroing circuitry, with Cp equal to 32fF. The chip micrograph is shown in Figure 14. Figure 15 shows the layout of a single DAZ sense amplifier. The layout of the output capacitor is shown on the left side of Figure 15, consuming an area of 2.97 μm × 3.9 μm. The sense amplifier and the charge pump layout are shown on the right side of Figure 15, consuming an area of 4.39 μm × 5.29 μm. The supply voltage is set to 0.6V during measurements to mitigate the effect of noise on the measured results. The control signals are supplied to the auto-zeroing circuit at 1 MHz. Figure 16 shows the measured offset distribution of both banks. The positive terminal of the SAs is connected to 0.45 V. The negative terminal of the SAs is swept from 0.3 V to 0.6 V in increments of 5 mV. The SAs are enabled during each increment, and measurements of the SAs outputs are recorded. This information is then used to construct the SAs offset distribution in Figure 16. The measured mean (μ) and standard deviation (σ) of the uncompensated SA banks is −31 mV and 45 mV respectively. The auto-zeroing circuitry reduced the value of μ to −13mV and lowered σ to 9.3 mV. This indicates an 80% improvement in σ. The scheme limits the absolute value of the maximum offset to 50 mV.

Figure 14. Chip Micrograph.

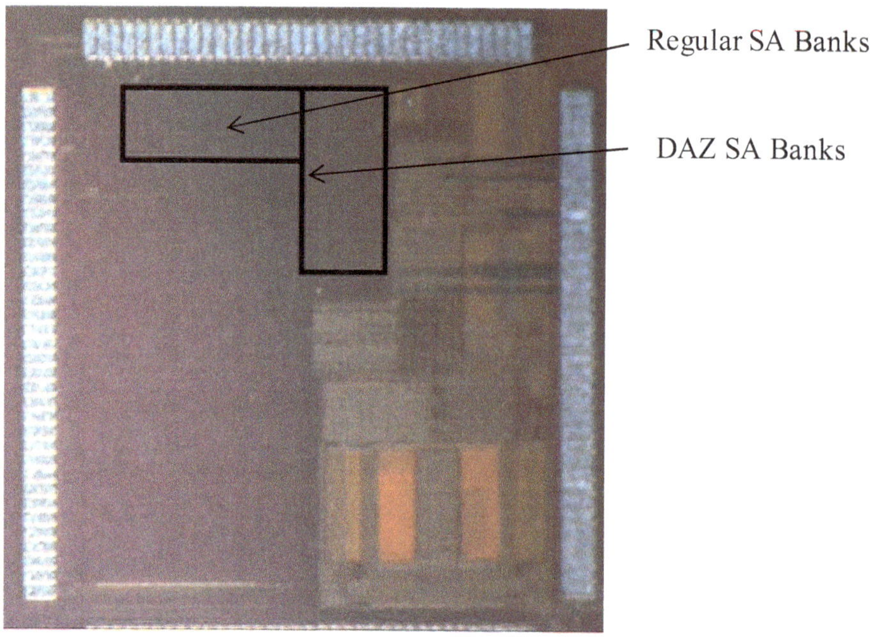

Figure 15. DAZ Sense Amplifier Layout.

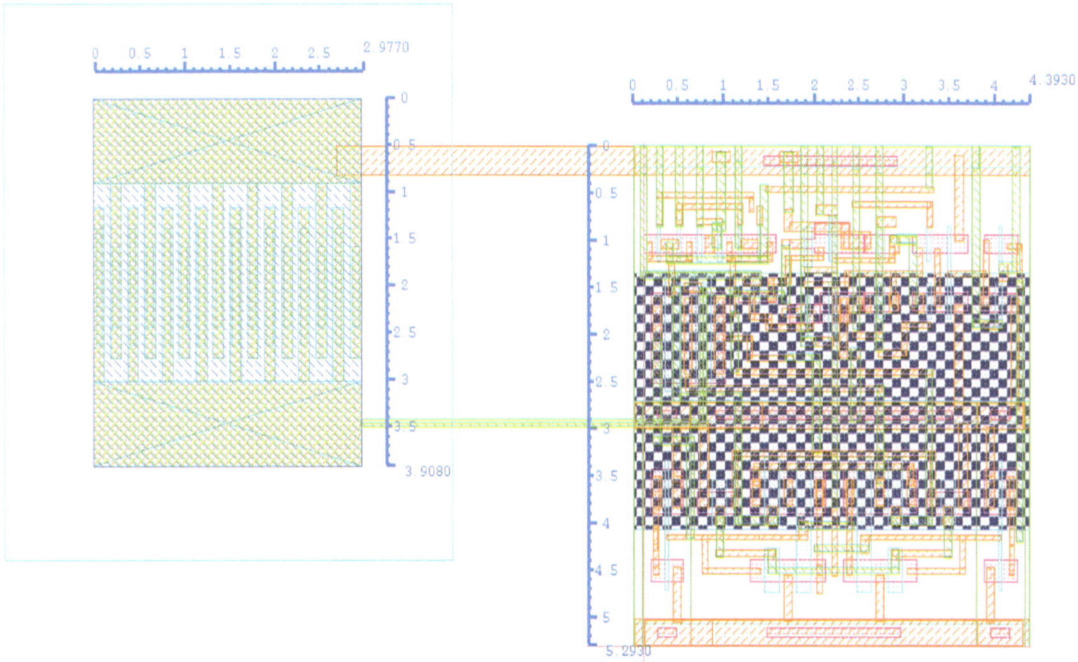

Figure 16. Measured offset voltage distribution (**a**) regular SA; (**b**) SA with auto-zeroing circuitry [7].

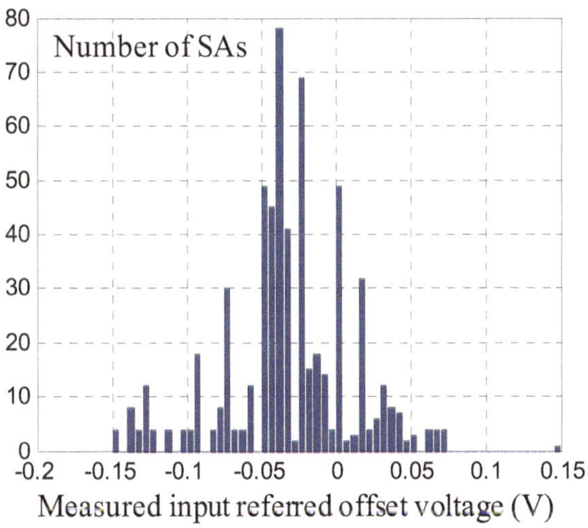

(**a**)

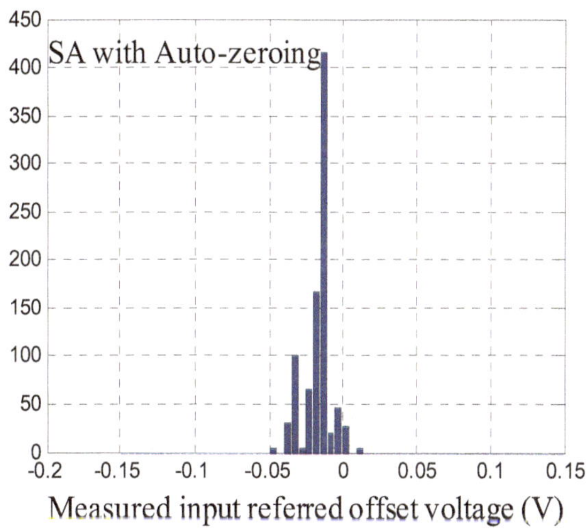

(**b**)

To verify the offset sensitivity to split phases, the offset of a sample DAZ SA is measured for different split phase frequencies. Figure 17 shows the offset voltage values for different split phase frequency.

Figure 17. Measured offset voltage *vs.* split phase frequency [7].

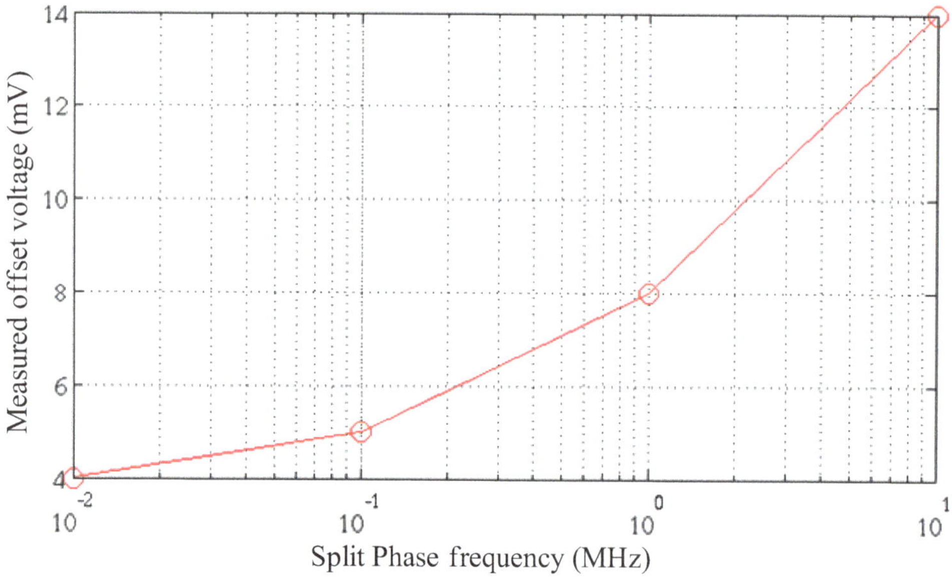

11. Conclusions

We proposed a circuit that is capable of improving sense-amp offset to near zero, which is valuable for sub-threshold operation due to the heightened effect of mismatch. Simulations of the design (0.5 V, 1 MHz) show a compensated offset voltage of 1mV, settling time of 37 μs, and total power consumption of 12 nW. Measurements from a test chip fabricated in 45 nm technology showed the circuit's ability to improve σ of the offset voltage by 80% and limited the absolute maximum value of the offset voltage to 50 mV using a 1 MHz split phase frequency and 32fF output capacitance. Using the circuit in a 16 kB SRAM showed a reduction in the total energy and delay of 10% and 15% respectively. The trade-off between the sense amp compensated offset and power consumption is demonstrated. This makes the circuit able to provide the offset/power values that can generate the optimal SRAM design.

References

1. Ryan, J.F.; Calhoun, B.H. Minimizing Offset for Latching Voltage-Mode Sense Amplifiers for Sub-Threshold Operation. In Proceedings of the 9th International Symposium on Quality Electronic Design, San Jose, California, USA, 17–19 March 2008; pp. 127–132.
2. Wang, A.; Chandrakasan, A.P.; Kosonocky, S.V. Optimal Supply and Threshold Scaling for Subthreshold CMOS Circuits. In Proceedings of the IEEE Computer Society Annual Symposium on VLSI, Pittsburgh, PA, USA, 25–26 April 2002; pp. 5–9.
3. Verma, N.; Chandrakasan, A.P. A 256 kb 65 nm 8T subthreshold SRAM employing sense-amplifier redundancy. *IEEE J. Solid State Circ.* 2008, *43*, 141–149.
4. Pileggi, L.; Keskin, G.; Li, X.; Mai, K.; Proesel, J. Mismatch Analysis and Statistical Design at 65 nm and Below. In Proceedings of the IEEE Custom Integrated Circuits Conference, San Jose, California, USA, 21–24 September 2008; pp. 9–12.

5. Bhargava, M.; McCartney, M.P.; Hoefler, A.; Mai, K. Low-overhead, Digital Offset Compensated, SRAM Sense Amplifiers. In Proceedings of the IEEE Custom Integrated Circuits Conference, San Jose, California, USA, 13–16 September 2009; pp. 705–708.
6. Sachdev, M.; Sharifkhani, M.; Shah, J.S.; Rennie, D. Sense-amplification with Offset Cancellation for Static Random Access Memories. U.S. Patent Application 12/757,033, 8 April 2010.
7. Beshay, P.; Calhoun, B.H.; Ryan, J.F. Sub-threshold Sense Amplifier Compensation Using Auto-zeroing Circuitry. In Proceedings of the 2012 IEEE Subthreshold Microelectronics Conference, Waltham, Massachusetts, USA, 9–10 October 2012; pp. 1–3.

2

Reconfigurable Threshold Logic Gates using Memristive Devices

Adrian Rothenbuhler, Thanh Tran, Elisa H. Barney Smith *, Vishal Saxena * and Kristy A. Campbell *

Department of Electrical and Computer Engineering, Boise State University, 1910 University Drive, Boise, ID 83725, USA

* Authors to whom correspondence should be addressed;
E-Mails: EBarneySmith@boisestate.edu (E.H.B.S.); VishalSaxena@boisestate.edu (V.S.); KrisCampbell@boisestate.edu (K.A.C.)

Abstract: We present our design exploration of reconfigurable Threshold Logic Gates (TLG) implemented using silver–chalcogenide memristive devices combined with CMOS circuits. Results from simulations and physical circuits are shown. A variety of linearly separable logic functions including AND, OR, NAND, NOR have been realized in discrete hardware using a single-layer TLG. The functionality can be changed between these operations by reprogramming the resistance of the memristive devices.

Keywords: memristors; threshold logic gates; reconfigurable circuits

---

1. Introduction

Moore's Law helped the semiconductor industry increase the computational power of computers while shrinking their size and power consumption. This has enabled the development of many new commercial products, which enables changes in daily life as well as new research breakthroughs. However, it seems continuing Moore's Law for much longer may not be possible. To continue to make sizable jumps in computer development, a different architecture might be needed.

One new technology that could be part of this uses the memristor. Recent progress in memristive devices has spurred renewed interest in reconfigurable and neuromorphic computing architectures [1]. The memristive devices, integrated with conventional CMOS, are expected to realize low-power circuits with increased reconfigurability and smaller physical layout area [2]. Threshold logic gates (TLG) are an interesting candidate for logic implementation using hybrid memristive-CMOS circuits [3]. We present our early results on TLG design using Silver–Chalcogenide (Ag–Ch) memristive devices developed at Boise State University [4–10]. This work is an expansion of the work presented in [11].

TLGs built with memristors can lead to changes in reconfigurable processing architectures. Memristor based circuits could be used to replace larger combinations of transistors, and through proper application of signals, the logic operations in the foundational architecture could be easily changed among AND, OR, NAND and NOR to create the logic necessary to solve each individual's computational problem. We present work in designing a circuit using memristors to implement a reprogrammable Threshold Logic Gate with memristors in discrete circuitry. The circuit is designed and modeled first in Cadence with a MATLAB controller, and then is built with discrete components and is controlled by an FPGA (field programmable gate array). While several papers have proposed using memristors [12–16], most are just simulation [17], often without a circuit [18]. A few papers go beyond simulation to design circuits, at least for a portion of the system [19,20], and even fewer design the complete circuit and build a working prototype [21–24]. Papers on physical memristors usually evaluate them in isolation using a signal analyzer [25,26]. This work reports on using physical memristors in a complete circuit application.

Section 2 describes the Threshold Logic Gates simulated in this paper. The memristive devices and their programming are described in Section 3. Section 4 describes circuits used to implement these TLGs. Experiments and results of the co-simulation are described in Section 6. The conclusions are portrayed in Section 7.

2. Threshold Logic Gate Design

Threshold Logic Gates are single node Artificial Neural Networks (ANNs) designed specifically to implement a logic operation, $g(X)$. The threshold logic gates produce a binary output by passing a weighted summation of several binary inputs ($X = [X_1 X_2 \cdots X_n]$) through a hard-limiting activation function, f,

$$Y = g(X) = f\left(\sum_{i=1}^{N} w_i \cdot X_i + w_0\right) \tag{1}$$

where w_0 is a bias or the threshold level [3]. This is conceptualized in Figure 1 for $N = 2$ with a bias. In this paper we are considering the two input logic operations AND, OR, NAND and NOR. These logic operations are all linearly separable and can be realized by this architecture. To implement a logic operation (or any function) that is not linearly separable, such as XOR and XNOR, the logic must be implemented using a two-layer TLG network.

Figure 1. Neuron with two input synapses and one bias synapse.

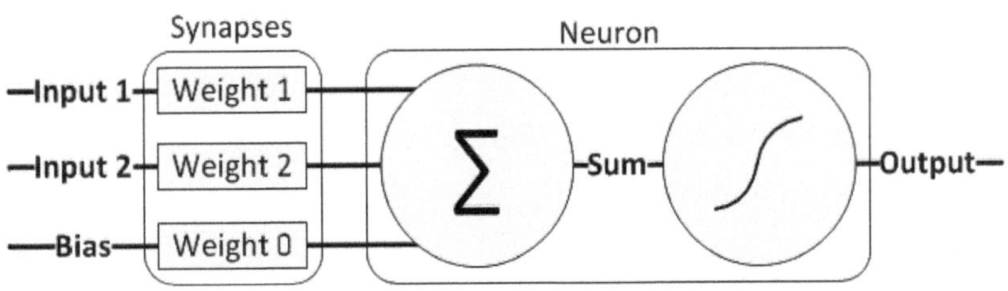

The same network structure can be used for multiple logic operations, simply by changing the weights, w_i. The weights necessary to implement any given TLG are not unique. Figure 2 shows in weight space the values that can be used to realize the four two-input logic operations AND, OR, NAND and NOR, over the normalized range $w_i \in [-1, 1]$. From these possible weight values, one target weight value for each logic operation was selected, Table 1. Note that all but the OR operation require negative weights.

Figure 2. Weight values necessary to realize the logic operations AND, OR, NAND and NOR.

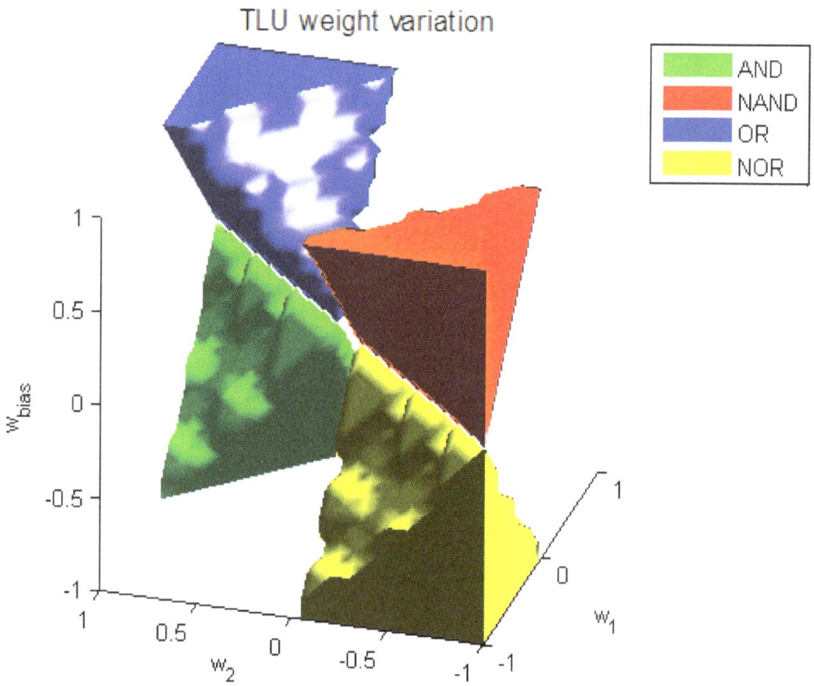

Table 1. Weight values chosen for linearly separable logic operations.

Logic operation	w_1	w_2	w_0
NAND	−0.5	−0.7	0.5
NOR	−0.5	−0.7	−0.5
AND	0.3	0.6	−0.5
OR	0.3	0.6	0.5

3. Memristive Devices

The Ag–Ch memristive device acts as a programmable two-terminal resistor, which can be programmed to assume a range of continuous resistance states. The resistance of the memristive device is decreased ("program" operation) and increased ("erase" operation) by applying positive and negative voltage pulses, greater than the device threshold voltage, respectively. The memristive device acts as a linear resistor when it receives an input below the threshold voltage. This behavior makes memristive devices suitable for use in reconfigurable hardware.

In our early design exploration using memristive devices, we are implementing neuromorphic circuits using discrete hardware. However, the true potential of these devices will be realized in the second-phase of our research where a fully-integrated platform with memristive devices fabricated in the back-end-of-the-line (BEOL) of a CMOS process will form compact reconfigurable circuit blocks.

A Verilog-A model for the memristive device has been developed and used with CMOS circuits for simulation in Spectre [27]. A memristive device model, similar to [28], is used to emulate the characteristics of the Boise State memristive device reported in [4–10], with appropriate device parameters. The Boise State memristors have been widely referenced by other researchers in their work [28–33]. This memristive model has a threshold voltage of 150 mV, while the read pulse voltage for the actual memristive devices fabricated and packaged at Boise State University should not exceed 20 mV for repeatable operation. This threshold voltage will be increased to the desired range using a modified material stack in the device. In our simulations, the initial state of the devices can be set to the desired resistance value. However, the algorithms assume random initial states, and all subsequent resistance changes are made through the application of voltage pulses.

4. Circuit Realization

The TLG defined in Equation (1) can be realized in discrete hardware using a summing op-amp circuit with memristive devices implementing the weights (w_i). The target logic function is then realized by re-programming the weights (*i.e.*, resistance) of the memristive devices.

To achieve the needed negative weights, the resistors (or memristive devices) are supplemented so the current coming out of the op-amp has the opposite direction of the input, and its combination with the variable current from the memristive device can take on both positive and negative values, Figure 3. The two resistors, R_N and R_{FS}, determine the gain of the circuit that sets the range of weights possible in the circuit realization. Adjusting the resistance of the memristive device R_M between 1 kΩ and 100 kΩ moves the weights in this range. Table 2 shows a set of target resistances for the memristors if $R_N = 1.98$ kΩ and $R_{FS} = 2.02$ kΩ.

Figure 3. Circuit implementation with memristive devices with circuitry for negative weights. In our circuit implementation the common mode voltage $V_{CM} = 2.5$ V.

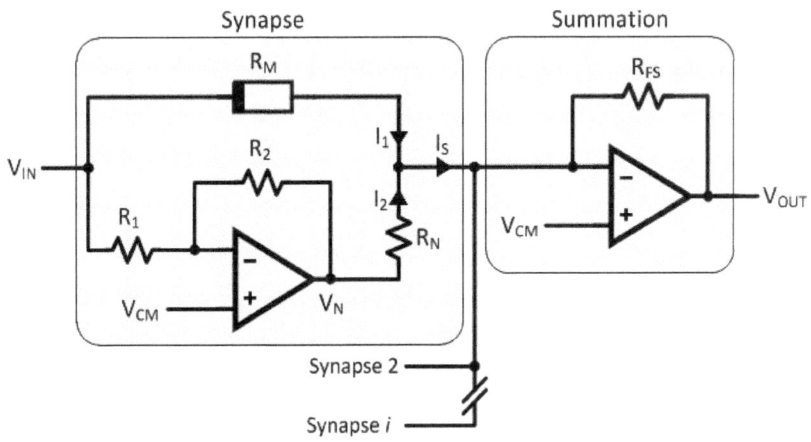

Table 2. Resistance values corresponding to theoretical weights in Table 1.

Logic operation	R_1	R_2	R_0
NAND	1.33 kΩ	1.17 kΩ	3.88 kΩ
NOR	1.33 kΩ	1.17 kΩ	1.33 kΩ
AND	2.81 kΩ	4.81 kΩ	1.33 kΩ
OR	2.81 kΩ	4.81 kΩ	3.88 kΩ

5. Memristive Device Characteristics

The memristive devices have been modeled in Cadence-Spectre and fabricated as physical devices. Programming the devices in both forms has the same basic approach. This approach and the necessary changes to that approach and observations about the physical devices are described next.

5.1. Simulations

The memristor model used in this paper is based on "A memristor device model" by C. Yakopcic [28] and is written in Verilog-A. The "state input" pin "xd" is used to sneak into the memristor state, but all memristor programming is done through the application of voltage pulses. The state input value is inversely proportional to the memristor resistance. A characterization of how the memristor state in the memristor model responds to the writing and erasing processes is shown in Figure 4. A series of positive and negative pulses with an amplitude of 200 mV were applied across the memristor. A pulse width of 1 μs was chosen so that the device is not under a constant stress. Applying a positive pulse causes the memristor to write, which decreases the memristor resistance, meaning that the "state" will be increasing. In contrast, applying a negative pulse causes the memristor to erase, which increases the memristor resistance, meaning that the "state" will be decreasing. Figure 5 shows the schematic of the programming circuit [34].

To read the current value of the memristor, the same circuit will be used, however, with a pulse amplitude that is much smaller than the threshold voltage. For the memristor model used in this paper, a (100 mV, 1 μs) pulse is chosen to be sufficiently small in order to avoid disturbing the current state of the device while reading it. The memristor resistance is then determined by measuring the output voltage of the circuit and applying Ohm's law

$$R_M = \frac{V_{read}}{V_{out} - V_{read}} R_{f1}. \tag{2}$$

The programming circuit in Figure 5 was built in Cadence first and then it was connected to Simulink using the Cadence coupler block. A MATLAB script was written to provide the voltage inputs to automatically drive the memristor to a desired resistance. The script generates programming pulses with amplitudes proportional to the difference between the current resistance R_M of the memristor and the desired value R_d. If the difference is large, it will send a bigger amplitude pulse and vice-versa. This amplitude gets smaller and smaller as the memristor approaches the desired resistance. The erasing process is more sensitive than the writing process; therefore, the gain in the erasing process is a factor of

1/3 smaller than the gain in the writing process. The programming pulse is limited below V_{max} so that it will not destroy the device or cause clipping in the output voltage of this programming circuit.

Figure 4. Movement of the memristor state in response to the (**a**) writing and (**b**) erasing processes. Writing decreases the device resistance, while erasing increases the device resistance. The input pulses are shown in red (top window) and the current through the device is shown in blue (middle window). The third window shows the movement of the device state "x" ($x_0 = 0.1$).

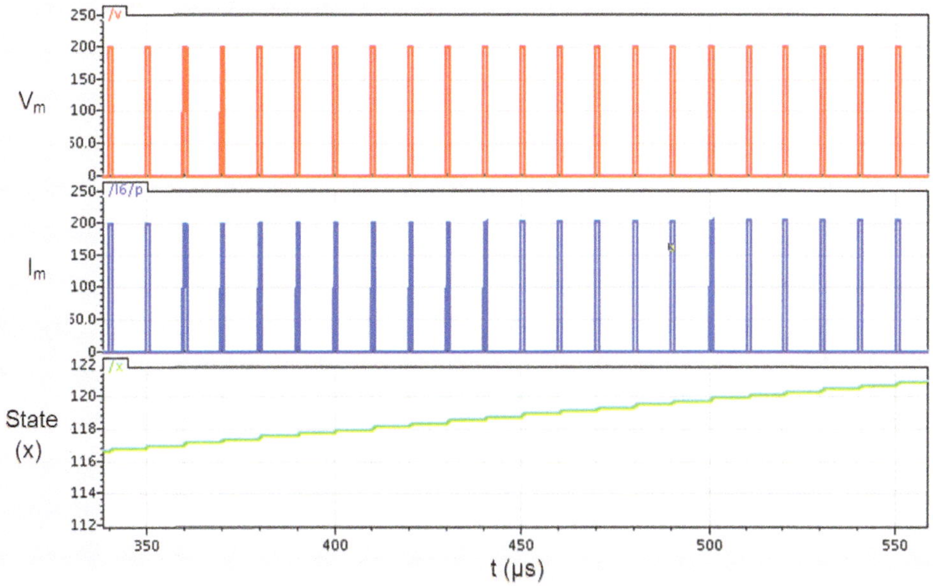

(**a**)

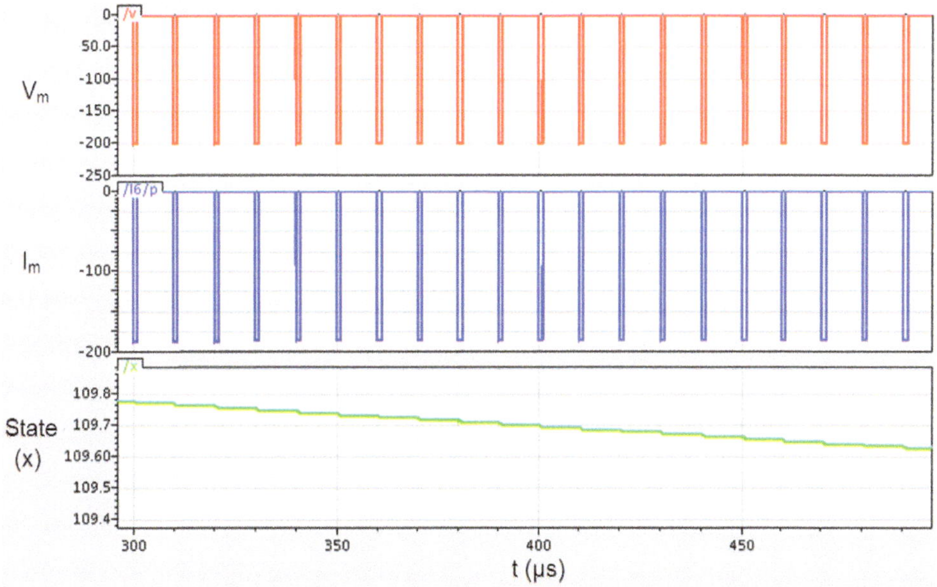

(**b**)

After each programming step, the MATLAB script will generate a reading pulse to read the current resistance of the device. In short, the programming circuit will keep sending pulses with an adapted amplitude to program, then read the memristor until its resistance is within a certain tolerance of the desired value. This behaves like a state machine in hardware. The operating range of the memristor in this paper is between 10 kΩ and 100 kΩ. The device was initialized at 40 kΩ, the desired value for the writing mode was 10 kΩ, and for the erasing mode was 100 kΩ. Figure 6 shows the amplitudes of the applied programming pulses and the measured memristor resistance after each programming cycle. The target resistance is highlighted in red. The script was able to drive the device to the desired resistance within a tolerance of 4 kΩ within an average of 10 programming cycles.

Figure 5. Memristor programming circuit.

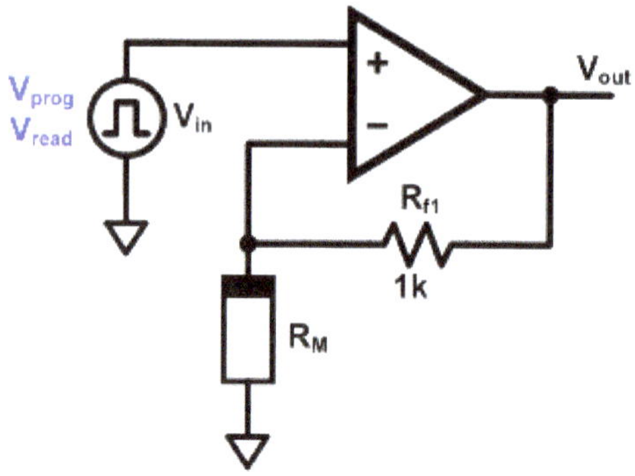

Figure 6. Programming a single memristor. (**a**) Programming from 40 kΩ to 10 kΩ; (**b**) Programming from 40 kΩ to 100 kΩ.

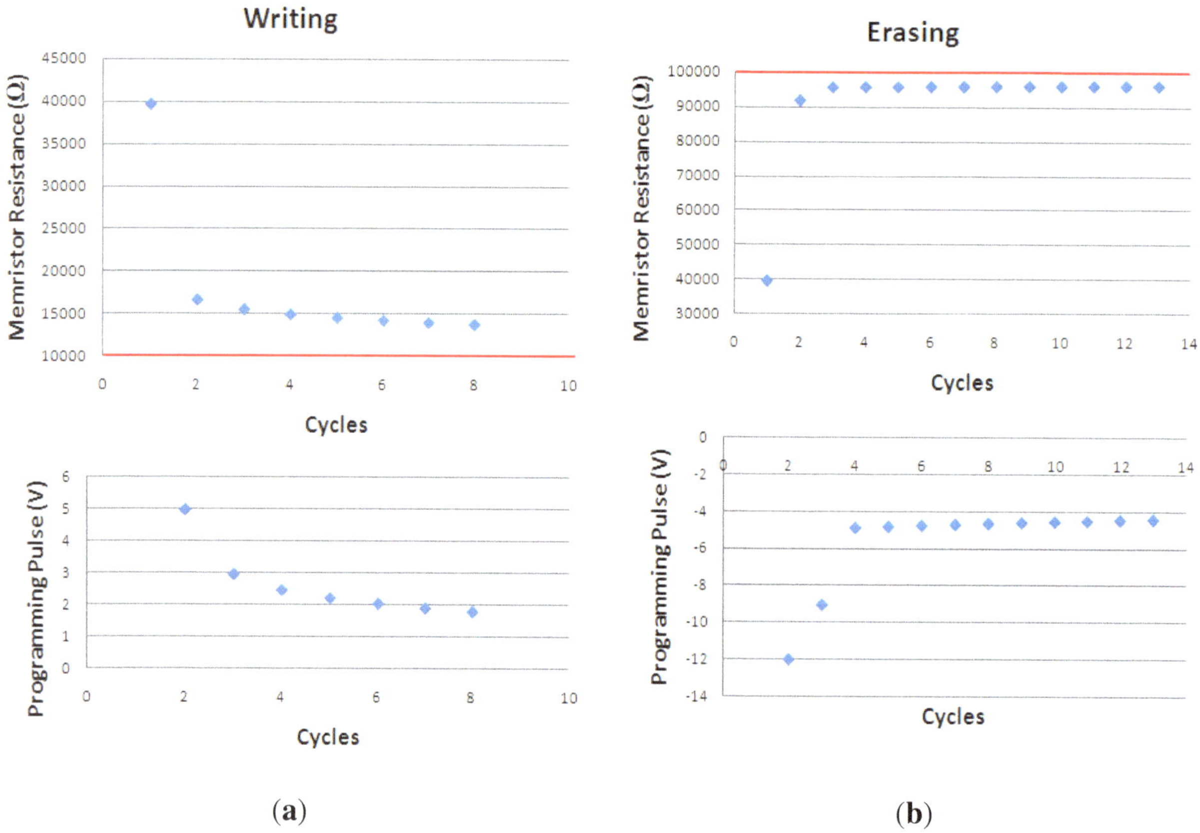

5.2. Hardware

While our simulations used a non-idealized Verilog-A memristor model, when working with physical memristors the behavior of the devices is not fully predicted by the modeling equations. Experience with the characterization of these devices is employed to make circuits robust in the presence of these deviations from the simplistic understanding of memristive devices. The memristor model was designed as a curve fit to the data obtained from our physical memristors [28], but the full characteristics of the physical memristors are still being studied. Most studies are with a semiconductor parameter analyzer (SPA). Some experiments with physical memristors are shown here to understand how the memristors function in a circuit environment.

At this point there is no accurate model that foresees how much the memristor's state will move if a voltage pulse of duration x and voltage y is applied. The writing and erasing voltages have to exceed a certain threshold for the memristor to change its state. The device has both a writing and an erasing threshold, which are not necessarily at the same absolute voltage level. Furthermore, these thresholds can change as the device's resistance is altered, and the threshold of a device can change during its lifetime and was found to exhibit significant fluctuations between devices, making it even more difficult to make an accurate prediction of where the state is going to end up. It is therefore important to read the device's state after each writing/erasing pulse so that programming error can be detected and compensated for by applying subsequent pulses. After programming the memristance can relax or drift, but certain programming pulses are less susceptible to the post-programming drift. For our application the amount of decay will not affect the final application response. For other applications that require finer memristance granularity, there would usually be a sequence of programming pulses and the subsequent pulses would "correct" any resulting drift. A typical writing/erasing algorithm will therefore need to send multiple pulses interspersed with reading pulses to the memristor.

Table 3 shows the threshold voltages that were found to work best for the devices under study. Note that these values are only valid for DC operation (slow pulses) and are not representative of the AC behavior of these devices. The following general rules have been developed to reach the desired state without damaging the memristor:

- The reading voltage should be as low as possible to prevent unintentional ion movement. For the Ag–Ch devices used for this work, the reading voltage should not exceed 40 mV.
- If a large reduction in resistance has been observed after applying a writing pulse, the device should be fully erased before applying another writing pulse. A large ΔR indicates hard device switching, meaning that most of the silver from the top electrode is embedded in the amorphous insulation layer. Applying another writing pulse might irreversibly damage the device.
- Erasing pulses of very short duration and high enough amplitude can cause the device to go into negative differential resistance (NDR) mode or even damage the device. Repeated mild erase pulses will not cause long-term damage.

Table 3. DC threshold levels for Ag–Ch memristors.

Programming operation	Threshold (volts)
Writing	0.23 V
Erasing	−0.58 V

Figure 7 shows erasing and writing pulses applied to a silver-based device. The device is initially in low resistance state. Applying a low erasing pulse does not change its state, unless the erasing threshold is reached. At 15 ms, the device switches to a higher resistance because the erasing threshold listed in Table 3 is exceeded. The same is true for programming the device back to a lower resistance at 25 ms. The three programming pulses between 30 and 40 ms are bad examples as they violate one of the programming rules. If a larger ΔR is observed when writing, no more subsequent writing pulses should be applied as it could potentially harm the device. Note that the resistance in Figure 7 was only slightly reduced by the first pulse at 25 ms, indicating that the device has been sufficiently written. At 50 ms, the amplitude needed to erase the device is significantly higher than previously reported. This shows how the threshold levels can change while programming the device. The programming algorithm in this work was designed to accommodate these non-ideal characteristics.

Figure 7. Experimental response of the device to program and erase pulses. The program and erase threshold can be seen to move between pulses. The moving threshold can be observed by the change in memristance relative to the program and erase pulse heights.

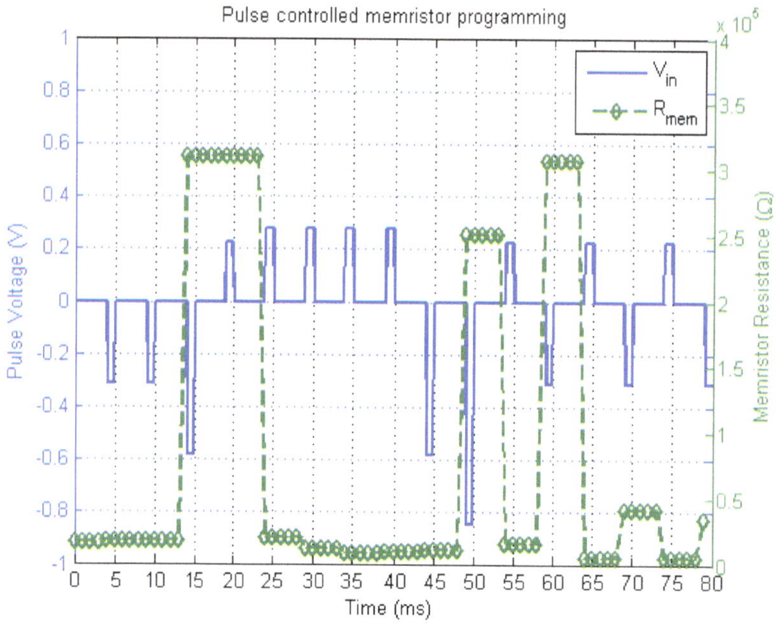

The device used for this experiment performed well for several cycles, indicating that its integrity was not decreased. In fact, the "moving threshold" phenomenon was observed in all devices under study. The threshold levels listed in Table 3 were found to generally be accurate, but exhibited a deviation of up to 50% over a memristor's lifetime. The moving threshold was observed because as every instance a voltage pulse is applied to change the resistance, there is an activation energy involved. At every step to move the Ag atoms to change the state, a higher pulse is required to affect the change. This gives the appearance of a moving threshold voltage. This variable threshold phenomenon has also been observed in other memristive devices [35].

In the Cadence memristor model a pulse train of writing pulses continues to decrease a memristor's resistance, while an erasing pulse train continually increased its resistance, Figure 4. This is based on a simulation model, which does not have to follow the memristor programming rules as outlined above. The physical devices look moderately different compared with a simulated environment. Experiments on the devices confirmed this indication and it was found that continuously increasing or decreasing the resistance with fine control can be challenging to achieve in practice. This is especially true for continuous writing, as the programming rules prohibit us from continually sending writing pulses. It is therefore better to write the device and gradually erase it. In most cases, the devices would perform hard switching (*i.e.*, jump from a low resistance to a high resistance and vice-versa). We chose a programming pulse that was suited to our application to enable us to jump in a single programming pulse between a high and low resistance value. Figure 8 shows the result of applying a pulse train of writing and erasing pulses with pulse amplitudes that are only slightly above the silver threshold levels (see Table 3) to a silver device. As can be seen, the resistance jumps up and down by at least two orders of magnitude in response to applied programming and erase pulses. The standard deviation was found to be around 45% for both high and low states. While this is a pretty significant variation, it enables the memristors to be used as a digital memory when toggling between the two states (high and low). The average high state resistance was found to be at 392 kΩ and the average low state resistance at 7.3 kΩ.

Figure 8. Memristor characterization plot showing toggling between high and low memristance states.

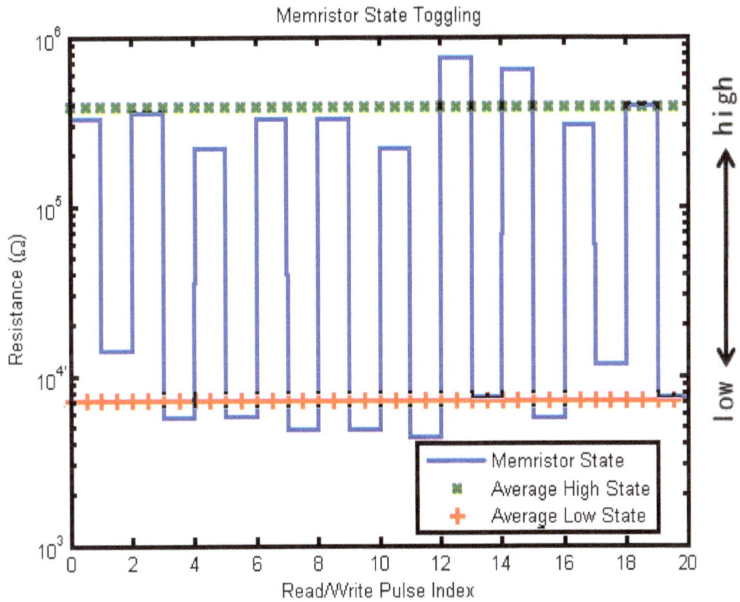

Due to their intrinsic device structure and material characteristics, silver-based devices are expected to exhibit a certain hard-switching behavior, making them suitable to reprogrammable TLG use as the weights are polarized. It has been proposed that memristors would be suitable for use in ANNs, but these need finely variable resistances. Further device development and alternate circuitry and control methodologies may be necessary for these applications. The programmer used for this work is purely voltage controlled. While the presence of a threshold voltage suggests that the programmer should be voltage controlled, it is possible that the device should be programmed in current mode once the threshold voltage is reached. In fact, literature suggests that memristors are both a voltage and current controlled device [1,4], but no solutions are presented as to how a memristor is best controlled in a practical application such as an TLG or ANN programming circuit. The work in this paper starts to develop this.

Figure 8 shows the device has clear high and low states, but they vary significantly. It may seem as though the deviation of the values representing these high and low states is random and does not suggest that a relationship between previous state (resistance) and the change in resistance exists. However, when the change in resistance is analyzed numerically, a relationship between the change in resistance and the previous resistance is found to exist, but only when writing. Figure 9 shows data points from selected datasets that exhibited nice hard-switching behavior of the silver-based devices. When writing the devices to a lower resistance, the change in resistance is dependent on the previous state, but is independent of the pulse amplitude (the pulse amplitudes used here were all the same).

Figure 9. Relationship between resistance change and previous state.

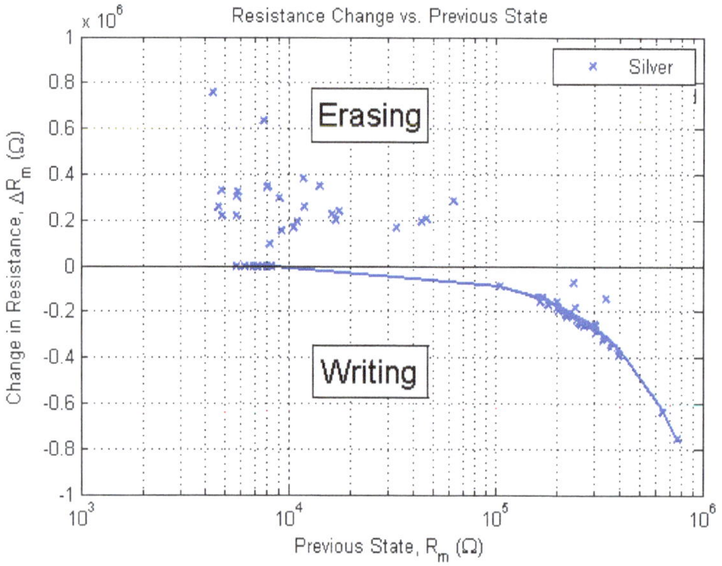

The resistance change can be said to be proportional to $-1/R^2$, where R is the resistance of the current state (before applying the pulse). This is consistent with literature investigating spike-timing-dependent-plasticity (STDP), which governs how synapses learn in biology [36]. This relationship between resistance change and previous state only holds for programming (*i.e.*, reducing the resistance). When erasing, there seems to be no apparent pattern between the change and previous state. Given that the relationship is proportional to $-1/R^2$, it makes sense that this only applies when writing a memristor to a lower state, as the change in resistance will always be a negative number, indicating a reduction in resistance. Further, in Figure 8 it can be observed that with the selected program/erase voltage pulses, we could achieve a binary switch functionality desired for our 2-input Boolean logic application. For TLGs with a larger fan-in ($N > 2$), a more continuous range of weight values will be needed. A different scheme of adaptive program/erase voltage pulses would be applied to achieve the desired functionality.

6. Experiments and Results

The circuit described in Section 4 was created in a Cadence Simulation and also built as a physical circuit with discrete devices. In both cases the memristors were programmed to change the functionality of the TLG circuit alternately to AND, OR, NAND and NOR. Descriptions of these experiments and their results are described next.

6.1. Simulations

A feedback-based adaptive programming circuit was developed to program the individual memristive devices to predetermined resistance values to create each logic operation. The programming circuit in Figure 10 was built in Cadence and connected to Simulink using the Cadence coupler block. The memristor has to alternately be isolated for the circuit to apply the programming via the circuit in Figure 5 and be connected with the other memristors to the negative weight and summing circuitry in Figure 3. This is accomplished from switches shown in Figure 10.

Figure 10. Circuit implementation for programming memristive devices. In our circuit implementation the common mode voltage $V_{CM} = 2.5$ V.

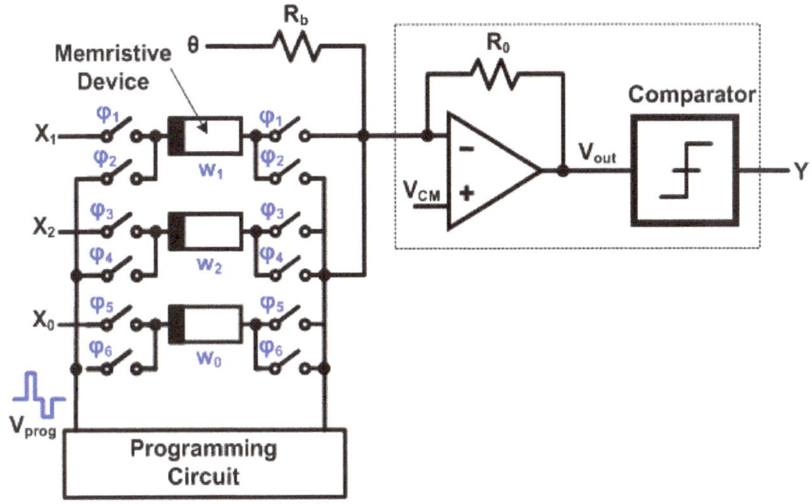

A MATLAB-Simulink/Cadence co-simulation was run to realize four threshold logic gates: AND, OR, NAND, and NOR. The memristive devices were programmed by providing the voltage inputs through a charge-scaling DAC (digital to analog converter) with an iterative search process, using the device programming method described in Section 3 to automatically drive the memristor to reach the resistance values in Table 2 within a tolerance of 100 Ω. The program-and-erase operation is performed by applying positive and negative pulses (1 μs pulse width). After the memristive devices were programmed to match all the target resistances, a set of feed forward voltages was applied to the network to confirm the logic outputs for all four AND, OR, NAND, and NOR operations.

The results are shown in Figure 11. The waveform in the top frame of the plots shows the clock signal of the circuit, and the second and third frames show the two binary input signals. The amplitudes of these inputs are set within a range of 100 mV to not disturb the current state of the memristive device. The fourth waveform shows the output voltage superimposed on the 2.5 V virtual ground. The logic output of the comparator is displayed in the fifth window. The results show that the circuit built with memristive devices can be programmed and re-programmed to create the four desired logic operators. It is confirmed that the circuit design and programming procedure work correctly.

Simulations are often based on ideal behavior of devices. The MATLAB-Simulink/Cadence co-simulation tried to account for hardware-related phenomena as much as possible, and used the most realistic memristor model available, but there are still a lot of things that a simulation cannot account for. Dynamic range, noise and signal integrity are only a few examples. In order to build the simulated TLG entirely in hardware, a new programming circuit was developed to program and read the memristor devices.

Figure 11. Results of TLG circuit when programmed to each of the four logic operations. (**a**) AND; (**b**) OR; (**c**) NAND; (**d**) NOR. The output is valid only when the clock is high.

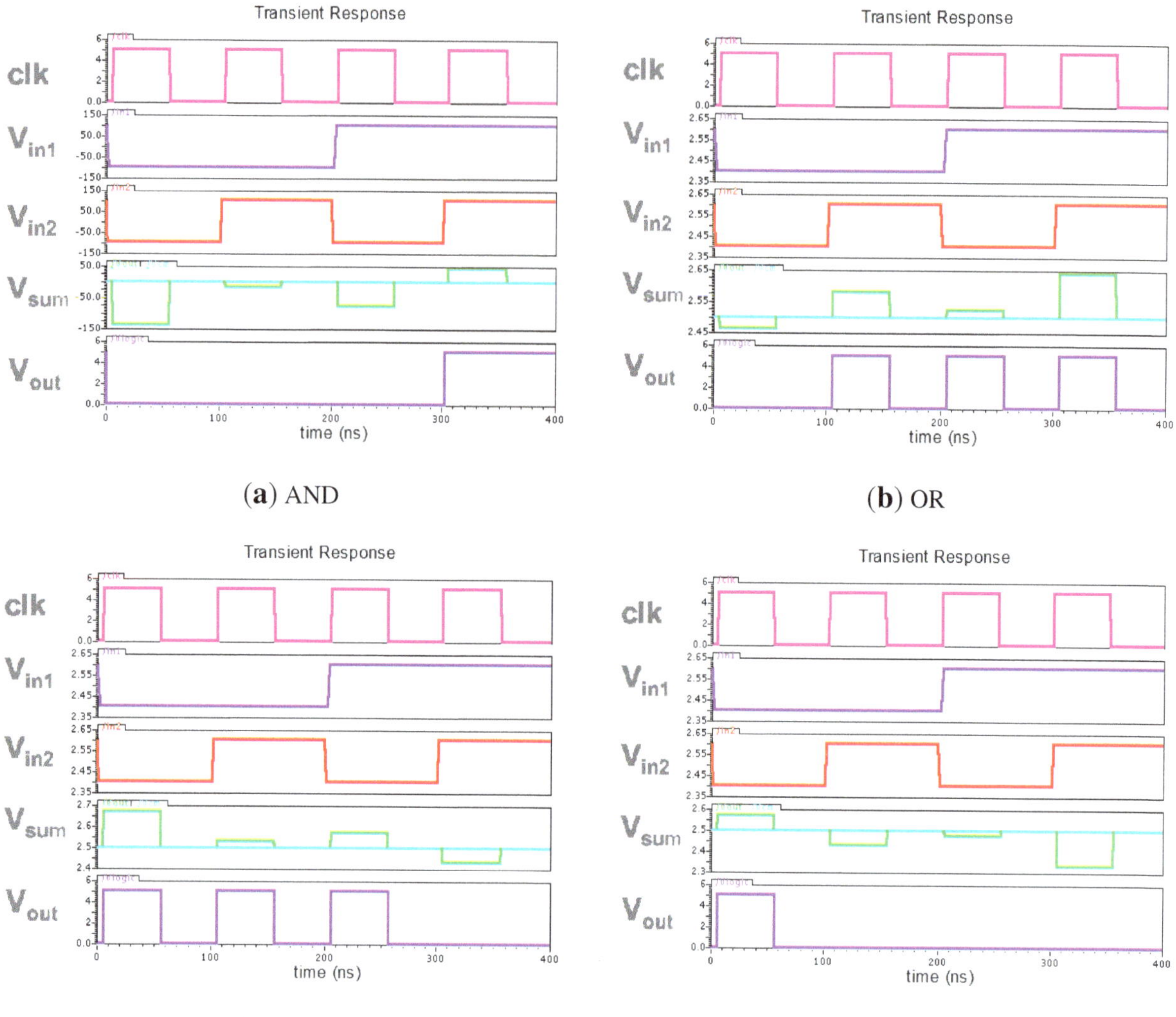

(**a**) AND (**b**) OR

(**c**) NAND (**d**) NOR

The programming setup used in the simulation is based on an op-amp in non-inverting configuration. The MATLAB-Simulink programming algorithm sends a pulse width and a pulse amplitude value to the Cadence simulation, which applies the programming pulse to the op-amp. The basic programming circuit used is shown in Figure 5. While this works fine in simulation, to build this circuit in hardware an op-amp supporting high speeds, high gains, and a wide dynamic range is needed and is hard to find. Given a fixed feedback resistor R_{f1}, the resistance value of the memristor can easily be determined by applying a pulse with a known pulse height to the positive input of the op-amp. The output V_{OUT} is then a function of R_M. However, depending on the value of R_M, the gain of the op-amp can range from very small values to very large values. This dynamic range issue is not a big deal for simulations, but it is for hardware-based applications. In order to improve the dynamic range, the op-amp in Figure 5 is turned into a programmable gain amplifier (PGA) by adding different feedback resistors in parallel to the existing feedback resistor R_{f1}. These additional resistors can be connected with switches to change the effective feedback resistance and therefore the gain of the PGA, Figure 12.

Figure 12. Programmable gain amplifier used to program memristors.

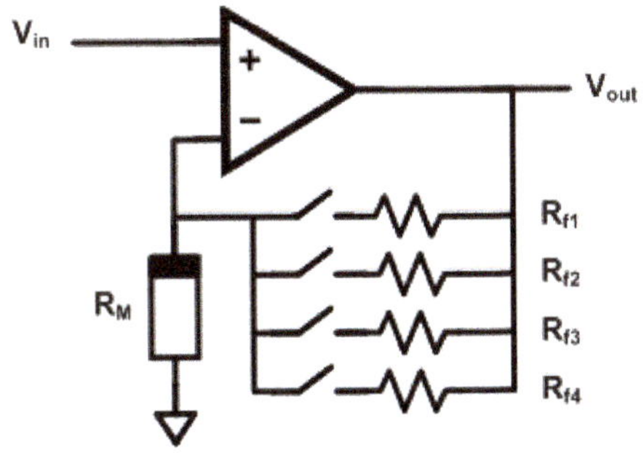

The output of the PGA for different memristor resistance values is shown in Figure 13. Note that there are four feedback resistors in Figure 12, but Figure 13 only shows three curves. This is because the first of the four resistors (R_{f1}) is only used for programming and not reading on the memristor device. We limited our circuit to a range from 1 kΩ to 200 kΩ such that when programming, the gain of the PGA is set sufficiently small. With a 1 kΩ feedback resistance, the programming current is also sufficiently limited such that the device will not be destroyed when its resistance is drastically reduced.

Figure 13. Output of programming/reading circuit with programmable gain.

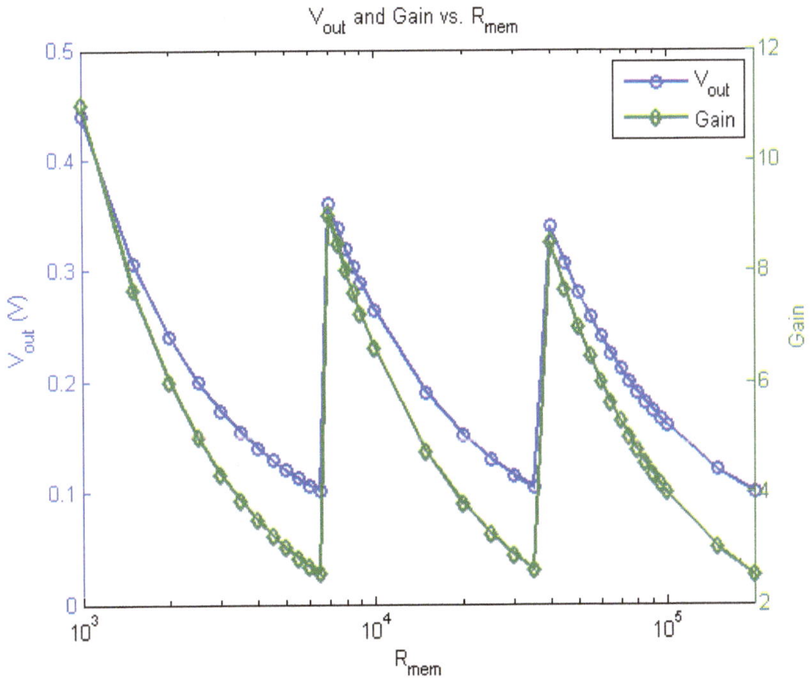

An FPGA (field programmable gate array) and a high-speed DSP (digital signal processing) board are used to generate the programming pulses and to read the response (V_{OUT}) of the programming circuit. A host computer communicates with a Nios-II soft-core processor implemented on the FPGA, controlling the pulse generation module also implemented on the FPGA. The pulse generation module generates both programming and reading pulses through a high-speed DAC, controls the switches for the PGA and reads the response of the memristor using an ADC (analog to digital converter). When reading a memristive device, a small reading pulse (20–40 mV) is applied and the response of the PGA is read back by the DSP board's DAC and is further processed by the FPGA. For the Ag–Ch devices, a pulse with fixed pulse width of 10 μs and variable amplitude was used.

For the simulation where greater control of the memristor is possible, exact target resistances were calculated in Table 1. Since the zones of acceptable weight values are large, binary toggling of the device was utilized for the hardware implementation. The weight values in Table 1 are therefore quantized to produce the binary states in Table 4. These correspond to the four corners of the weight cube in Figure 2 where a logic operation of interest is located.

Table 4. Binary memristor states for each logic function.

Logic operation	w_1	w_2	w_0
AND	high kΩ	high kΩ	low kΩ
OR	high kΩ	high kΩ	high kΩ
NAND	low kΩ	low kΩ	high kΩ
NOR	low kΩ	low kΩ	low kΩ

Figure 10 shows the fully connected TLG with quantizer (activation function of the neuron) and programming circuit attached. The switches are needed to disconnect each memristor individually and connect it to the programmer for reconfiguration. This circuit was built from discrete components. Memristors were packaged and connected. The circuit board is shown in Figure 14.

Figure 14. TLG programming circuit.

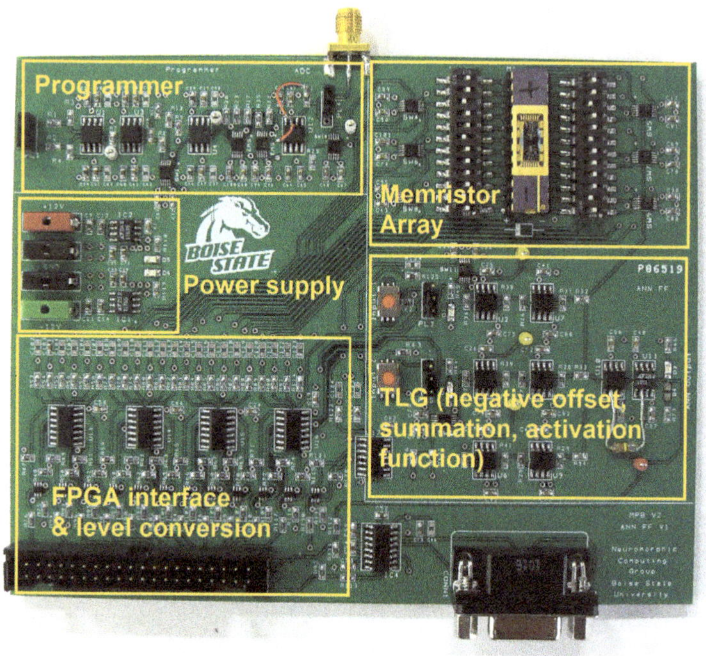

The performance of the TLG was measured by the four basic logic operations (AND, OR, NAND, NOR), but there are a total of 16 two-input logic operations possible, 14 of which could be achieved by this circuit. The remaining 12 cases for the purposes of this paper are described as invalid or "other" logic operations. In Figure 15 the output of the TLG through several cycles of programming to take on different logic operations is shown. The top two frames show the input signals cycling through 4 input options, then resting while voltages are applied to change the memristor's state. The resulting output signals after each reprogramming cycle are shown in frame 3. Frames 4 and 5 show additional results that were generated after the first set of results. They use the same functional inputs as frames 1 and 2. This clearly shows that the TLG can be reconfigured to many more possible functions. Since we used memristors, it is non-volatile and retains the settings even when we remove the power source.

The programming intervals in Figure 15 indicate where a single memristor was programmed according to Table 4. As Table 4 shows, going from an AND to a NOR configuration requires two memristors to change (memristor 1 and 2 need to change from high to low; memristor 3 stays low). Therefore, it required two programming cycles—one for each memristor—to reconfigure the TLG. In order to reconfigure the TLG from an AND to an OR or a NAND to a NOR configuration, only one programming cycle is required, as the bias weight (memristor 3) is the only one that needs to be changed.

Figure 15. TLG inputs and outputs.

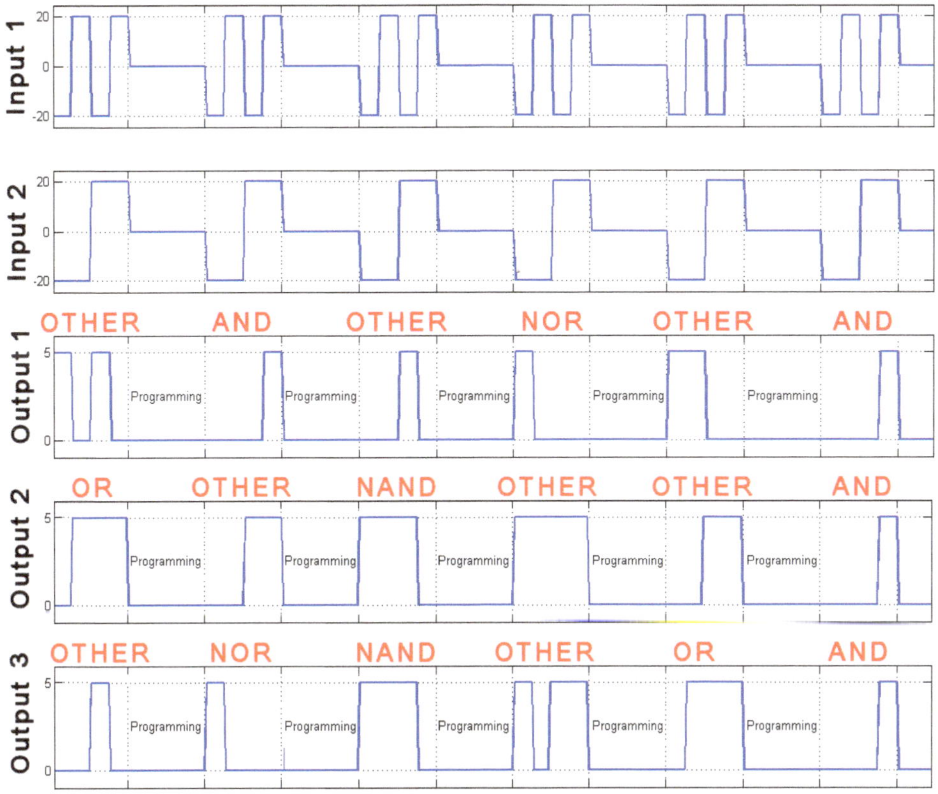

Figure 16a shows the weight space with a possible programming path for reconfiguring the TLG from NOR to OR. In this case it will take a total of three programming cycles as all three memristors need to be changed. As Figure 16a shows, the path from NAND to OR leads through a corner that is not one of the four functions used for this work. Figure 16b shows the programming paths required to obtain all the results in Figure 15 in the order presented. As can be seen, the TLG is reconfigured to operate only in the corner regions. Four of the eight corners are valid configurations (AND, OR, NAND, NOR), while the rest are unused configurations and are simply used as way-points to reconfigure the TLG from one function to another.

Figure 16. (**a**) Weight cube indicating the programming path required to go from NOR to OR and (**b**) all programming paths required to obtain the results seen in Figure 15.

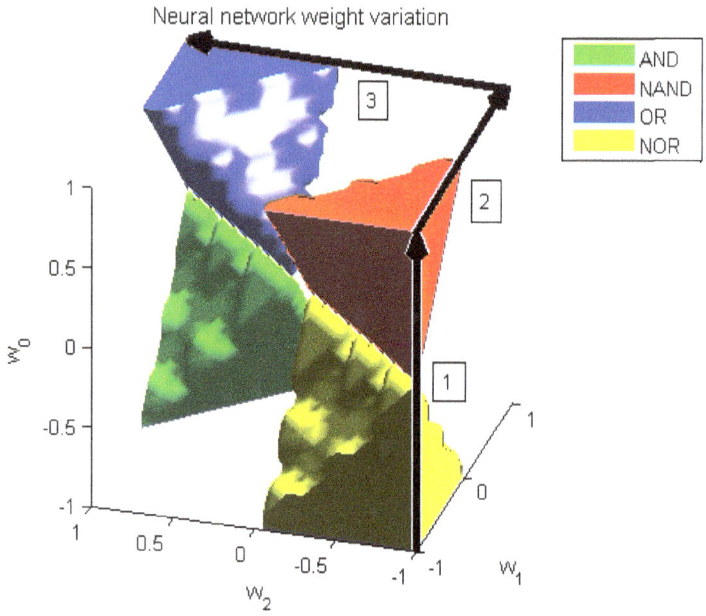

(**a**)

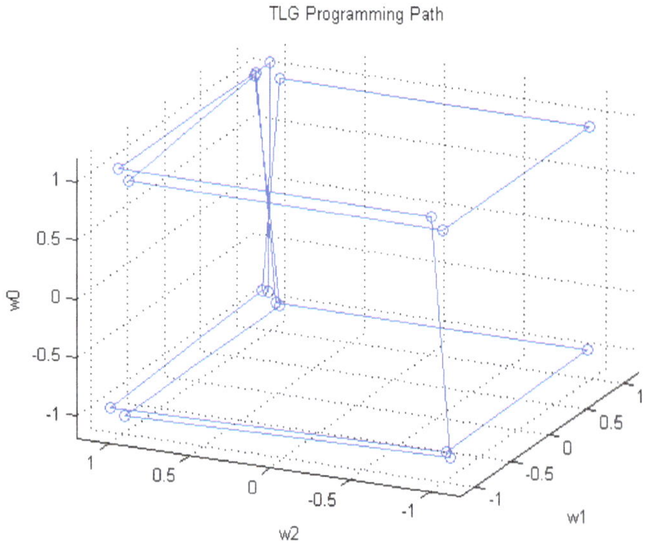

(**b**)

7. Conclusions

We have developed a circuit design capable of realizing four different logic operations by changing the resistance of the memristive devices. The resistance is changed through an iterative procedure through the application of voltage pulses to reach resistances determined outside the circuit. This was implemented first in simulation, and this same circuit with only minor modifications was then physically realized on a printed circuit board (PCB). The memristors in this circuit were reprogrammed based on voltage pulses determined by an FPGA so the circuit could take on the four logic operations of interest.

We believe that this is the first reported experimental demonstration of a reprogrammable TLG circuit with physical memristors. This work developed a greater understanding of the constraints involved when using memristors in complete circuits, instead of in signal analyzers. Techniques necessary to work with these memristors were also developed.

This programming procedure can be replaced by a learning algorithm, such as back propagation, to determine the resistance values based on input-output pairs. Memristors and a control circuitry that can attain finer granularity in resistances will be needed. The switching circuitry introduced a high level of overhead. Future work should look at alternate approaches to do both programming and use of the memristors without this.

The prototype TLG circuit, presented in this paper, used discrete circuit components to demonstrate the hybrid memristor and circuit application. Due to the limitations set by the available discrete op-amps, a 5 V supply was used. Eventually, the memristive devices will be fabricated in the back-end-of-the-line (BEOL) on a CMOS chip, with a compact TLG circuit implementation and comprised of the peripheral circuits for program and read operations. A lower supply voltage and smaller device geometries in nanometer CMOS processes would result in significant power savings. Further, the integrated version of the threshold logic gate would use a compact transistor level implementation instead of a power hungry op-amp. Our work thus far has explored feasibility of two-input TLGs with Ag–Ch memristive devices. In the future, hybrid CMOS-memristor integrated TLGs are expected to have fan-in of >10, switching speed >1 GHz and dissipate lower static power than a corresponding pure-CMOS logic implementation. Such high density of integration will allow development of several new applications.

Acknowledgments

The authors gratefully acknowledge Kolton Drake and Stephen Stauts for invaluable support with testing and packaging of memristors and Jeanette Brooks for assistance with PCB assembly.

References

1. Strukov, D.B.; Snider, G.S.; Stewart, D.R.; Williams, R.S. The missing memristor found. *Nature* **2008**, *453*, 80–83.
2. Strukov, D.B.; Likharev, K.K. A Reconfigurable Architecture for Hybrid CMOS/Nanodevice Circuits. In Proceedings of the 14th ACM/SIGDA International Symposium on Field Programmable Gate Arrays, Monterey, CA, USA, 22–24 February 2006; pp. 131–140.
3. Beiu, V.; Quintana, J.M.; Avedillo, M.J. VLSI implementations of threshold logic—A comprehensive survey. *IEEE Trans. Neural Netw.* **2003**, *14* 1217–1243.
4. Oblea, A.S.; Timilsina, A.; Moore, D.; Campbell, K.A. Silver Chalcogenide Based Memristor Devices. In Proceedings of the International Joint Conference on Neural Networks (IJCNN), Barcelona, Spain, 18–23 July 2010; pp. 1–3.
5. Campbell, K.A. Variable integrated analog resistor. U.S. Patent No. 8238146, August 2012.
6. Campbell, K.A. SnSe-based limited reprogrammable cell. U.S. Patent No. 8101936, January 2012.

7. Campbell, K.A.; Moore, J.T. Silver-selenide/chalcogenide glass stack for resistance variable memory. U.S. Patent No. 8080816, December 2011.
8. Campbell, K.A. Resistance variable memory device and method of fabrication. U.S. Patent No. 7868310, January 2011.
9. Campbell, K.A. Differential negative resistance memory. U.S. Patent No. 7745808, June 2010.
10. Campbell, K.A.; Moore, J.T. Silver-selenide/chalcogenide glass stack for resistance variable memory. U.S. Patent No. 7646007, January 2010.
11. Tran, T.; Rothenbuhler, A.; Barney Smith, E.H.; Saxena, V.; Campbell, K.A. Reconfigurable Threshold Logic Gates Using Memristive Devices. In Proceedings of the IEEE Subthreshold Microelectronics Conference, Waltham, MA, USA, 9–10 October 2012.
12. Merrikh-Bayat, F.; Shouraki, S.B. Memristor-based circuits for performing basic arithmetic operations. *Procedia Comput. Sci.* **2011**, *3*, 128–132.
13. Di Ventra, M.; Pershin, Y.V.; Chua, L.O. Circuit elements with memory: Memristors, memcapacitors, and meminductors. *Proc. IEEE* **2009**, *97*, 1717–1724.
14. Yuriy, P.V.; Di Ventra, M. Practical approach to programmable analog circuits with memristors. *IEEE Trans. Circuits Syst. Regul. Pap.* **2010**, *57*, 1857–1864.
15. Talukdar, A.; Radwan, A.G.; Salama, K.N. Generalized model for Memristor-based Wien family oscillators. *Microelectron. J.* **2011**, *42*, 1032–1038.
16. Muthuswamy, B. Implementing memristor based chaotic circuits. *Int. J. Bifurc. Chaos* **2010**, *20*, 1335–1350.
17. Rose, G.S.; Rajendran, J.; Manem, H.; Karri, R.; Pino, R.E. Leveraging memristive systems in the construction of digital logic circuits. *Proc. IEEE* **2012**, *100*, 2033–2049.
18. Victor, E.; Howard, G.D.; Adamatzky, A. Organic memristor devices for logic elements with memory international. *J. Bifurc. Chaos* **2012**, *22*, 1250283; doi: 10.1142/S0218127412502835.
19. Qiangfei, X.; Robinett, W.; Cumbie, M.W.; Banerjee, N.; Cardinali, T.J.; Yang, J.J.; Wu, W.; Li, X.; Tong, W.M.; Strukov, D.B.; Snider, G.S.; Medeiros-Ribeiro, G.; Williams, R.S.; Memristor? CMOS hybrid integrated circuits for reconfigurable logic. *Nano Lett.* **2009**, *9*, 3640–3645.
20. Robinett, W.; Pickett, M., Borghetti, J.; Qiangfei, X.; Snider, G.S.; Medeiros-Ribeiro, G.; Williams, R.S. A memristor-based nonvolatile latch circuit. *Nanotechnology* **2010**, *21*, doi:10.1088/0957-4484/21/23/235203.
21. Snider, G.S. Self-organized computation with unreliable, memristive nanodevices. *Nanotechnology* **2007**, *18*, 1–13.
22. Adhikari, S.P.; Yang, C.; Kim, H.; Chua, L.O. Memristor bridge synapse-based neural network and its learning. *IEEE Trans. Neural Netw. Learn. Syst.* **2012**, *23*, 1426–1435.
23. Ziegler, M.; Soni, R.; Patelczyk, T.; Ignatov, M.; Bartsch, T.; Meuffels, P.; Kohlstedt, H. An electronic version of Pavlov's dog. *Adv. Funct. Mater.* **2012**, *22*, 2744–2749.
24. Bichler, O.; Zhao, W.; Alibart, F.; Pleutin, S.; Lenfant, S.; Vuillaume, D.; Gamrat, C. Pavlov's dog associative learning demonstrated on synaptic-like organic transistors. *Neural Comput.* **2013**, *25*, 549–566.

25. Yu, S.; Wu, Y.; Jeyasingh, R.; Kuzum, D.; Wong, H.-S.P. An electronic synapse device based on metal oxide resistive switching memory for neuromorphic computation. *IEEE Trans. Electron Devices* **2011**, *58*, 2729–2737.
26. Jo, S.H.; Chang, T.; Ebong, I.; Bhadviya, B.B.; Mazumder, P.; Lu, W. Nanoscale memristor device as synapse in neuromorphic systems. *Nano Lett.* **2010**, *10*, 1297–1301.
27. Pino, R.; Bohl, J.W.; McDonald, N.; Wysocki, B.; Rozwood, P.; Campbell, K.A.; Oblea, A.; Timilsina, A. Compact Method for Modeling and Simulation of Memristor Devices: Ion Conductor Chalcogenide Based Memristor Devices. In Proceedings of the 2010 IEEE/ACM International Symposium on Nanoscale Architectures (NANOARCH), Anaheim, CA, USA, 17–18 June 2010; pp. 1–4.
28. Yakopcic, C.; Taha, T.M.; Subramanyam, G.; Pino, R.E.; Rogers, S. A memristor device model. *IEEE Electron Device* **2011**, *32*, 1436–1438.
29. Yakopcic, C.; Taha , T.M.; Subramanyam, G.; Shin, E.; Murray, P.T.; Rogers, S. Memristor-based Pattern Recognition for Image Processing: An Adaptive Coded Aperture Imaging and Sensing Opportunity. In Proceedings of the Adaptive Coded Aperture Imaging, Non-Imaging, and Unconventional Imaging Sensor Systems II, San Diego, CA, USA, 1 August 2010; Volume 7818, 78180E.
30. Yakopcic, C.; Sarangan, A.; Gao, J.; Taha , T.M.; Subramanyam, G.; Rogers, S. TiO_2 Memristor Devices. In Proceedings of the 2011 IEEE National Aerospace and Electronics Conference (NAECON), Dayton, OH, USA, 20–22 July 2011; pp. 101–104.
31. Yakopcic, C.; Taha, T.M.; Subramanyam, G.; Pino, R.E.; Rogers, S. Analysis of a Memristor Based 1T1M Crossbar Architecture. In Proceedings of the 2011 International Joint Conference on Neural Networks (IJCNN), San Jose, CA, USA, 31 July–5 August 2011; pp. 3243–3247.
32. Yakopcic, C.; Taha, T.M.; Subramanyam, G.; Rogers, S. Memristor-based unit cell for a detector readout circuit. *SPIE Optical Eng. Appl.* **2011**, 8165, 81651F.
33. Yakopcic, C.; Taha , T.M.; Subramanyam, G.; Pino, R.E. Memristor SPICE modeling. In *Advances in Neuromorphic Memristor Science and Applications*, Springer: Berlin, Germany, 2012; pp. 212–244.
34. Drake, K.; Campbell, K.A. Chalcogenide-Based Memristive Device Control of a Lego Mindstorms NXT Servo Motor. In Proceedings of the AIAA Infotech at Aerospace Conference and Exhibit, St. Louis, MO, USA, 29–31 March 2011.
35. Jo, S.H.; Kim, K.-H.; Lu, W. Programmable resistance switching in nanoscale two-terminal devices. *Nano Lett.* **2008**, *9*, 496–500.
36. Zamarreño-Ramos, C.; Camuñas-Mesa, L.A.; Pérez-Carrasco, J.A.; Masquelier, T.; Serrano-Gotarredona, T.; Linares-Barranco, B. On spike-timing-dependent-plasticity, memristive devices, and building a self-learning visual cortex. *Front. Neurosci.* **2011**, *5*(26), doi:10.3389/fnins.2011.00026.

3

Bias-Flip Technique for Frequency Tuning of Piezo-Electric Energy Harvesting Devices

Jianying Zhao [1], Yogesh Ramadass [2], Jeffrey Lang [3], Jianguo Ma [1] and Dennis Buss [2,3,*]

[1] Tianjin University, School of Electronic Information Engineering, 92 Weijin Rd, Nankai Dist., Tianjin, China; E-Mails: jyzhao@tju.edu.cn (J.Z.); majg@tju.edu.cn (J.M.)
[2] Texas Instruments, Inc., 12500 TI Blvd., Dallas, TX 75243, USA; E-Mail: Yogesh.Ramadass@ti.com
[3] Massachusetts Institute of Technology, EECS Dept, 77 Mass Ave, Cambridge, MA 02139, USA; E-Mail: lang@mit.edu

* Author to whom correspondence should be addressed; E-Mail: buss@ti.com

---

Abstract: Devices that harvest electrical energy from mechanical vibrations have the problem that the frequency of the source vibration is often not matched to the resonant frequency of the energy harvesting device. Manufacturing tolerances make it difficult to match the Energy Harvesting Device (EHD) resonant frequency to the source vibration frequency, and the source vibration frequency may vary with time. Previous work has recognized that it is possible to tune the resonant frequency of an EHD using a tunable, reactive impedance at the output of the device. The present paper develops the theory of electrical tuning, and proposes the Bias-Flip (BF) technique, to implement this tunable, reactive impedance.

Keywords: energy harvesting; bias-flip; piezo-electric

---

1. Introduction

Figure 1 shows a schematic of a Piezoelectric (PZ) Energy Harvesting Device (EHD) that is the subject of this research. This structure is referred to as a cantilever structure, and is used to amplify the amplitude of the source vibration [1]. Previous work has shown that maximum output power is achieved when the cantilever has a high Q resonance at the frequency of the source vibration. However, the frequency of the source vibration is not usually matched to the resonant frequency of the EHD. The source vibration may vary with time. This paper addresses the problem of electronically tuning the PZ EHD to achieve maximum power in situations where the source vibration is not a stable frequency matched to the mechanical resonant frequency of the EHD.

Figure 1. Schematic of a Piezo-Electric (PZ) Energy Harvesting Device (EHD) based on the Cantilever Beam structure.

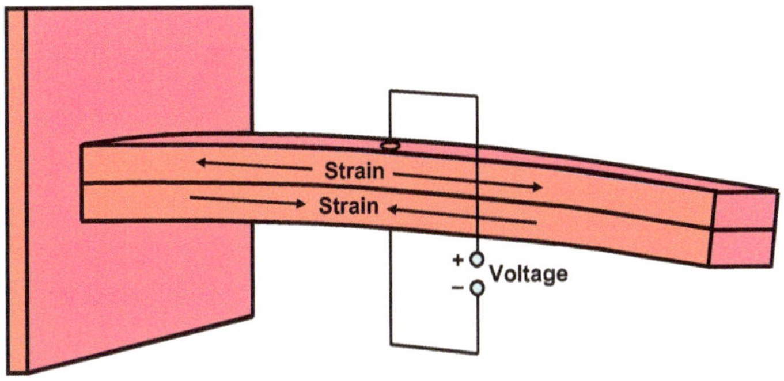

In the interest of simplicity, we will analyze the structure in Figure 2. The results achieved through analysis of this structure can be generalized to the cantilever structure through the addition of geometrical constants.

Figure 2. Schematic of the simplified EHD that is analyzed in this paper. A_p is the area of the PZ capacitor, and t_p is the thickness. Z is the complex amplitude of the source vibration, and X is the complex amplitude of the mechanical displacement of the mass M. This simplified model illustrates the concepts of electronic tuning that apply to the cantilever structure of Figure 1.

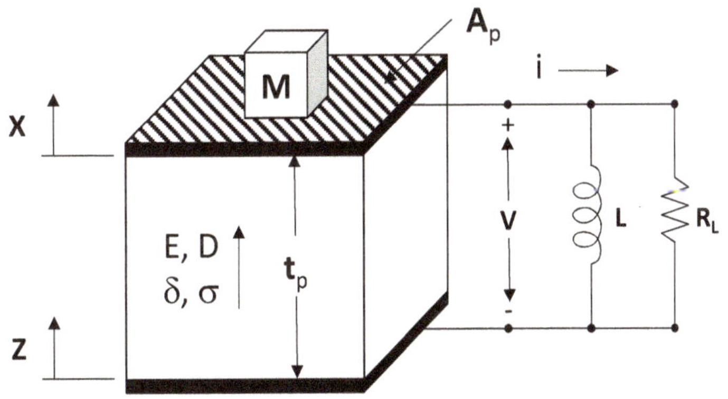

This paper describes three concepts for electrically tuning of PZ EHDs.

1. Use of voltage amplitude to tune the mechanical stiffness of the EHD;
2. Coupling of the mechanical resonator to an electrical RLC tank circuit;
3. Bias-Flip (BF) technique to emulate the large tunable inductor that is required for the RLC tank circuit.

These three concepts were introduced in summary form in [2]. In the succeeding sections of this paper, these concepts will be presented in more detail. Section 6 shows that BF can be used to effectively optimize the power output from a PZ EHD. In this paper, we have changed some of the notation that we used in [2], in order to conform to generally accepted usage.

In this paper, we will analyze the PZ EHD. However, many of the results and conclusions are equally applicable to electromagnetic and electrostatic EHDs. Cammarano *et al.* [3] have described concepts very similar to #1 and #2 above in the context of electro-magnetic EHDs.

2. Frequency Tuning by Voltage

The material equations for PZ material can be written as follows [1].

$$\delta = \frac{\sigma}{Y} + dE \tag{1}$$

$$D = \varepsilon E + d\sigma \tag{2}$$

The parameters are defined below.

$\delta =$ mechanical strain (displacement/length)
$\sigma =$ mechanical stress (force/area)
$Y =$ Young's Modulus (force/area)
$d =$ piezo-electric (PZ) coefficient (m/volt)
$\kappa^2 = \frac{Y}{\varepsilon} d^2$
$\rho = \frac{\kappa^2}{1-\kappa^2} =$ PZ coupling constant
$E =$ electric field (volt/m)
$D =$ electrical displacement (coulomb/m^2)
$\varepsilon =$ dielectric constant (coul/volt-m)
$k_m =$ mechanical (short-circuit) spring constant
$\omega_m = \sqrt{k_m / m} =$ mechanical (short-circuit) resonant frequency
$C_e = \varepsilon A_p / t_p =$ electrical ($\sigma = 0$) capacitance
$C_{mc} = C_e /(1+\rho) =$ motion constrained ($\delta = 0$) capacitance.
$L =$ inductance
$\omega_{mc} = 1/\sqrt{LC_{mc}} =$ motion constrained resonant frequency
$\eta =$ mechanical damping factor (force/velocity)
$Q_m = \frac{m\omega_m}{\eta} =$ mechanical Q-factor
$G_L = 1/R_L =$ load conductance
$G_L^N = \frac{G_L}{\omega_m C_{mc}} =$ normalized load conductance
$G_{in} = 1/R_{in} =$ internal conductance

$$G_{in}^N = \frac{G_{in}}{\omega_m C_{mc}} = \rho Q_m = \text{normalized internal conductance}$$

Refer to the device of Figure 2. When the output is shorted, E = 0, and the mechanical stiffness is given by Young's modulus, $k_m = A_p Y / t_p$. The short-circuit, resonance frequency is given by $\omega_m^2 = k_m / m$. However, when the output is in the open circuit condition, D = 0; the open-circuit stiffness is given by $k_{oc} = k_m(1+\rho)$, where ρ is the dimensionless PZ coupling constant, defined in Table 1. From this, it can be shown that the open-circuit and short-circuit resonant frequencies ω_{oc} and ω_m are related by the equation $\omega_{oc}^2 = \omega_m^2(1+\rho)$. This relationship is well-known, and has been used to experimentally determine the coupling constant ρ [1].

Similarly, we define the electrical capacitance $C_e = \varepsilon A_p / t_p$ for the case when there is no stress $(\sigma = 0)$; and we define the motion constrained capacitance $C_{mc} = C_e/(1+\rho)$ for the case when $\delta = 0$.

The equations for the PZ EHD shown in Figure 2 are given below

$$F = ma \Rightarrow F = -k_m X - k_m dV - j\omega\eta X = -\omega^2 m(X+Z) \tag{3}$$

$$i = -\frac{dQ}{dt} \Rightarrow -j\omega Q = -j\omega C_{mc} V + j\omega k_m dX = \left(\frac{1}{j\omega L} + \frac{1}{R_L}\right)V \tag{4}$$

These equations can be solved for $V(\omega)$ and $X(\omega)$ as shown below.

$$V = \frac{-\rho\omega^4 Z / d}{\left(\omega_m^2 - \omega^2 + \frac{j\omega\omega_m}{Q_m}\right)\left(\omega_{mc}^2 - \omega^2 + j\omega\omega_m G_L^N\right) - \rho\omega^2\omega_m^2} \tag{5}$$

$$X = \frac{\left(\omega_{mc}^2 - \omega^2 + j\omega\omega_m G_L^N\right)\omega^2 Z}{\left(\omega_m^2 - \omega^2 + \frac{j\omega\omega_m}{Q_m}\right)\left(\omega_{mc}^2 - \omega^2 + j\omega\omega_m G_L^N\right) - \rho\omega^2\omega_m^2} \tag{6}$$

When the source vibration frequency ω equals the mechanical resonant frequency ω_m, Equation (5) for output voltage reduces to a familiar form.

$$V = \frac{I_p}{G_L + G_{in} + j\omega_m C_{mc} + (j\omega_m L)^{-1}} = \frac{\rho Q_m Z / d}{G_L^N + G_{in}^N + j\left(1 - \frac{\omega_{mc}^2}{\omega_m^2}\right)} \tag{7}$$

where $G_{in} = \rho Q_m \omega_m C_{mc}$ and $I_p = G_{in} Z / d$.

This results in the familiar circuit model for the PZ EHD, shown at the left of Figure 3.

Figure 3. The circuit within the dashed box is the equivalent circuit that applies to a PZ EHD when $\omega = \omega_m$. At this frequency, the PZ current I_p is independent of the load. For $\omega \neq \omega_m$ this equivalent circuit does not accurately model device behavior, because the PZ current I_p changes as the load changes. The circuit at the right includes an inductor to cancel the capacitive reactance of the EHD.

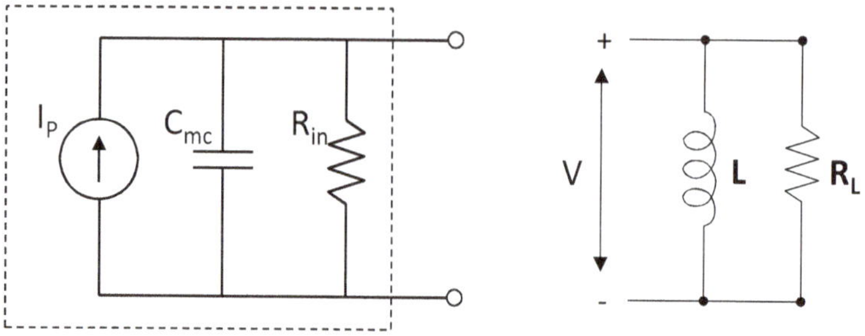

If a purely resistive load is connected to the EHD, the device capacitance C_{mc} degrades output power. As a result, an inductor (or an effective inductor) is added to the output circuit for the purpose of cancelling the capacitive admittance and achieving maximum average power to the load.

$$P_{\max}^{av} = \frac{1}{8}\left(\frac{Z}{d}\right)^2 G_{in} \tag{8}$$

In succeeding sections, simulations of voltage and output power are shown as a function of frequency. In [2] simulations were shown for $\rho = 0.2$; $Q_m = 50$; and $G_{in}^N = \rho Q_m = 10$. Throughout this paper, simulations are shown for $\rho = 0.05$; $Q_m = 20$; and $G_{in}^N = \rho Q_m = 1.0$. These values are more representative of today's commercial devices.

Figure 4. Voltage magnitude for the case of no inductor. Voltage is normalized to Z/d, the open-circuit voltage at $w=1$.

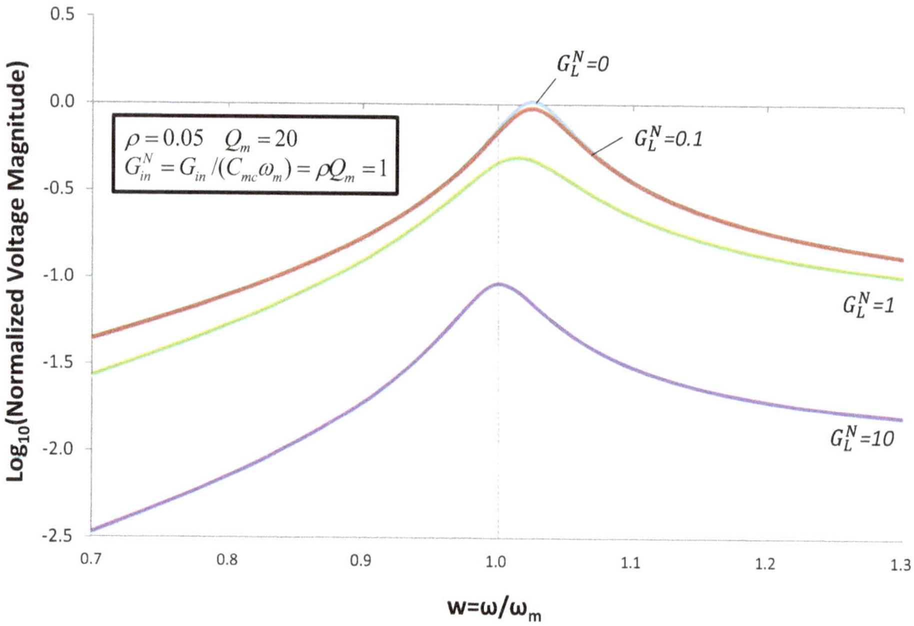

Figure 4 shows the magnitude of the output voltage as a function of frequency, calculated using Equation (5), for the case of no inductor ($\omega_{mc} = 0$). On the vertical axis, voltage is normalized to the maximum open-circuit voltage at the mechanical resonant frequency $V_{oc}^{\max} = Z/d$. Frequency, on the horizontal axis, is normalized to ω_m. For $w = \omega/\omega_m = 1$, Figure 4 confirms the predictions of Equation (7). For $G_L^N = 0$, the normalized voltage magnitude is 0.707; and for $G_L^N = G_{in}^N$, the normalized voltage magnitude is 0.447. However, for $w \neq 1$, Figure 4 shows interesting behavior, especially for $G_L^N \leq G_{in}^N$. For $G_L^N = 0$, the voltage shows a resonance at $\omega_{oc} = \omega_m\sqrt{1+\rho} = 1.025\omega_m$.

Figure 4 shows that, for large G_L^N, the voltage peaks at the mechanical or short-circuit resonance frequency ω_m. However, for small G_L^N, the device characteristics change. When $G_L^N = 0$, the peak voltage occurs at $\omega_{oc} = \omega_m\sqrt{1+\rho} = 1.025\omega_m$. This can be understood as follows. For large G_L^N, the electric field is effectively shorted. As G_L^N decreases, the electric field in the EHD increases, and alters the effective spring constant of the cantilever beam.

It is tempting to assume from the above that frequency tuning is possible only in the narrow range $\omega_m \leq \omega \leq \omega_{oc}$ [4]. However, as we will see below, the addition of a resonant electrical circuit allows the voltage to swing below zero and above V_{oc}, thereby enabling a wider tuning range.

3. Coupled Oscillators

In the simulations in the previous section, we did not attempt to cancel the reactive admittance of the PZ capacitor, and we observed the degradation in output voltage. Since $G_L^N = G_L/(\omega_m C_{mc}) = 1$ in the simulations above, the degradation is not large. However, in some cases, the PZ EHD has a large capacitance, which can substantially degrade output power at $\omega = \omega_m$. In Figure 5, we simulate Equation (5) for the case $\omega_{mc} = \omega_m$. This is equivalent to adding an inductor that cancels the reactive admittance of the capacitor at $\omega = \omega_m$. The inductor performs as expected at $\omega = \omega_m$. The output voltage equals $V_{oc}^{\max}$. Moreover, for large G_L^N, the output voltage continues to show a resonant peak at $\omega = \omega_m$; and the voltage falls away sharply away from $\omega = \omega_m$.

However, for $\omega \neq \omega_m$, Figure 5 shows a surprising result when G_L^N is small. Two peaks occur at $w_+ = \omega_+/\omega_m = 1.118$ and $w_- = \omega_-/\omega_m = 0.894$. These resonances result from the poles in the denominator of Equation (5) when $G_L^N = 0$.

$$\left(\omega_m^2 - \omega^2\right)\left(\omega_{mc}^2 - \omega^2\right) - \rho\omega^2\omega_m^2 = 0 \tag{9}$$

Solution to the pole-splitting equation above, for the case $\omega_{mc} = \omega_m$ and $\rho = 0.05$ gives the values of w_+ and w_- above. Note that, when $\omega_{mc} = 0$ (case of no inductor), the solutions of Equation (9) are $w_- = 0$ and $w_+ = \sqrt{1+\rho} = 1.025$. Pole splitting, which describes coupled modes of the mechanical and electrical resonators, occurs only for small G_L^N. When G_L^N is large, the electric field in the PZ material is screened, and coupling is suppressed.

Figure 5. Voltage magnitude for the case when an inductor of value $L = (\omega_m^2 C_{mc})^{-1}$ is added to the circuit. Voltage is normalized as in Figure 4.

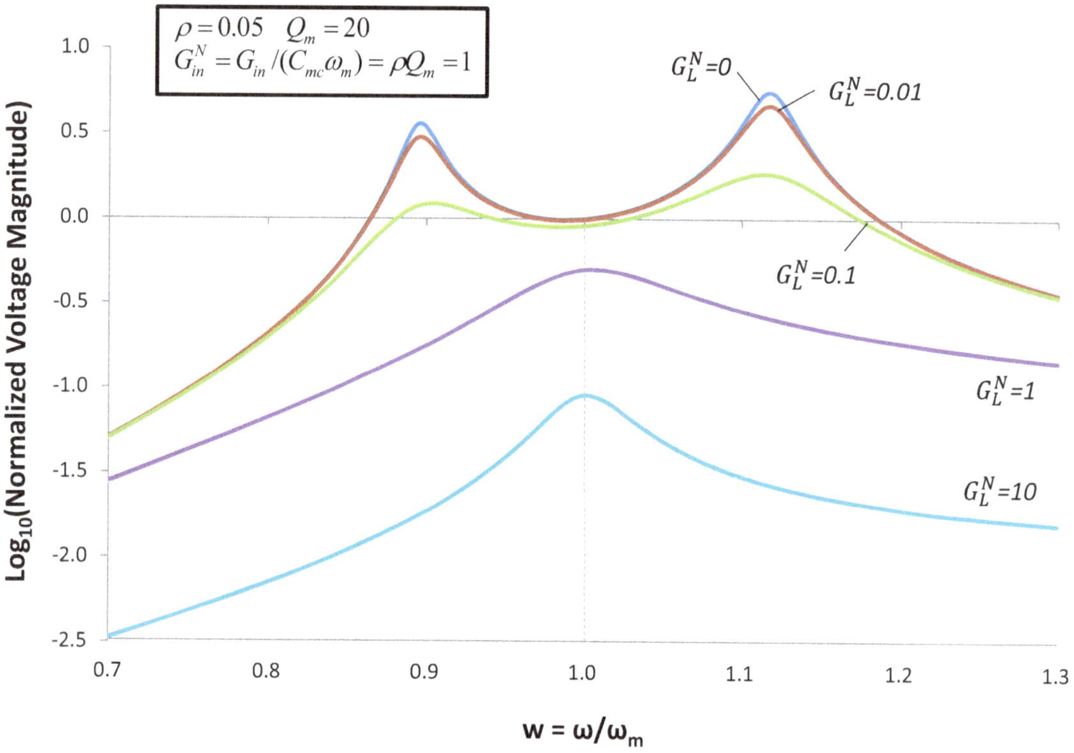

Figure 6. Roots of the Pole-splitting Equation (9). The normalized pole frequencies $w_+ = \omega_+ / \omega_m$ and $w_- = \omega_- / \omega_m$ are plotted *vs.* $w_{mc} = \omega_{mc} / \omega_m$.

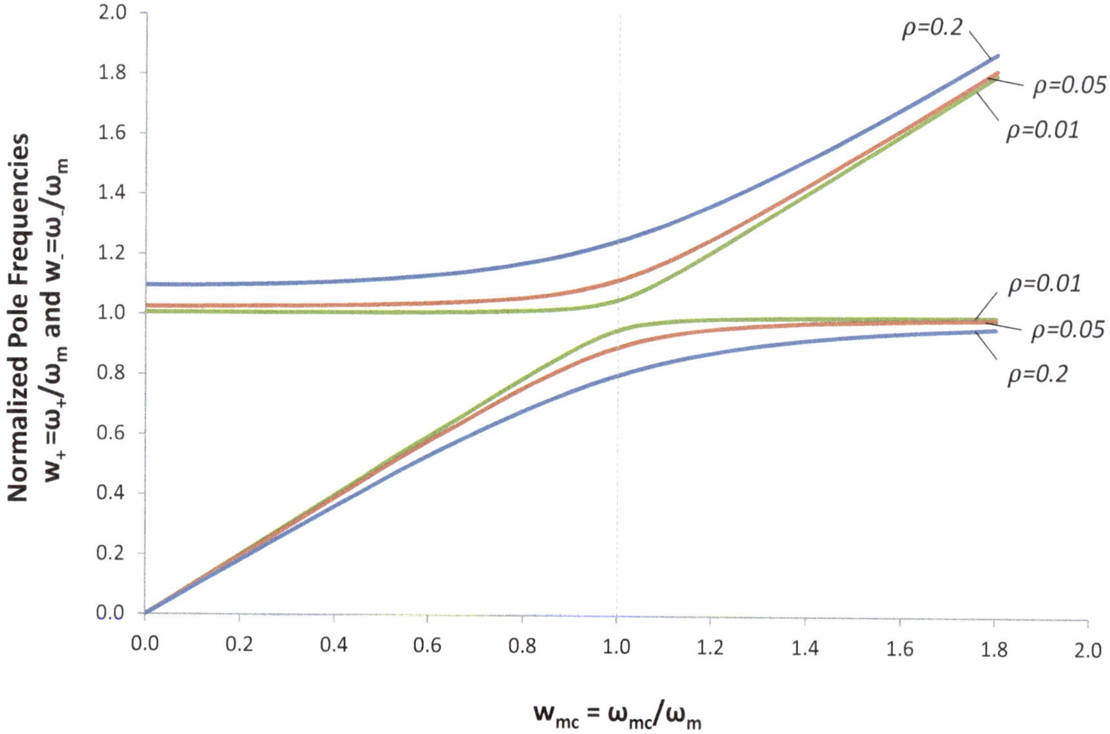

The roots of the pole-splitting equation are shown in Figure 6 for several values of coupling constant ρ. In Figure 5, we selected $\omega_{mc} = \omega_m$, in order to optimize output power at $\omega = \omega_m$. But, we discovered that, if we increased the load resistance, we could also optimize output voltage at w_+ and w_-. (We will show in the next section that output power is also optimized at these frequencies). This analysis suggests that we can tune the EHD resonant frequency by varying ω_{mc}.

We can gain further insight into the pole-splitting by returning to Equations (5) and (6). For small G_L^N, the following relationship holds between $X(\omega)$ and $V(\omega)$.

$$dV = \frac{-\rho\omega^2 X}{\omega_{mc}^2 - \omega^2 + j\omega\omega_{mc}G_L^N} \approx \frac{-\rho\omega^2 X}{\omega_{mc}^2 - \omega^2} \tag{10}$$

Using Equation (10) the force of the spring can be written as

$$F_{spring} = -k_m X - k_m dV = -k_m\left(1 - \frac{\rho\omega^2}{\omega_{mc}^2 - \omega^2}\right)X \tag{11}$$

The above equation shows that, by tuning L, we can vary ω_{mc} and vary the effective spring constant. When $\omega < \omega_{mc}$, $V(\omega)$ has a phase of 180° relative to $X(\omega)$; and the voltage reduces the effective spring constant. When $\omega > \omega_{mc}$, $V(\omega)$ has a phase of 0° relative to $X(\omega)$; and the voltage increases the effective spring constant. When $\omega = \omega_{mc}$, the magnitude of $X(\omega)$ is zero. If we adjust ω_{mc} such that the corresponding root of Equation (9) equals the source vibration frequency, then Equation (11) reduces to

$$F_{spring} = -k_m\left(\frac{\omega^2}{\omega_m^2}\right)X \tag{12}$$

Equation (12) shows that the effective spring constant can be tuned over a wide range.

It is also somewhat surprising that the peak voltages at ω_- and ω_+ are 3.5× to 5.5× higher than V_{oc}^{max}. The simulations in this paper assume that the source magnitude Z is held constant as the frequency changes. As a result, the input acceleration increases in proportion to ω^2, and $|V(\omega_+)| > |V(\omega_-)|$. The important result is that the peak voltages away from resonance can be somewhat higher than V_{oc}^{max}, and the higher voltage enables frequency tuning.

4. Optimizing Output Power

In the previous section, we showed that voltage can be made to peak at frequencies ω_- and ω_+, which are different from ω_m. Figure 7 shows that power is also maximized at these frequencies. The curve for $G_L^N = G_{in}^N$ indicates that output power peaks at $\omega = \omega_m$, at the power P_{max}^{av} given by Equation (8). The curve for $G_L^N = 10G_{in}^N$ shows a peak at $\omega = \omega_m$ at a degraded power level. The curve for $G_L^N = 0.1G_{in}^N$ shows two peaks at ω_- and ω_+ that have output power comparable to P_{max}^{av}. When G_L^N is further reduced to $G_L^N = 0.01G_{in}^N$, the output power at ω_- and ω_+ decreases.

Figure 7. Normalized average output power for the case when an inductor of value $L = (\omega_m^2 C_{mc})^{-1}$ is added to the circuit. Average power is normalized by $P_{\max}^{av}$ Equation (8).

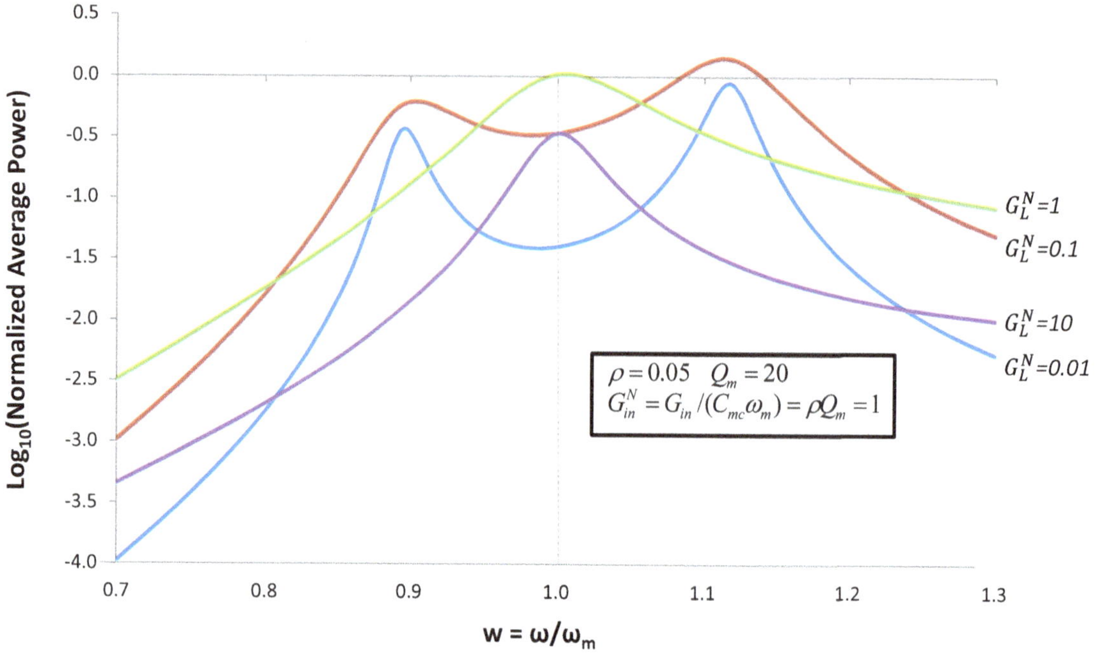

Figure 7 suggests that output power can be optimized at frequencies different from ω_m by adjusting the external inductor (or effective inductor). Equation (5) provides a general expression for $V(\omega)$. Varying the parameter $\omega_{mc} = (LC_{mc})^{-1}$ in Equation (5) is equivalent to varying the inductor. Maximizing power with respect to ω_{mc} gives the expression

$$\omega_{mc}^2 = \omega^2 + \frac{(\omega_m^2 - \omega^2)\omega^2\omega_m^2\rho}{(\omega_m^2 - \omega^2)^2 + \dfrac{\omega^2\omega_m^2}{Q_m^2}} \tag{13}$$

When we use Equation (13) to determine the value of ω_{mc} that optimizes power at each frequency, the resulting voltage and power are shown in Figures 8 and 9.

Equation (13) suggests that we need two strategies for optimizing power, depending on the source frequency ω. For frequencies near ω_m, (Region 2), Equation (13) reduces to

$$\omega_{mc}^2 = \omega^2 + (\omega_m^2 - \omega^2)\rho Q_m^2 \tag{14}$$

Note that, when $\omega = \omega_m$, Equation (14) reduces further to $\omega_{mc} = \omega_m$, which is equivalent to matching the capacitive admittance (refer to Figure 3). When ω is above or below ω_m (Regions 1 and 3), Equation (13) reduces to the pole-splitting Equation (9). Far from ω_m, output power is maximized at the pole frequencies [roots of Equation (9)]. However, as ω approaches ω_m, interaction between the poles shifts the max-power frequency [given by Equation (13)] slightly away from the pole frequency.

Figure 8. Normalized voltage magnitude for the case when the reactive admittance is optimized to give maximum output power using Equation (13). Plots of Voltage *vs.* frequency are given for different values of normalized load conductance G_L^N. These plots can be thought of as envelopes of curves such as Figure 5 for different values of ω_{mc}.

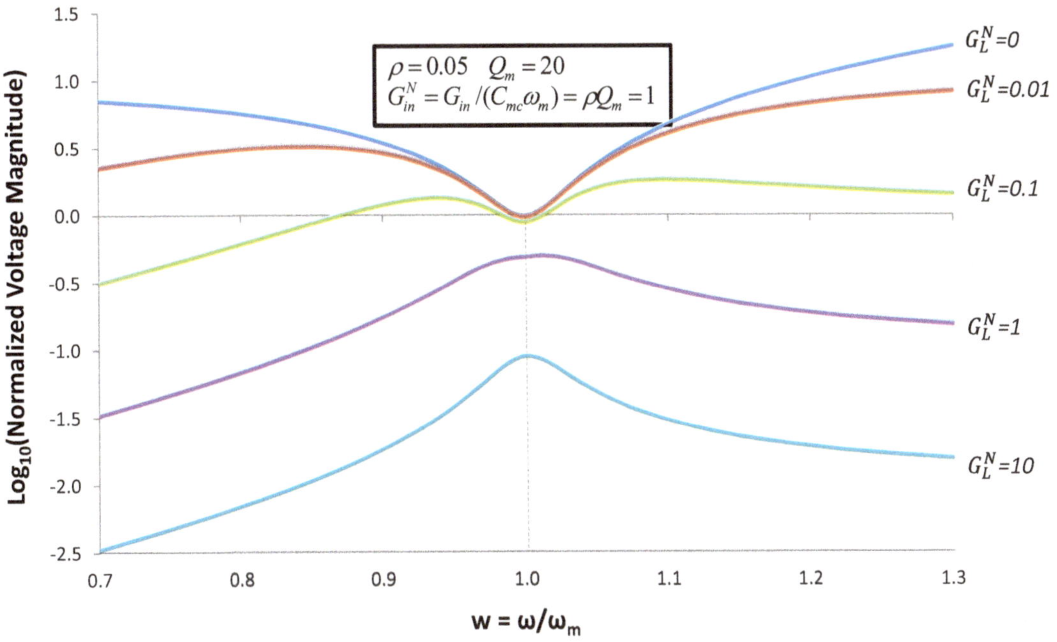

Figure 9. Normalized average power for the case when the reactive admittance is optimized to give maximum output power using Equation (13). Plots of power *vs.* frequency are given for different values of normalized load conductance G_L^N. Power increases with increasing ω because source acceleration increases, thereby increasing input power.

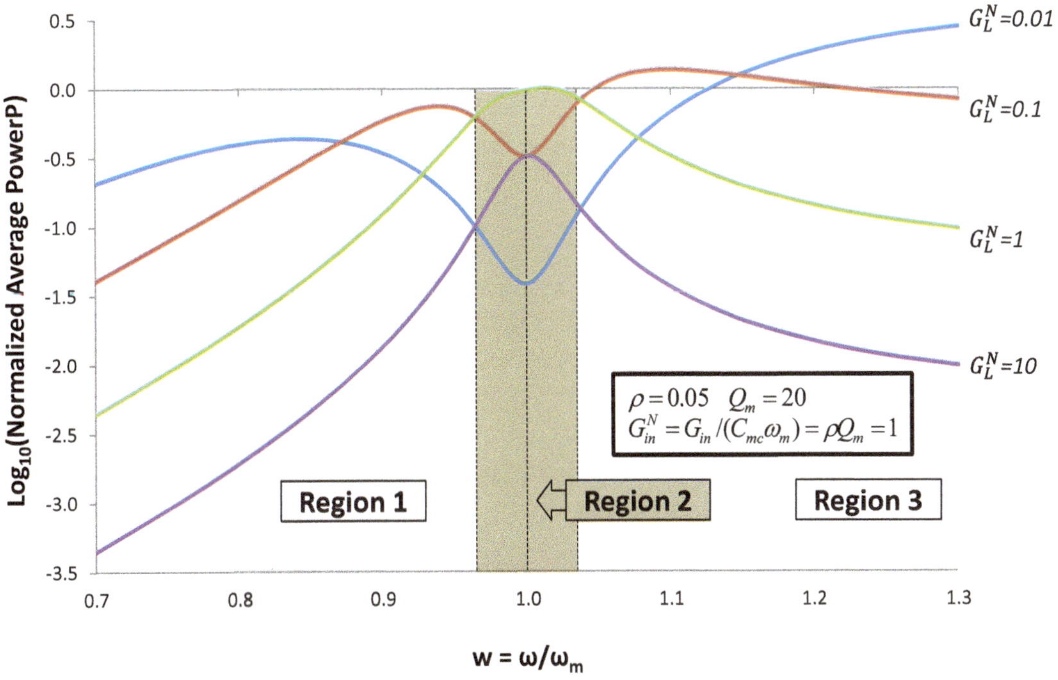

Within the region $\Delta w \approx \pm(2Q_m)^{-1}$ around the mechanical resonant frequency (Region 2), output power can be optimized by using (14) for reactive admittance and $G_L \approx G_{in}$. In Regions 1 and 3, power is optimized by using the pole splitting Equation (9) for reactive admittance and $G_L << G_{in}$. For the parameters used in this example ($Q_m = 20$), $\Delta w \approx 0.025$.

Cammarano *et al.* [3] have derived an equation very similar to Equation (13): Equation (8) in [3]. They observe that the power conditioning system at the output of the EHD can be used to synthesize the complex load impedance required by Equation (13), and they comment on the challenge of reducing the power of such systems. Chang *et al.* [5] have implemented a switch-mode power conditioning system to the output of a PZ EHD, and have demonstrated the ability to harvest energy from two sources simultaneously: $\omega = \omega_m$ and $\omega = 1.2\omega_m$.

5. Bias-Flip Technique

For a typical, discrete EHD, $C_{mc} \approx 100nF$, and the inductor required to match this reactance at 100 Hz is impractically large: $L \approx 25H$. However, it has been shown that the Bias-Flip technique can be used to synthesize a reactive impedance for effective impedance matching [6,7]. This technique is suitable to ULP miniaturization. It utilizes a very small inductor together with ULP microelectronics to emulate an inductor that is large and tunable. The BF technique has been shown to be effective in maximizing the output power of PZ EHDs at $\omega = \omega_m$ [7]. In this section, we will describe the Bias-Flip technique in the context of the equivalent circuit of Figure 3 describing a PZ EHD operating at the mechanical resonance frequency.

Figure 10. Operation of a Bias Flip (BF) Inductor. (**a**) A small inductor is connected to the output through ideal switches; (**b**) When the switchers are closed, the LC tank circuit begins to oscillate, when the switches are opened, half a period later, the sign of the voltage has been adiabatically "flipped"; (**c**) To achieve maximum power to the load, the bias is flipped when Ip changes sign; (**d**) The resulting voltage waveform is "in-phase" with the current.

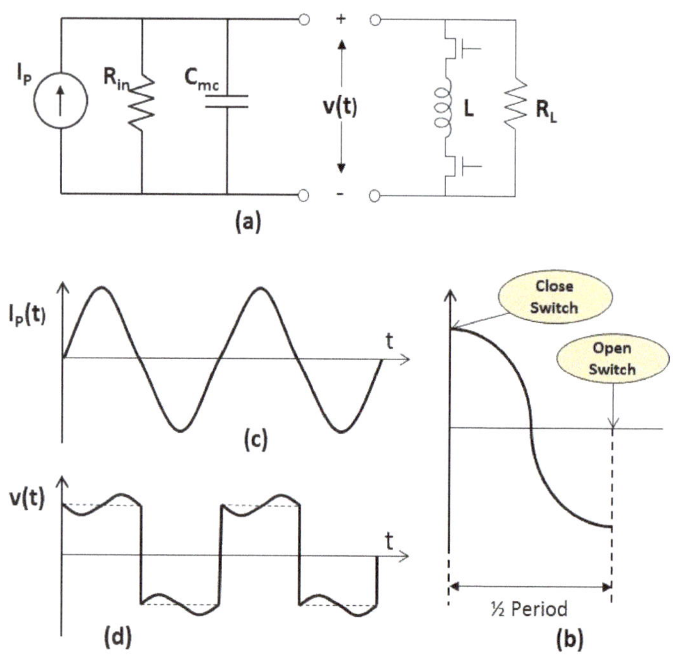

The BF technique is illustrated in Figure 10. In the BF circuit, the large inductor is replaced by a small inductor, connected by MOS switches. When the switches are closed, a high frequency tank circuit is formed. After ½ period of oscillation of this tank circuit, the switches are opened, and the voltage on the capacitor has "flipped" adiabatically from +V to −V. In this paper, the switches are assumed to be ideal and lossless.

Refer to Figure 3. When an ideal inductor is used together with a matched resistive load $R_L = R_{in}$, the maximum average output power is given by $P_{\max}^{av} = I_p^2/(8G_{in})$. [See Equation (8)] In Figure 11, we show how effective the ideal Bias-Flip circuit is in achieving maximum power.

Figure 11. Normalized Average Power delivered to the load at $\omega = \omega_m$. The equivalent circuit of Fig 3 is used with $R_L = R_{in}$. The dashed line shows the case of no inductor. The solid line shows the improvement in output power that is achievable using the Bias-Flip inductor. Both curves are normalized to the average power obtained using an ideal tuned inductor $P_{\max}^{av} = I_p^2/(8G_{in})$.

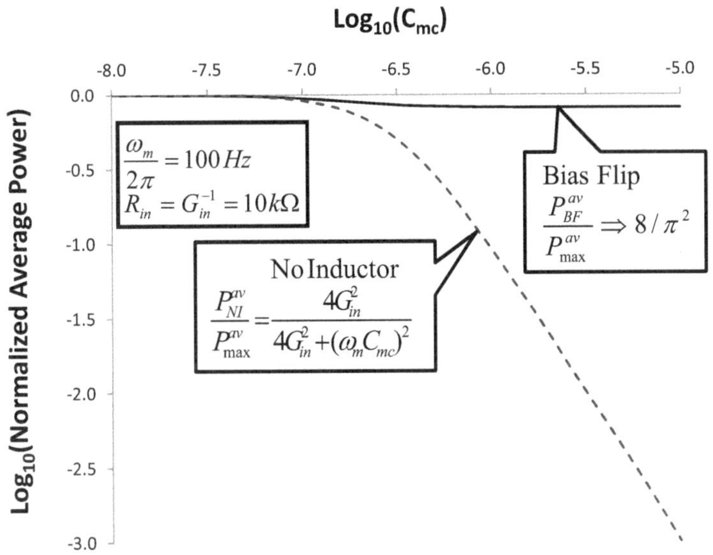

In the worst case of very large C_{mc}, the output power is degraded by several orders of magnitude, when no inductor is used. However, the Bias-Flip approach delivers power $P_{BF}^{av} = I_p^2/(\pi^2 G_{in})$, which is $8/\pi^2 - 81\%$ of the max power obtained using an ideal inductor. This illustrates the effectiveness of Bias-Flip circuits to achieve high output power when C_{mc} is large [7].

So far in this paper, we have discussed the case in which AC power is delivered to a resistive load. We have done this because the analysis can be performed in closed form. However, in many energy harvesting applications, it is necessary to rectify the AC power and store it in a battery or super-capacitor. The Bias-Flip technique is especially applicable to this case, as shown in [7].

The rectification circuit analyzed in [7] is shown in Figure 12. For simplicity, we assumed that the EHD is operating at the mechanical resonance frequency, and we use the equivalent circuit of Figure 3. The output AC voltage $v(t)$ is rectified in the diode bridge and stored on the capacitor C_{RECT} that is maintained at voltage V_{RECT} by the Energy Management Circuit. The analysis below assumes ideal diodes.

Figure 12. Circuit to rectify and store the AC power being generated by the EHD. It is assumed that the EHD is operating at the mechanical resonance frequency.

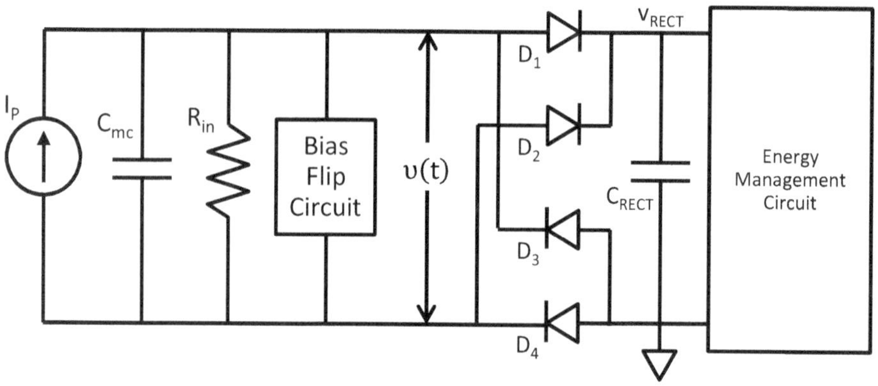

Operation of the Bias-Flip rectifier is described with reference to Figure 13.

Figure 13. (**a**) Voltage waveform $v(t)$ in Figure 12 for the case $C_{mc} = 0$; in this case, the Bias-Flip circuit is not required; (**b**) Voltage waveform for the case of large C_{mc} without Bias-Flip compensation; when the current becomes positive, there is a large negative voltage on the capacitor C_{mc}, this must be discharged before the voltage can swing positive; (**c**) When the current becomes positive, the polarity of the voltage $v(t)$ is "flipped". This reduces the time to diode turn-on and increases power transferred to C_{RECT}.

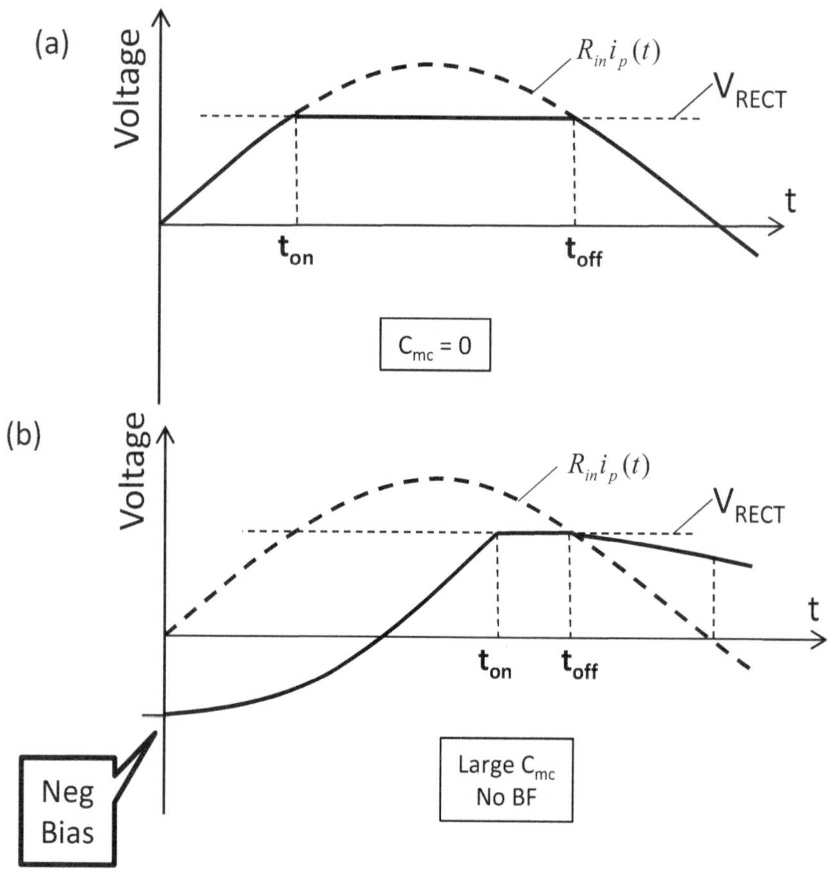

Figure 13. *Cont.*

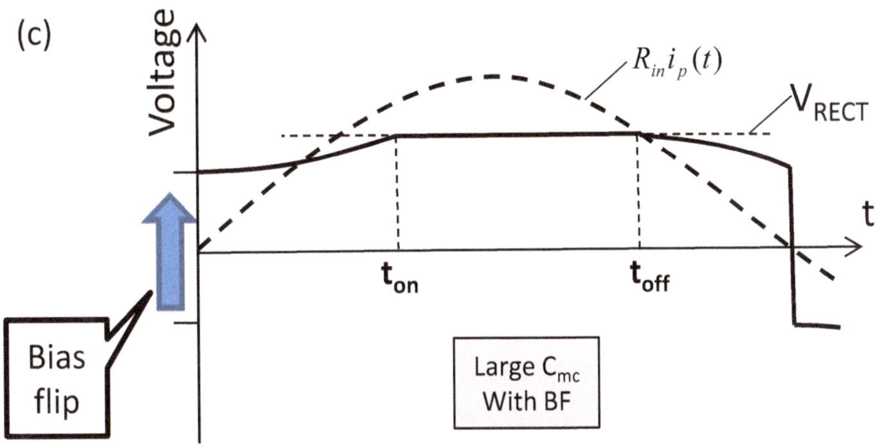

When the capacitance is zero, as shown in Figure 13a, Bias-Flip is not required. $v(t) = I_p R_{in} \sin(\omega t)$ until $t = t_{on}$; at which time, $v(t)$ becomes clamped at V_{RECT}. Between t_{on} and t_{off} power is supplied to the storage unit. At $t = t_{off}$, the diodes turn off, and $v(t)$ returns to zero following the curve $v(t) = I_p R_{in} \sin(\omega t)$. The presence of non-zero C_{mc} degrades transferred power: Figure 13b. When the current turns positive, there is a negative bias on C_{mc} that must be discharged before the voltage can swing positive. This delays diode turn-on, and forces a reduction in V_{RECT}, both of which degrade transferred power. This degradation can be corrected by adiabatically flipping the bias on C_{mc} when the current changes sign, as illustrated in Figure 13c.

Figure 14. Power transferred to the storage capacitor C_{RECT} as a function of V_{RECT} for different values of C_{mc}. Power is normalized to $P_{\max}^{av}$ (see Equation (8)). The curve for $C_{mc} = 0$ is negligibly different from the curve for $C_{mc} = 10nF$, and is not shown. For the case of no capacitor, the max power transfer of 0.92 occurs at $V_{RECT} = 0.4 I_p R_{in}$.

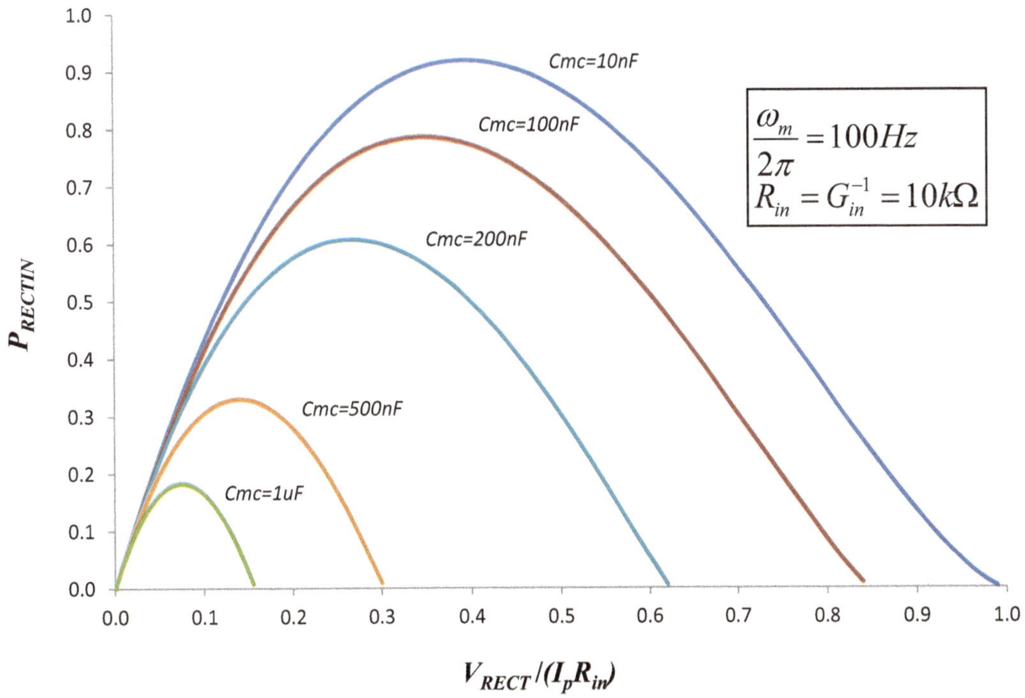

The energy transferred per cycle depends on V_{RECT}. When V_{RECT} is low, the power transfer interval $t_{off} - t_{on}$ is long, but the power is low. When V_{RECT} is high, the power transfer is high, but the transfer interval is short. In fact, for V_{RECT} above a maximum value the diodes do not turn on, and no power is transferred. Figure 14 shows the power transfer as a function of V_{RECT}, for various values of C_{mc}. This simulation is made using the values $R_{in} = 10k\Omega$ and $\omega_m/(2\pi) = 100Hz$.

Figure 15. Shows the rectified power as a function of C_{mc}. For each point on the curve, V_{RECT} was selected to give the max power transfer. Power is normalized to $P_{\max}^{av}$, and capacitance has units Farads.

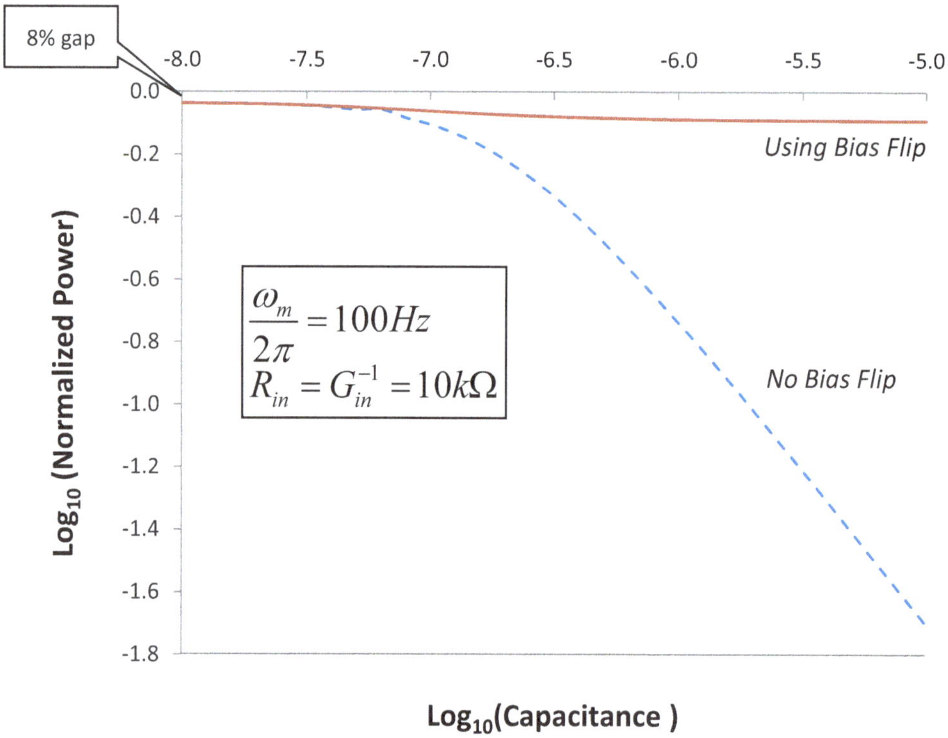

Figure 15 shows that, for large C_{mc}, output is severely degraded. However, the Bias-Flip circuit is effective in recovering most of the lost output power. This analysis confirms the conclusions of Ramadass and Chandrakasan [7] that for an EHD, operating at resonance, the Bias-Flip circuit is effective in canceling the reactive impedance of the device, and achieving near-optimum output power. In the next section, we will demonstrate that the Bias-Flip technique can be used to form an effective LC tank circuit that, when coupled to the EHD can tune the resonant frequency.

6. Bias Flip for Frequency Tuning

In Section 5, we confirmed the effectiveness of the BF technique for power optimization at the mechanical resonant frequency. When $\omega = \omega_m$, and the equivalent circuit of Figure 3 applies, we can optimize power to the load by "canceling" or "matching" the reactive admittance of the capacitor with an inductor. We select an inductor value such that $\omega_{mc} = (LC_{mc})^{-1/2} = \omega_m$. Matching the reactive admittance aligns the phase of the voltage across the load with the phase of the current. This works because the current source is assumed to be ideal. Varying the reactive load does not change the current from the source. We confirmed the finding of [7] that the BF inductor is effective in canceling capacitive admittance at ω_m. In this section, we will demonstrate that the BF inductor can also be used to tune the resonant frequency of the EHD and optimize power at frequencies substantially different from ω_m.

In order to maximize output power at any frequency, we need to maximize the input power delivered from the mechanical source to the EHD. In other words, we need to align the phase of the force with the phase of the source velocity.

In the following analysis, we assume the phase of $z(t)$ to be zero. $z(t) = Z\cos(\omega t)$, and the velocity of the source is $dz/dt = -\omega Z\sin(\omega t)$. The source velocity has a phase of +90°. The force acting on the EHD is given by Equation (3). Our goal is to maximize.

$$P_i^{av} = \frac{\omega}{2\pi}\int F dz = \frac{1}{2}\omega F_I Z \tag{15}$$

Where F_I is the imaginary part of F. $F(t) = F_R\cos(\omega t) - F_I\sin(\omega t)$. From this, we conclude that input average power is maximized when F has phase +90°, matched to source velocity, and X has phase −90°.

Figure 16. Phase of mechanical displacement $X(\omega)$ for three cases: (1) No inductor; (2) Inductor optimized using (13) and (3) Inductor optimized using the pole-splitting Equation (9). In all three cases, $G_L = 0$.

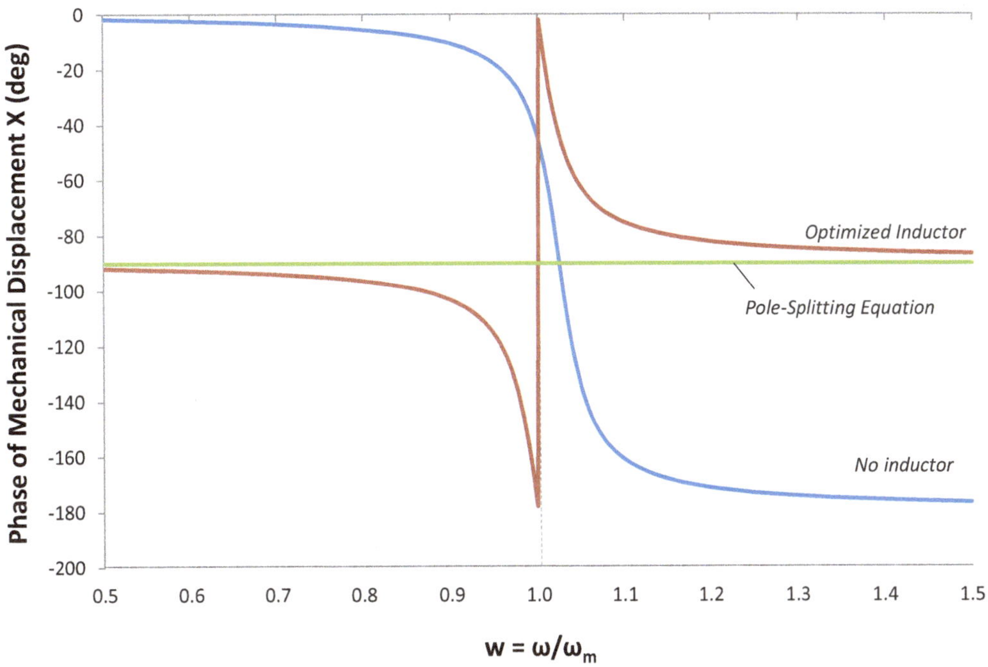

Additional insight into maximization of output power is seen in Figure 16. Here, we compare the phase of mechanical displacement $X(\omega)$ for three conditions

1. No inductor. The phase is −90° only at $\omega = \omega_{oc} = \omega_m\sqrt{1+\rho}$. Below ω_m, the phase is ≈0°, and above ω_m, the phase is ≈ −180°. Only at $\omega = \omega_{oc}$ is the force in phase with the source velocity;
2. Inductor, optimized using Equation (13). Note that the phase approaches −90° above and below $\omega = \omega_m$;

3. Inductor optimized using the pole-splitting Equation (9). The phase is −90° for all frequencies.

The improvement in power at frequencies above and below ω_m results from phase alignment between force and source velocity.

Figure 17 shows output power for 3 conditions.

1. No inductor: $G_L^N = G_{in}^N = 1$;
2. Inductor optimized using Equation (13): $G_L^N = G_{in}^N = 1$;
3. Inductor optimized using Equation (13): $G_L^N = 0.1$.

Figure 17. Normalized Average Output Power for three cases. (1) No inductor & $G_L^N = G_{in}^N = 1$; (2) Inductor optimized using Equation (13) & $G_L^N = G_{in}^N = 1$; (3) Inductor optimized using Equation (13) & $G_L^N = 0.1$.

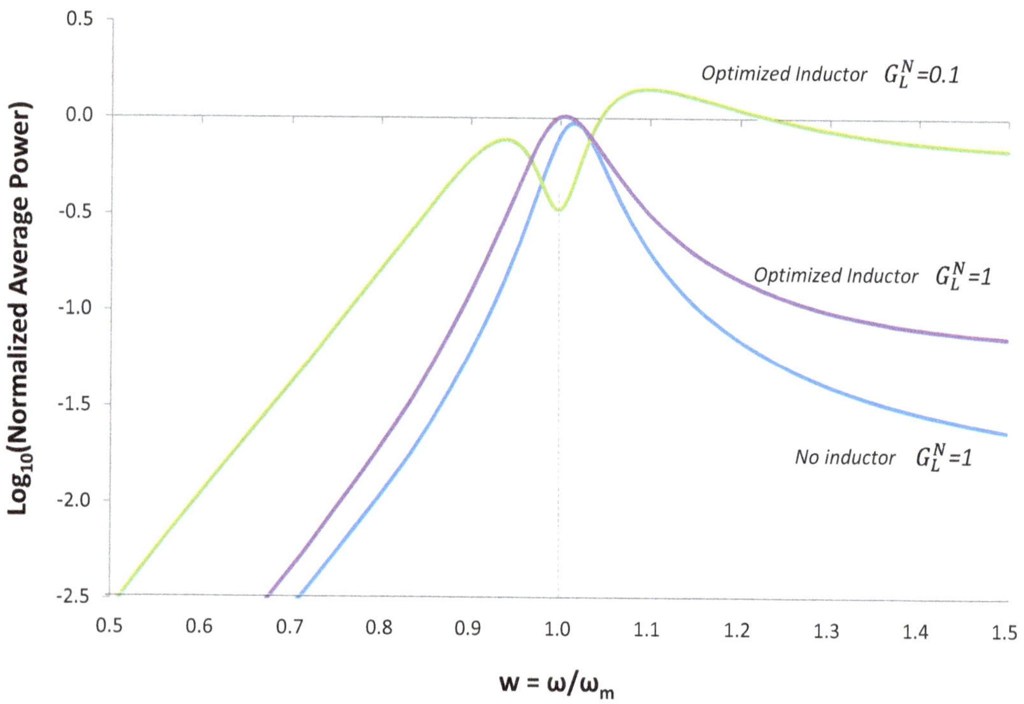

Case #2 illustrates the case where the reactive admittance is chosen to optimize output power, but the load conductance G_L^N is kept at $G_L^N = G_{in}^N = 1$. Very little improvement is achieved, because the voltage is kept low by the high load conductance, and the voltage is ineffective in modulating the cantilever spring constant. Case #3 shows power improvement of ~50X compared to case #1. When we compensate for the increase in acceleration with frequency, case #3 demonstrates that it is possible to achieve output power at $\omega \neq \omega_m$ that is comparable to the maximum power at ω_m.

Additional insight into the mechanism of frequency tuning can be obtained by transforming the mechanical equations of motion to an equivalent circuit [1,8,9], as shown in Figure 18.

Figure 18. (**a**) Equivalent circuit model for a PZ EHD in which $L_m \Leftrightarrow m$, $R_m \Leftrightarrow \eta$, $C_m = k_m^{-1}$, $V_F \Leftrightarrow m\omega^2 Z$, $I_s(\omega) \Leftrightarrow S(\omega) = j\omega X(\omega)$, and $A \Leftrightarrow k_m d$ The subscript m denotes the mechanical circuit. $S(\omega)$ is used for velocity to avoid confusion with voltage; (**b**) Equivalent circuit model of Figure 18a, in which the electronic circuit is replaced by an electrical impedance Z_e.

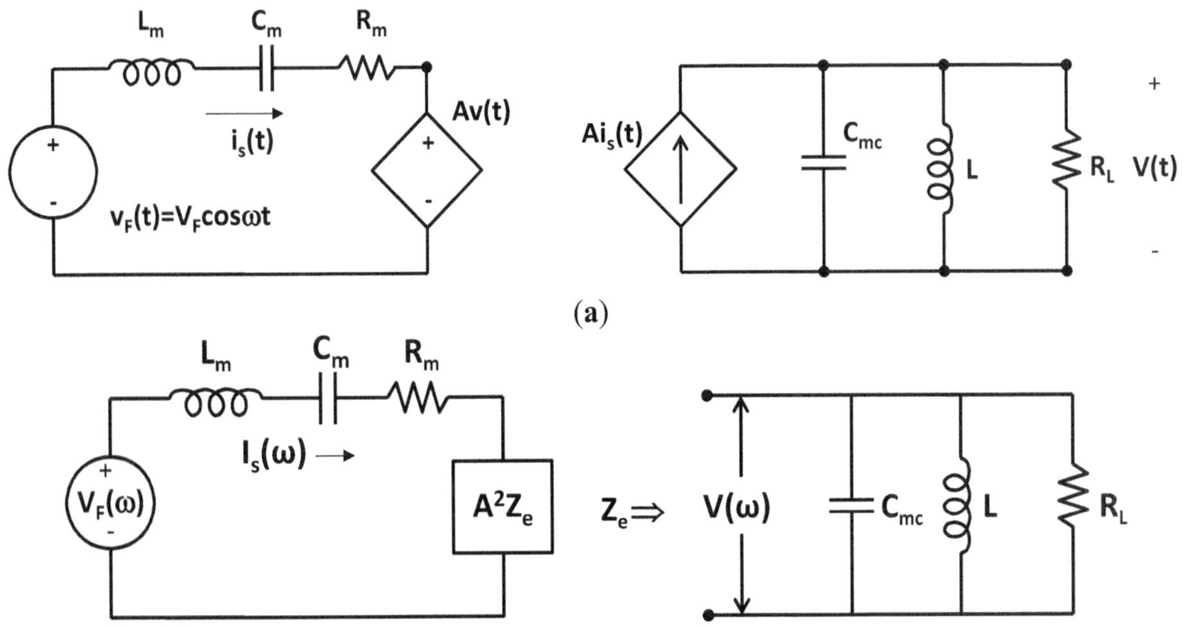

(**b**)

The equations for the mechanical portion of the equivalent circuit are shown below.

$$V_F = \left(j\omega L_m + \frac{1}{j\omega C_m} + R_m + A^2 Z_e(\omega) \right) I_s \Leftrightarrow m\omega^2 Z = \left(j\omega m + \frac{k_m}{j\omega} + \eta \right)(j\omega X) + k_m dV \qquad (16)$$

where Z_e, defined in Figure 18, is given by

$$Z_e = \frac{1}{j\omega C_{mc} + 1/(j\omega L) + 1/R_L} = \frac{j\omega L}{1 - \omega^2/\omega_{mc}^2 + j\omega L/R_L} \qquad (17)$$

Define Z_m to be the impedance seen by the voltage source $V_F(\omega)$ in Figure 18b. Setting $\mathrm{Im}(Z_m) = 0$ gives the pole-splitting equation, equivalent to Equation (9), in the limit $R_L >> \omega L$.

$$\mathrm{Im}(Z_m) = \omega L_m - \frac{1}{\omega C_m} + \frac{\omega A^2 L\left(1 - \omega^2/\omega_{mc}^2\right)}{\left(1 - \omega^2/\omega_{mc}^2\right)^2 + \left(\omega L/R_L\right)^2} = 0 \qquad (18)$$

The last term in the above equation can be used to tune the resonance frequency above or below the mechanical resonance frequency. The last term takes the form

$$\frac{j\omega A^2 L(1 - \omega^2/\omega_{mc}^2)}{(1 - \omega^2/\omega_{mc}^2)^2 + (\omega L/R_L)^2} = j\omega L_{eff} \qquad (19)$$

where L_{eff} can be made positive or negative to add or subtract from L_m. At the mechanical resonance frequency $\omega = \omega_m$, $\mathrm{Im}(Z_m) = \omega L_m - \frac{1}{\omega C_m} = 0$, and output power is optimized by setting $\omega = \omega_{mc}$. When $\omega < \omega_m$ the resonant frequency ω_- must be reduced. This can be achieved by increasing the effective inductance, which can be achieved by setting $\omega_{mc} > \omega$. When $\omega > \omega_m$, the resonant frequency ω_+ must be increased. This can be achieved by decreasing the effective inductance, which can be achieved by setting $\omega_{mc} < \omega$.

These results are summarized in Table 1.

Table 1. Strategy For Power Optimization in the Three Frequency Regions of Operation.

Frequency region	L_{eff}	ω_{mc}	Phase of voltage	Optimum power
Region 1: $\omega < \omega_m$	>0	$\omega_{mc} > \omega$	~ +90°	$R_L >> R_{in}$
Region 2: $\omega \approx \omega_m$	≈0	$\omega_{mc} = \omega \approx \omega_m$	~0°	$R_L = R_{in}$
Region 3: $\omega > \omega_m$	<0	$\omega_{mc} < \omega$	~ −90°	$R_L >> R_{in}$

Maximizing power in the three regions can be envisioned in term of an effective inductor. Alternatively, it can be envisioned in terms of setting the phase of the voltage $V(\omega)$. The phase in each of the three regions is given in the table. In Region 2, the voltage across the load $V(\omega)$ is in phase with $I_s(\omega)$, and output power is optimized by setting $R_L = R_{in}$. However, in Regions 1 and 3, max power occurs when $V(\omega)$ has a phase of +90° (Region 1) and −90° (Region 3) relative to the phase of $I_s(\omega)$, and optimum power is achieved by setting R_L to be substantially larger than R_{in}.

The foregoing analysis suggests that the Bias-Flip technique can be used to synthesize an inductor, by flipping the polarity of the voltage in such a way that $X(\omega)$ has a phase of −90° in all three regions of operation. Recall from Equation (10) that, when G_L^N is small, the phase of $X(\omega)$ is related to the phase of $V(\omega)$ in a simple way. By adjusting the phase of $V(\omega)$ as shown in Table 1, we also adjust the phase of $X(\omega)$ to −90°.

Simulations were performed starting from the differential equations for $v(t)$ and $x(t)$ that are comparable to Equations (3) and (4), for the case of no inductor. These equations were solved subject to the boundary conditions.

$$v(t = T/2) = +v(t = 0) \quad x(t = T/2) = -x(t = 0) \text{ and } \frac{dx}{dt}(t = T/2) = -\frac{dx}{dt}(t = 0) \tag{20}$$

In the case of no Bias-Flip, $v(t = T/2) = -v(t = 0)$. The effect of the BF inductor is to change the phase of $v(t)$ every half-period.

Using the voltage waveform, we calculated average output power. This is shown normalized to $P_{\max}^{av}$ in Figure 19. These simulations show that the BF technique is capable of achieving output power, comparable to the optimum power achievable with an optimized inductor. Moreover, the BF technique is self-tuning. If the bias is flipped whenever the source velocity crosses zero (as assumed in this simulation), the desired phase is maintained as the source frequency changes. No calculation is required to solve Equation (13).

Figure 19. Normalized average power as a function of frequency. In Regions 1 and 3, the Bias-Flip technique (red and blue dashed lines) improves output power by ~100X compared to the case of no inductor and no BF (green line). Moreover, it gives output power that is comparable to the maximum power achievable with an optimized inductor (red and blue solid lines).

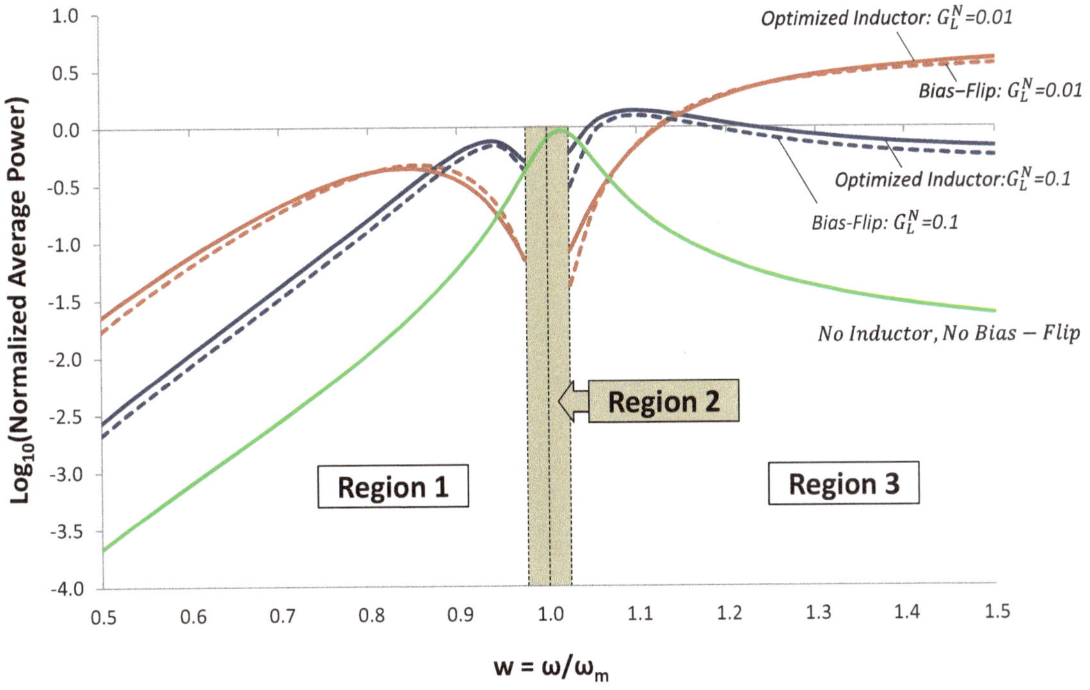

Analysis of the voltage waveforms reveals another aspect of self-tuning. In Region 1, the bias flips from negative to positive at $t = 0$ and from positive to negative at $t = T/2$, thereby emulating a +90° phase shift. In Region 3, the reverse happens. The bias flips from positive to negative at $t = 0$ and from negative to positive at $t = T/2$, thus emulating a −90° phase shift.

7. Conclusions

In the preceding sections, we have explained the principles for electrically tuning of PZ EHDs. These principles are summarized below.

Equation (11) shows that the effective spring constant of the mechanical resonator is a function of voltage. If the load conductance is large, the voltage is kept small, and the resonator responds only at the mechanical resonant frequency ω_m. However, for small load conductance G_L, the voltage can be used to tune the spring constant, and the resonant frequency of the mechanical oscillator.

In Regions 1 and 3, output power is maximized by maximizing input power (force x velocity), transferred from the source to the EHD. At frequencies below ω_m (Region 1), this occurs when the phase of the voltage is +90° relative to the source vibration, and at frequencies above ω_m (Region 3), output power is optimized when the phase of the voltage is -90° relative to the source vibration.

This optimum phase relationship can, in theory, be achieved using a tunable inductor, whose value can be obtained from Equation (13). A large tunable inductor is not generally practical. However, the Bias-Flip technique can be used to emulate a large, tunable inductor. Previous work has shown that the

BF technique can be used to optimize the output power at ω_m, by cancelling the capacitive admittance of the EHD. In this work, we have shown how the BF technique can be used to tune an EHD and harvest energy from frequencies other than the mechanical resonance frequency.

Acknowledgments

The authors acknowledge informative exchanges with Samuel Chang of MIT.

Conflict of Interest

The authors declare no conflict of interest.

References

1. Roundy, S.; Wright, P.K.; Rabaey, J.M. *Energy Scavenging for Wireless Sensor Networks, with Special Focus on Vibrations*; Kluwer Academic Publishers: Norwell, MA, USA, 2004.
2. Zhao, J.; Ramadass, Y.; Buss, D.; Ma, J. Microelectronic Techniques for Frequency Tuning of Piezo-Electric Energy Harvesting Devices. In Proceedings of the 2012 IEEE Subthreshold Microelectronics Conference (SubVt), Waltham, MA, USA, 9–10 October 2012.
3. Cammarano, A.; Burrow, S.G.; Barton, D.A.; Carella, A.; Clare, L.R. Tuning a resonant energy harvester using a generalized electrical load. *Smart Mater. Struct.* **2010**, *19*, 1–7.
4. Wu, W.-J.; Chen, Y.-Y.; Lee, B.-S.; He, J.-J.; Peng, Y.-T. Tunable Resonant Frequency Power Harvesting Devices. In *Smart Structures and Materials*: *Damping and Isolation*; Clark, W.W., Ahmadian, M., Lumsdaine, A., Eds.; SPIE: San Diego, CA, USA, 2006; Volume 6169, pp. 61690A-1–61690A-8.
5. Chang, S.C.; Ocalan, M.; Pabon, J.; Lang, J.H. Harvesting Energy from Multi-Harmonic Vibrations with a Tunable Electric Load. In proceedings of PowerMEMS, Atlanta, GA, USA, 2–5 December 2012; pp. 72–75.
6. Guyomar, D.; Badel, A.; Lefeuvre, E.; Richard, C. Toward energy harvesting using active materials and conversion improvement by nonlinear processing. *IEEE Trans. Ultrason. Ferroelectr. Freq. Control* **2005**, *52*, 584–595.
7. Ramadass, Y.; Chandrakasan, A. An Efficient Piezoelectric Energy Harvesting Interface Circuit Using a Bias-Flip Rectifier and Shared Inductor. In Proceedings of the IEEE International Solid-State Circuits Conference, San Francisco, CA, USA, 8–12 February 2009.
8. Toh, T.T; Mitcheson, P.D.; Dussud, L.; Wright, S.W.; Holmes, A.S. Electronic Resonant Frequency Tuning of a Marine Energy Harvester. In Proceedings of the PowerMEMS, Seoul, Korea, 15–18 November, 2011; pp. 383–386.
9. Chang, S.C. Harvesting Energy from Non-Ideal Vibration Sources. Ph.D. Thesis, Massachusetts Institute of Technology, Cambridge, MA, USA, May 2013.

4

Exploration of Charge Recycling DC-DC Conversion Using a Switched Capacitor Regulator

Kaushik Mazumdar * and Mircea R. Stan

Charles. L. Brown Department of Electrical and Computer Engineering, University of Virginia, Charlottesville, VA 22904, USA; E-Mail: mircea@virginia.edu

* Author to whom correspondence should be addressed; E-Mail: km3sj@virginia.edu

Abstract: The increasing popularity of DVFS (dynamic voltage frequency scaling) schemes for portable low power applications demands highly efficient on-chip DC-DC converters. The primary aim of this work is to enable increased efficiency of on-chip DC-DC conversion for near-threshold operation of multicore chips. The idea is to supply nominal (high) off-chip voltage to the cores which are then "voltage-stacked" to generate the near-threshold (low) voltages based on Kirchhoff's voltage law through charge recycling. However, the effectiveness of this implicit down-conversion is affected by the current imbalance among the cores. The paper presents a design methodology and optimization strategy for highly efficient charge recycling on-chip regulation using a push-pull switched capacitor (SC) circuit. A dual-boundary hysteretic feedback control circuit has been designed for stacked loads. A stacked-voltage domain with its self-regulation capability combined with a SC converter has shown average efficiency of 78%–93% for 2:1 down-conversion with I_{Load} (max) of 200 mA and workload imbalance varying from 0–100%.

Keywords: voltage stacking; charge recycling; DC-DC converter; switched capacitor; energy efficiency; near-threshold; hysteresis feedback; ripple; phase interleaving

1. Introduction

With the industry focusing more and more on SoC (system on chip) solutions for the low power portable applications, the role of efficient power management has become very crucial. An emerging trend for lowering energy is to scale the supply voltage V_{dd} to the near threshold region which brings not only quadratic dynamic energy savings, but also super-linearly reduced leakage currents. This shift in design trend, popularly known as near threshold computing (NTC) has been proposed to improve energy efficiency of the system, but this comes at the cost of severe degradation in performance [1]. Therefore multicore computing has been combined with NTC design to exploit parallelism and improve throughput while lowering energy/operation. However with the supply voltage being scaled down and the chip power density increasing with the number of cores, unsustainable increases in current density become a major design challenge. This leads to higher IR drop and I^2R power loss in the board/package resistance negating the effective energy gains.

To balance performance/energy, on-chip DVFS has been widely used to provide just-as-needed V_{dd}. With this approach, lower V_{dd} can be used for slower parts of the die, and higher V_{dd} for the high performance section of the chip. By bringing the DC-DC converter module closer to the processor and with boosted external voltage and local on-chip regulation, the current in the off-chip package and board parasitic can be reduced. However large variations in performance requirements and workload characteristics make on-chip DC-DC converter design very challenging. Also, while making claims about energy savings in the cores, often the energy overheads to generate such low V_{dd} power supply voltages are neglected [2]. While this can be a somewhat fair assumption for relatively narrow ranges, e.g., between full V_{dd} to 2/3 V_{dd} (as efficiency can be more than 90% in such cases), down conversion from V_{dd} to 1/2 V_{dd} or less can consume considerable overhead, which reduces the potential overall energy savings—under certain conditions, the overall power consumption might even increase.

Existing highly efficient regulator structures (like Buck-Boost architecture [3]) are not a good option for on-chip regulation as they need big inductors (also with small losses) which cannot be integrated fully on-chip. Low dropout (LDO) regulators, which are more commonly used for on-chip DC-DC conversion, enable fast voltage transition for multiple voltage domains. However LDOs suffer from low conversion efficiency, restricting their use for high step-down ratios. Switched capacitor (SC) converters are easier to integrate on-chip, but their output impedance varies with load and are not highly efficient with varying workloads [4]. Thus the benefits of wide range DVFS are negated by the lack of efficient conversion techniques. This bottleneck has been referred to as the *on-chip power regulation efficiency wall* (where the relatively poor efficiencies achievable with on-chip regulators limit the effectiveness of many low power schemes that depend on on-chip regulation) [5].

1.1. Voltage Stacking Idea

Voltage stacking simply refers to the power delivery arrangement of two or more circuit blocks such that the ground of one block becomes the power connection for the next, thus the blocks being connected as a *series* stack for power delivery, with all of them sharing the same current (thus the *charge being recycled* in the stack), while their V_{dd} values are added. With this internal recycling, for almost the same power consumption, the current drawn through the supply will be reduced to $1/n$ of conventional parallel loads (n is the number of cores stacked) [5–7]. This will also reduce the off-chip I^2R power loss by a factor of n^2 and IR drop by a factor of n. The efficiency of this technique depends on the current mismatch among the stacked domains. In the best-case scenario when all the domains are perfectly balanced, IR noise is the lowest and regulation efficiency the highest, while the more the imbalance, the less the efficiency. Thus there is an implicit self-regulating loop that controls the internal voltage distribution and unless the current consumption across the layers is balanced, it can contribute to internal voltage noise [6].

In this work we explore voltage stacking as an alternative technique for on-chip DC-DC conversion for NTC. We discuss in depth the mechanism behind this highly efficient power delivery technique as well as the challenges. Optimization strategies were developed to improve the regulation performance and a novel dual-boundary hysteresis based feedback circuit was designed for increasing the efficiency. With simulations we show how performance can actually benefit from the absence of feedback circuitry for low power loads. We also compare our proposed push-pull SC converter with traditional SC regulators to demonstrate a performance improvement. The paper focuses on the simplest possible case of only two cores using voltage stacking, but the same methods and optimizations can be extended to a larger number of cores with voltage stacking as long as maximum voltage levels for the process are not exceeded by the circuit. A notable contribution of the work is the characterization of the loads. Finally, simulations are shown for various workload conditions to demonstrate the usefulness of voltage stacked power delivery.

2. DC-DC Conversion using Voltage Stacking

The idea of charge recycling through voltage stacking to generate on-chip low voltage at higher efficiency has also been considered before (Figure 1a) [7].

Figure 1. (**a**) Implicit DC-DC down conversion through voltage stacking [7]; (**b**) Implicit voltage conversion for Resistive/CMOS load.

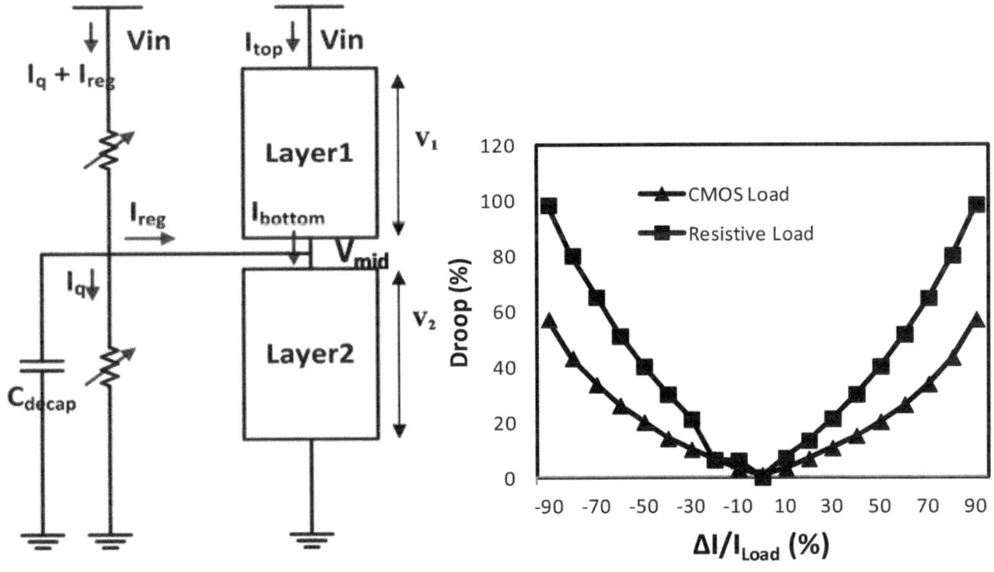

(a) (b)

The power efficiency of this technique depends on the mismatch between the stacked domains, which in turn depends on the activity of the circuits, the evaluation node capacitance and the voltage swing in the domains. The efficiency is given by:

$$\text{Efficiency} = \frac{\text{Power\_logic}}{\text{Power\_system}} = \frac{V_{IN}\, I_{top} + V_{INT}\, |I_{reg}|}{V_{IN}\, (I_{top} + I_{reg}| + I_q)} \tag{1}$$

where $|I_{reg}|$ is the difference between the top and bottom stack and I_q is the quiescent current of the regulator [7]. Different works in the literature have handled this imbalance in different ways. In [7], charge balance was maintained between the domains through active regulation of the intermediate node using a push-pull linear regulator. However if the top core has larger current requirements than the bottom core, then the closed-loop regulator will force the excess current to ground, thus wasting power. To compensate for this loss, granules were shifted between the domains using switching logic—this came with a power and area overhead needed for the switching logic. In [8], a shunt regulator was used, and, to balance the different domains, software scheduling was done to distribute the workload at runtime. In our own work, we recycled the imbalance current among the stacked domains using an explicit regulator to improve the efficiency of this technique and maintain the output node within a certain tolerance limit [9]. By using a SC regulator, the achievable efficiency can be more than LDO. This idea is an extension of our work on GALS-based stacked cores which allow the intermediate node to implicitly track the workload of the different cores [5].

2.1. Voltage Stacking: Dependency on Nature of the Load

The idea of implicit down-conversion through voltage stacking is very intuitive if we think of the loads as identical resistors stacked upon each other. However, real load may have behavior very different from an ideal resistor (Figure 1b). Even resistive load of different magnitudes will act as an imbalance, leading to internal voltage noise. Thus this notion of implicit down-conversion will vary not only on the workload difference, but also on the nature of the load. To demonstrate this load dependency, we modeled two different kinds of load, CMOS load (multiple blocks of ring oscillators) and resistor load, drawing the same amount of current.

From first order analysis, $V_{CMOS}\ \alpha \sqrt{I_{CMOS}}$ while $V_{Resistive}\ \alpha\ I_{Resisitve}$. Therefore, load current imbalance will cause a larger variation in resistive load than CMOS load. This quadratic versus linear dependency between voltage and current is also evident from Figure 1b where mid-voltage droop due to resistive load variation is larger than due to CMOS load variation.

2.2. Positive/Negative Regulation

In order to recycle the current imbalance between two voltage-stacked cores, current needs to be either *sourced* or *sunk* from the regulator, depending on which core consumes more current. When the bottom core consumes more, current needs to be sourced (*positive regulation*) which is similar to the conventional case (Figure 2a). However, when the top core consumes more, current needs to be *sunk* and we call this *negative regulation* (Figure 2b). To achieve this unique feature, we use a modified version of a conventional switched capacitor circuit [4]. The design is implemented with eight switches and two capacitors (Figure 3a). Unlike the case of a conventional push-pull design, with such an SC solution the *sunk* current is fed back to the top core, thus reducing power waste. The two fly-capacitors change roles periodically, providing the *source/sink* of charge.

Figure 2. (**a**) Positive Regulation (Similar to conventional regulator, sourcing I_{Load}); (**b**) Negative Regulation (Regulator absorbs current, sinking I_{Load}).

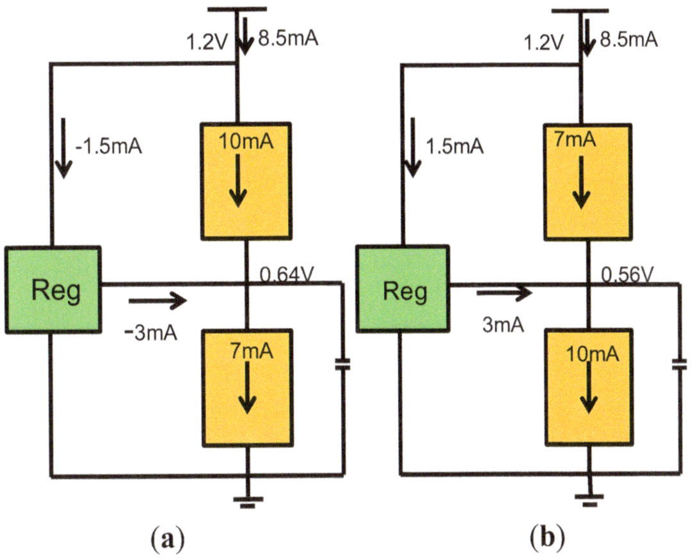

Figure 3. (**a**) Illustrates the push-pull switched capacitor designed to assist voltage stacking; (**b**) Fly-capacitors are swapped over the phases to regulate the *imbalance* between the stacks. Arrows indicates the direction from which charges are flowing to and from V_{out}.

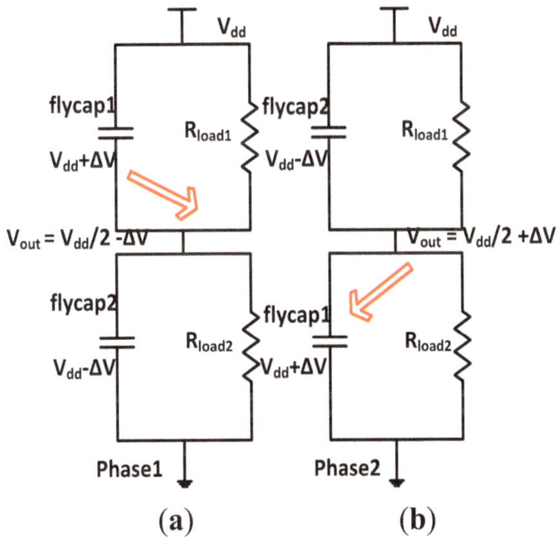

To understand how the 2:1 SC circuit works, consider an example of a slightly imbalanced workload with a supply voltage of V_{dd}, where the current offset pushes V_{out} to droop below 1/2 V_{dd} by ΔV (Figure 3b). In the first phase, as the voltage droops down at the load, flyCap1 begins charging to ΔV voltage above 1/2 V_{dd}, while the voltage on flyCap2 falls below 1/2 V_{dd} by ΔV. In the second phase, through the on-chip switches, flyCap1 and flyCap2 swap places. Since flyCap1 was charged to a higher voltage, it redirects this charge back onto the V_{out} node. This redirection of charge helps pull the load voltage ΔV above 1/2 Vdd. This ripple (2ΔV) is a manifest of the capacitor charging/discharging, and the faster the switching frequency, the lower the ripple.

2.3. Higher Efficiency in Voltage Stacked Regulation

One of the primary loss components in a switched capacitor regulator is the IR loss across the power switches. For a conventional DC-DC converter providing current to parallel cores, the *entire* current needs to flow through the regulator causing large IR drop across the switches. The traditional way of lowering this loss is to design efficient switches with low parasitic resistance. However beyond a limit, it is impossible to reduce the on-resistance of the switches and conductance loss starts to dominate. The uniqueness of voltage-stacked down-conversion lies in the fact that most of the current is reused from the top core and only the *difference* of the currents needs to flow through the regulator (instead of the sum of the currents) as shown in Figure 4b. This accounts for the *higher efficiency in stacked DC-DC conversion.*

Figure 4. (**a**) Conventional DC-DC converter sourcing currents (i1 + i2) to parallel loads; (**b**) Voltage Stacked DC-DC converter sourcing/sinking currents (|i1 − i2|) to stacked loads.

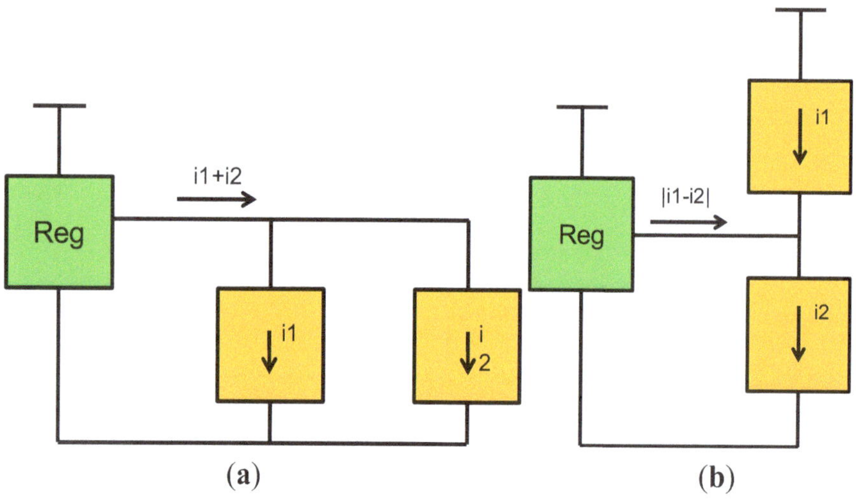

2.4. Power Loss Optimization for Switched Capacitor Circuit

As explained in [10], merely optimizing for the most energy efficient design can be misleading or impractical. Ideally the optimization should try to achieve the minimum energy point subject to some design constraint, for example by using the concept of *hardware intensity* [10]. In our work, we considered a design methodology to maximize the achievable efficiency by tuning different *sensitivity knobs* for a given area constraint [11,12].

In [9] we briefly discussed the design rules for this regulator. For a given current imbalance, while the fly-capacitor size determines the *amount of charge* that can be delivered, the switching frequency sets the *output ripple*. Taking maximum ripple as a design constraint, we chose the capacitor size and switch width as the two tuning knobs. As pointed out in [11], an energy efficient design is achieved when the costs (sensitivity ratios) of tuning the knobs are balanced. Each of the points on the energy-area design space represents percent power loss per percent area for an energy-efficient regulator design. In order to understand the design space, we developed analytical expressions (sensitivities) for all the tuning variables [12].

Switched capacitor circuit power loss can be categorized into two kinds, the series loss and the shunt loss [4]. Series loss consists of the intrinsic switched capacitor loss and the conductance loss of the switches:

$$P_{series} = P_{Cfly} + P_{Rsw} = I_{Load}\frac{\Delta V}{2} + m.I_{Load}{}^{2}\frac{R_{on}}{N.W_{sw}} \quad (2)$$

where R_{on} is the switch resistance density (Ω.m); W_{sw} is the size of each of the switches (we took all the switches to be equal sized); N the number of switches and m is a constant determined by the switched capacitor topology [4].

The additional shunt losses arise from switching the parasitic capacitance of the fly-capacitors and the power switches:

$$P_{shunt} = M_{bott.}V_0^2.C_{bott}.f_{sw} + V_{sw}^2.N.W_{sw}.C_{gate.}\frac{I_{Load}}{C.\Delta V} \quad (3)$$

where M_{bott} is a constant depending on SC topology; V_o and V_{sw} are the bottom plate-capacitor (plate-cap) and gate voltage swings; C refers to the fly-capacitor; C_{gate} refers to the gate oxide capacitance (gate-cap) of the switches and ΔV is the voltage droop.

The bottom plate-cap loss scales with only frequency while the gate-cap loss scales with both frequency and switch width. At higher load-currents the latter starts dominating, hence we neglected the bottom plate loss in our analysis. Also by using metal-insulator-metal (MIM) or trench-capacitors for the fly-capacitors, the bottom-plate loss can be minimized further. Thus total loss in switched capacitor regulator is given by:

$$P_{loss} = I_{Load}\cdot\frac{\Delta V}{2} + m.I_{Load}^2\cdot\frac{R_{on}}{N.W_{sw}} + V_{sw}^2.N.W_{sw}.C_{gate}.\frac{I_{Load}}{C.\Delta V} \quad (4)$$

2.5. Sensitivity Analysis

As mentioned before, we selected fly-capacitor and switch width as the two tuning variables for studying the power loss vs. area tradeoff-sensitivity analysis of the cost metrics with respect to the tuning variables and equalizing the cost across the design space provides us with optimum design points. We chose the power loss and area consumed by the switched capacitor circuit as the cost metrics:

$$A = A_cC + N.W_{sw} \quad (5)$$

where A is the total area and A_c is the area per unit capacitance—this value will depend on the capacitor technology. In our model we used the value assuming MIM cap density of 3 nF/mm^2. In [11] the sensitivity ratio for knob X is defined as:

$$S_x(X) = \frac{\frac{\partial P}{\partial X}}{\frac{\partial A}{\partial X}}\Big|\, x = X \quad (6)$$

where S_x represents the amount of energy that can be traded-off for area by tuning variable X. The sensitivity of power loss to area due to switch width and fly capacitor are given by:

$$\frac{\frac{\partial P}{\partial W_{sw}}}{\frac{\partial A}{\partial Wsw}} = \frac{V_{sw}^2 C_{gate.}}{C.\Delta V} I_{Load} - \frac{I_{Load}^2 R_{on}}{N^2 {W_{sw}}^2} \quad (7)$$

$$\frac{\frac{\partial P}{\partial C}}{\frac{\partial A}{\partial C}} = \frac{-N.V_{sw}^2.W_{sw} C_{gate}}{\Delta V.A_C C^2} I_{Load} \quad (8)$$

For optimizing the design in the energy/area design space, the costs across the tuning variables should be equal [11]:

$$\frac{\frac{\partial P}{\partial W_{sw}}}{\frac{\partial A}{\partial Wsw}} = \frac{\frac{\partial P}{\partial C}}{\frac{\partial A}{\partial C}} \quad (9)$$

This means that at optimal design points, any marginal gain in energy will result in the same percentage amount of loss in area. The values of R_{on} and C_{gate} are fixed, depending on the technology node, while fly-capacitor values are chosen in accordance with the maximum allowable droop ΔV and I_{Load} max. In this topology, I_{Load} max refers to the worst-case current imbalance between the domains. Plugging-in the values and solving for W_{sw} provides the optimum switch width for a given capacitor size at the minimum energy point. Thus this global optimization process can allow us to look at the entire design space while balancing out the different tuning variables depending on the weight of their individual cost. To verify our claim, we did a Monte Carlo simulation selecting a large number of arbitrary capacitor values and switch widths and plotting them against our result based on sensitivity optimization. As can be seen in Figure 5, our method indeed yields the lowest points in the design space, also known as the Pareto curve. Based on the above optimization, for a given fly-capacitor size, allowable droop, and maximum I_{Load}, we can find the optimum switch widths. In Figure 6, we plugged-in this optimal value of switch width and performed efficiency analysis for different load currents. Unlike for a conventional regulator, the optimization of a voltage-stacked regulator will depend not only on the I_{Load} but also on the difference of current between the domains [6]. In order to have a fair comparison, we kept one of the stacked loads at 200 mA while varied the other load from 0–200 mA. We have plotted this ratio of current imbalance to chip current consumption ($\Delta I/I_{Load}$) along the x-axis in Figure 6b. As can be seen from the plots, stacked regulation can yield higher efficiency, and since the bulk of the current comes from the off-chip supply, the switch widths and fly-capacitor area can be reduced to improve the power density of the circuit. Stacked regulation achieves 81%–95% efficiency (depending on imbalance between stacked domains) compared to 81% for non-stacked load of 200 mA for 1.2 V–0.6 V down conversion. Thus it is fair to say, even in the worst case, stacked regulation is comparable to the best case in conventional SC regulation [9].

2.6. *Open-Loop* versus *Closed-Loop*

In SC circuits, regulation is performed by modulating the output resistance of the converter in response to changes in load current [13]. The output of a SC converter is given by:

Figure 5. Monte Carlo Analysis of power optimization. Zoomed in view (right) shows that our sensitivity-based optimization gives the lowest points on the curve.

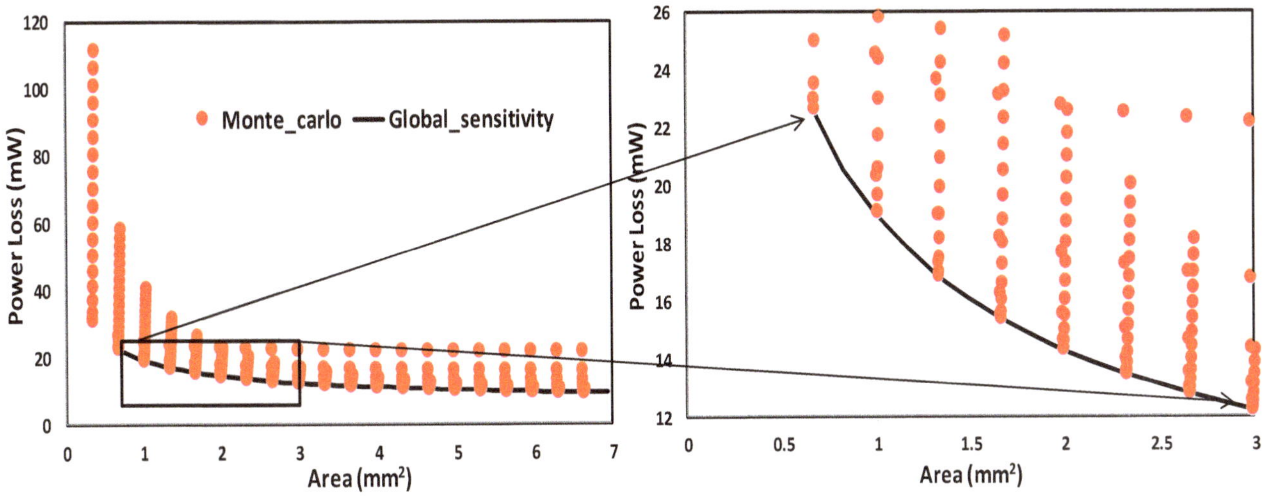

Figure 6. (**a**) Efficiency with varying conventional load (left); (**b**) Stacked load (right) with max I_{Load} = 200 mA. The X-axis indicates relative imbalance (%) between the domains.

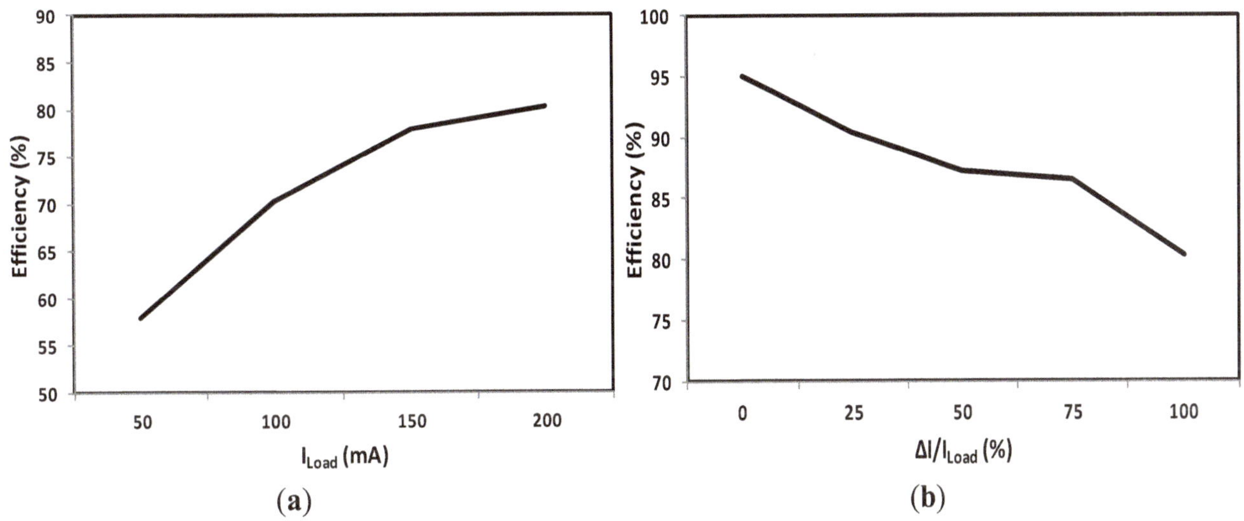

$$V_{out} = nV_{in} - i_{out} R_{out} (f_{sw}, D_i, G_i) \tag{10}$$

where n is the conversion ratio; f_{sw} the switching frequency; D_i the duty cycle of switching and G_i the conductance of the switches. Each of these variables can be used to control regulation. The conversion ratio (n) is fixed by the number of layers (cores) stacked. One of the main drawbacks of varying Di or Gi of the switches at constant frequency is reduced efficiency at lighter load. However, keeping duty cycle fixed at 50% and by modulating switching frequency with load current, higher efficiency can be achieved, especially at lighter load [4]. The downside of this is the high output voltage ripple as charge transfer is impulsive with slow switching limit of frequency [14]. Hybrid regulation, with two or three control variables together, will bring the highest efficiency, however complicated the control circuitry. In our work, we used modulation of switching frequency as the controlling variable with additional interleaving mechanism to reduce the output ripple. The traditional control method for a SC converter may include a linear feedback loop to control the switching frequency in terms of the output voltage. However obtaining stability and good transient response over varying load conditions using such a control method is difficult. A nonlinear control can provide superior results; hence we have used a hysteretic feedback scheme with lower and upper bounds to control the regulation (Figure 7) [15]. However our control circuit is different from traditional hysteretic control and the difference comes from the stacked loads as opposed to conventional parallel loads. As explained above, current needs to be either sourced or sunk in this type of load. Consequently the output can go both high and low depending on the current mismatch between the layers, hence higher switching frequency (Clk_high) is needed whenever either of the boundaries is crossed by output voltage while low frequency (Clk_low) can regulate the in-between state. Table 1 explains the states.

Figure 7. Dual-boundary hysteretic feedback control scheme for stacked load. Output clock is pulsed between high and low frequency depending on comparator detected trigger signal.

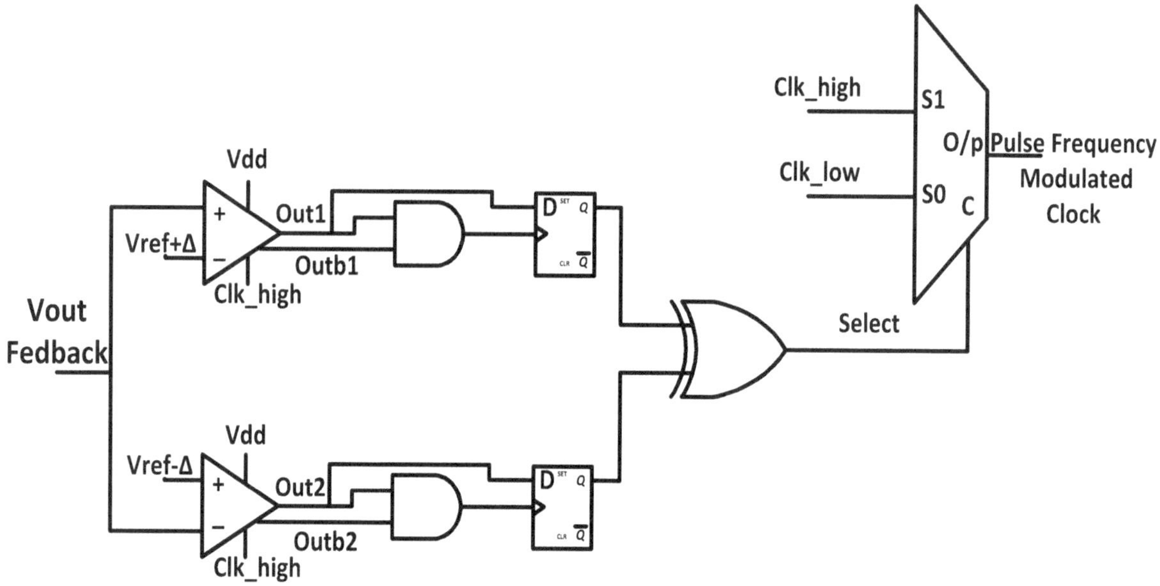

Table1. Different states for the feedback circuit to regulate V_{out}.

State of O/P	Out1	Out2	Select	O/P Clock
$V_{out} > V_{ref} + \Delta$	Toggle	Low	1	Clk_high
$V_{ref-\Delta} < V_{out} < V_{ref+\Delta}$	Low	Low	0	Clk_low
$V_{out} < V_{ref-\Delta}$	Low	Toggle	1	Clk_high

The feedback circuit consists of two comparators (along with latches, XOR gate and mux) that detect when the output voltage crosses the control boundary. By adding an edge-triggered latch we make sure that only rising edges are associated with a charge transfer [15]. If there were no latches, then the falling edge, which appears because of the clocked comparator and not as a trigger for boundary crossing, will cause an unwanted charge transfer. In order to account for comparator response time, we generated the clock to the latches out of the comparator itself—consequently comparator delay implicitly tracks across all PVT corners. Depending on the "Select" signal which triggers the output mux, low or high clock frequency is applied to the switches. A latch-based voltage sense amplifier was used to reduce power consumption [16]. The dual-comparator and logic gate controlled feedback circuit power consumption are critical, especially for low power operation. For 2 V–1 V stacked conversion for a load current of 200 mA with the comparator running at 1 GHz, feedback circuit power consumption is 1.2 mW. However this can be scaled down depending on the operational frequency of the load. For the low power (1.2 V–0.6 V, 10 mA) conversion, feedback power scales down to 0.2 mW.

This feedback circuit was used with the SC converter and we have done extensive simulations for 2:1 DC-DC conversion using both parallel loads and stacked loads. Near threshold circuits are typically operated 200 mV above their threshold value [17]. Hence here we considered 0.6 V as NTC V_{out} to be delivered. Conversion efficiency for both low power (0.5 mW–10 mW, 1.2 V–0.6 V) and high power (10 mW–400 mW, 2 V–1 V) loads have been shown for comparison.

Figure 8. (**a**) Feedback SC circuit applied to conventional load; (**b**) Comparisons between unregulated, open-loop and closed-loop stacked loads.

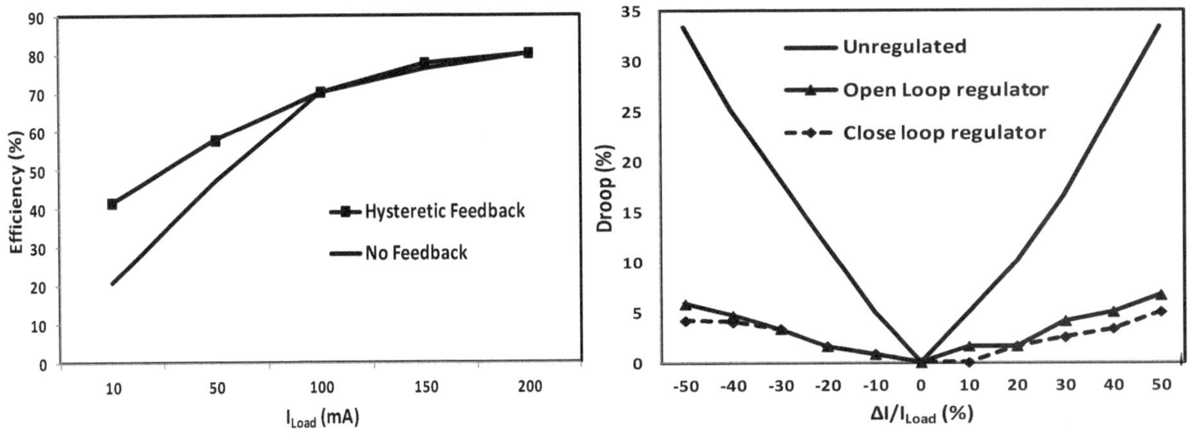

(**a**) (**b**)

As Figure 8a shows, the feedback circuit has considerable impact on efficiency for conventional load, especially at lower current. However stacked load has unique characteristics which are explained below with the help of Figure 1a [9].

In Section 2.1 we discussed how the nature of the load can affect voltage stacked DC-DC conversion. Here we discuss how the voltage headroom in each of the stacked cores is going to change with load or activity variation. CMOS load has been considered here, representing the core (Figure 1a).

By charge conservation:

$$I_{top} = I_{bottom} \tag{11}$$

Current consumption of the two cores is given by:

$$I_{top} = \alpha_{top} C_L \left(V_{dd} - V_{mid}\right) F_c \qquad I_{bottom} = \alpha_{bottom} C_L V_{mid} F_c \tag{12}$$

$$V_1 = V_{dd} - V_{mid} = \frac{\alpha_{bottom}}{\alpha_{bottom} + \alpha_{top}} V_{dd} \qquad V_2 = V_{mid} = \frac{\alpha_{top}}{\alpha_{top} + \alpha_{bottom}} \tag{13}$$

where α_{top} and α_{bottom} are the top and bottom core activity factors; F_c is the core frequency and V_{mid} is the implicit voltage generated by stacking two cores.

For

$$\alpha_{bottom} = \alpha_{top}, V_{mid} = 0.5V_{dd} \quad (14)$$

However, if $\alpha_{top} > \alpha_{bottom}$, $V_2 > V_1$; $\alpha_{top} < \alpha_{bottom}$, $V_1 > V_2$. This means that the inherent feedback of voltage stacking forces the voltage headroom to be lower for the core that demands *higher* current, thus acting against the idea of DVFS (that when computing demands are *lower* the voltage should be reduced). Therefore, this internal voltage buildup in voltage stacking can not only add noise to the stacked cores, but also oppose DVFS. This is shown in Figure 8b. The more the imbalance, the more the self-regulation of the system will force the V_{mid} node to go in the opposite direction as seen from the unregulated stacked mid rail. Thus for voltage stacking to work, we need to compensate for this natural feedback tendency and the push-pull scheme is therefore essential for maintaining the charge imbalance within bounds.

However whether adding a feedback loop over the SC converter can bring any benefits or not, is what we try to show in Figure 9a,b. As shown in some of the earlier plots, to demonstrate effectiveness of voltage stacking, efficiency needs to be plotted against the ratio of load current to load imbalance. Thus midpoint on the x-axis for both the plots indicates when the loads are perfectly balanced. Ideally the SC does not need to regulate at that point and it can lower its switching frequency to minimal value to improve energy efficiency. This is shown in Figure 9a,b where the feedback controller reduces the switching frequency around the balanced load condition (midpoint region ofx-axis) and increases the efficiency as compared to the open loop regulator. However when the loads are not in balance, the performance of the closed loop regulator is in fact worse compared to the open loop one for low power loads (Figure 9b). This is because for low power (NTC) stacked loads, the feedback circuit overhead reduces the efficiency. At an imbalance of 50%, closed-loop SC converter suffers an efficiency loss of 10.8% over an open-loop SC converter. Thus for low power DC-DC conversion, voltage stacking can provide an attractive alternative technique and by removing the additional losses in the feedback circuitry, the efficiency can be improved further [9]. However, for closely matched high power loads, hysteretic feedback can still provide higher efficiency as shown in Figure 9a.

Figure 9. (**a**) Comparison of open-loop/close-loop SC circuit for high power: 10 mW-400 mW, 2 V–1 V; (**b**) Low power: 0.5 mW–10 mW, 1.2 V–0.6 V.

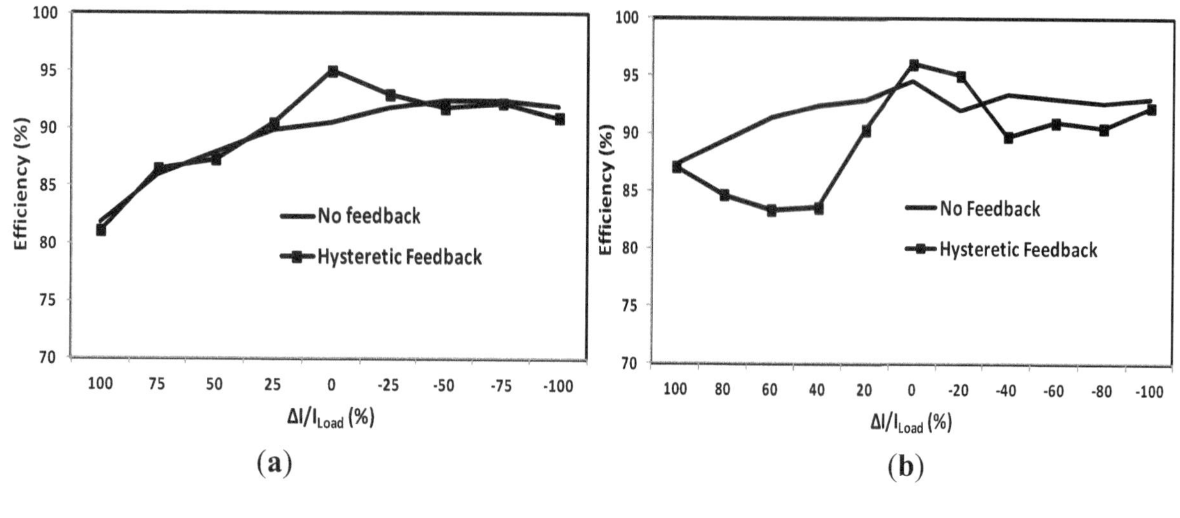

3. Practical Loads and Capacitor Models

While the SC circuits described above work well for DC current loads, their performance needs to be evaluated with respect to transient response. For this we used two different types of load to analyze the performance of the SC converter. The first load in Figure 10a,b models the big changes in load current during different power modes and performance states in real cores. The transient response in Figure 10c shows that the output remains within a 5% tolerance limit even in the worst case scenario (I_{min} to I_{max}). An additional startup circuitry can reduce the initial voltage droop. With actual applications running on the core, the current load hardly retains the steady state DC value as used in the model. To mimic a more real situation we generated different current profile by running a benchmark application (Dedup, a benchmark from the PARSEC suite) using an accurate full system multicore simulator M5 [18]. These current traces are then included in Cadence ADE as current sources, shown in Figure 11a,b while the output voltage generated from the stacked SC converter is shown in Figure 11c. Adopting the calculation as shown in Equation (1), 2 V–1 V conversion using this stacked approach yields efficiencies as high as 93%. The p-pripple on the V_{mid} (V_{out}) still remains within the tolerance limit. Further reduction in V_{mid} variation can be achieved by increasing the SC converter frequency or by increasing the on-chip decap, however this comes with tradeoffs in terms of efficiency or area.

Figure 10. Current load for different mode change. (**a**) upper stacked loads; (**b**) lower stacked loads; (**c**) 2:1 down converted V_{out}.

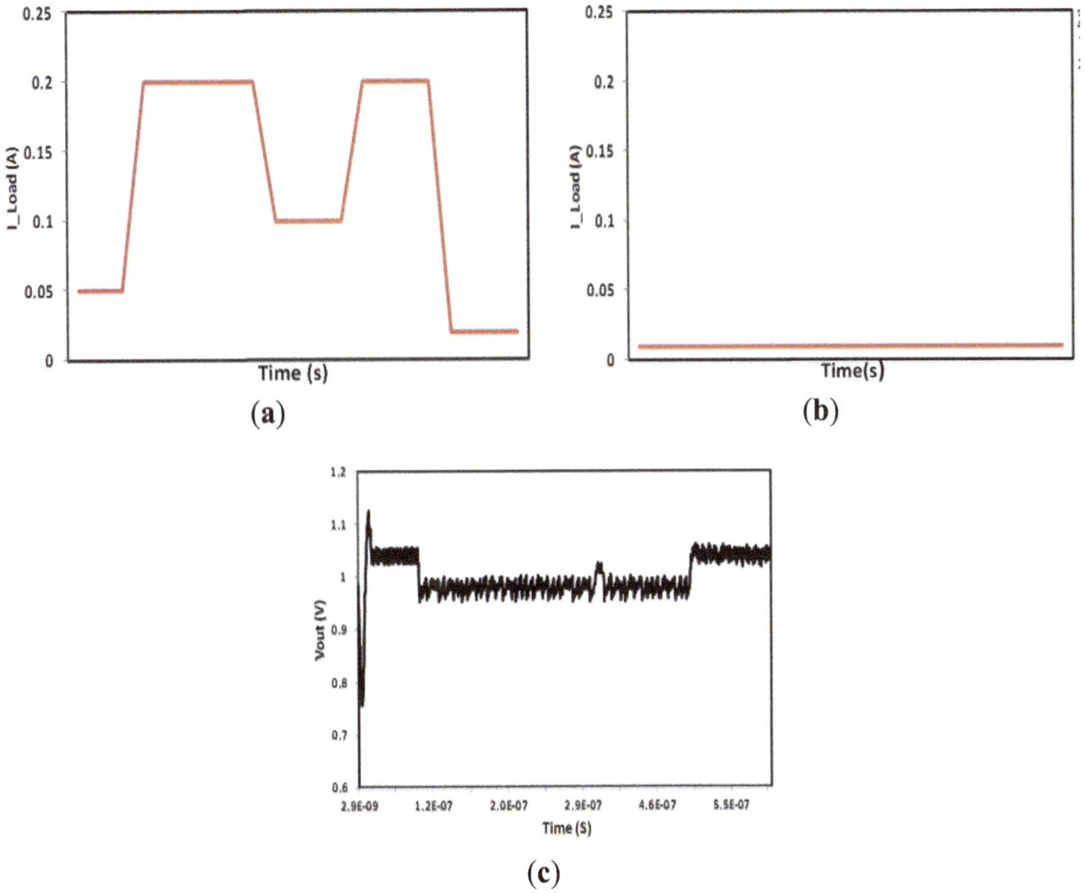

Figure 11. Current traces generated from running benchmark and including them in Cadence ADE as load. (**a**) upper stack load; (**b**) lower stacked loads; (**c**) 2:1 Down converted Vout (Vdd = 2 V, Vout ~ 1 V).

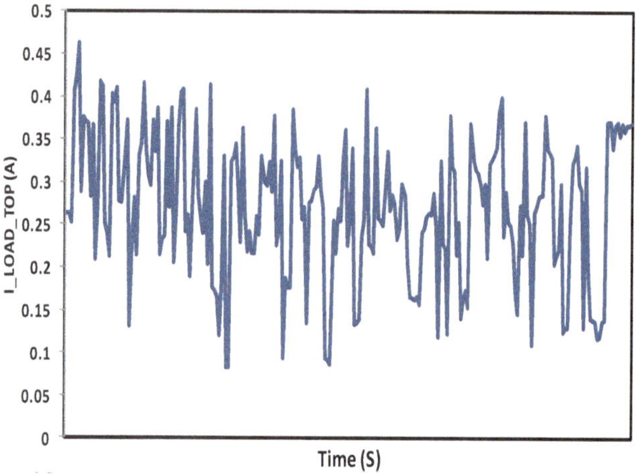

(**a**)

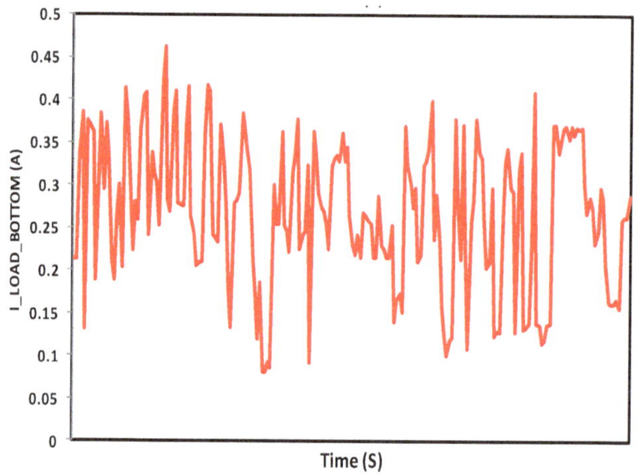

(**b**)

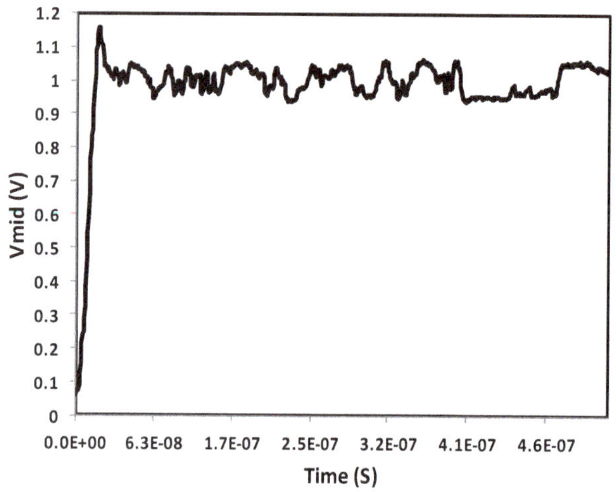

(**c**)

All the simulations so far have used ideal capacitors. However bottom plate parasitic and ESR of the capacitors can have a huge impact on the conversion efficiency and area constraint. Bulk CMOS technology with low ESR can provide capacitive density of up to 12 nF/mm^2, but bottom plate capacitance is highest in MOS Caps among the capacitor technology owing to proximity to substrate (5%–10%). MIM (Metal-insulator-Metal) capacitor with a lower bottom plate parasitic (1%–3%) can provide a good alternative to higher efficiency at the cost of area (3 nF/mm^2) and high ESR (Equivalent Series Resistance) [17]. To account for the parasitic losses, we used models of MIM cap and MOS cap in the simulation including the top and bottom plate capacitive parasitic as well as the contact and plate resistance (Figure 12a). Efficiency for 2:1 down-conversion using MIM cap ranges as high as 72%–91% for a load imbalance of 0–200 mA and maximum load being 400 mA (Figure 12b). Performance of the MOS cap is the worst due to the parasitic. However it gives the area advantage over MIM cap.

Figure 12. (**a**) MIM capacitor model, including all the parasitic; (**b**) Efficiency comparison between real (MIMcap and MOScap) and ideal capacitor.

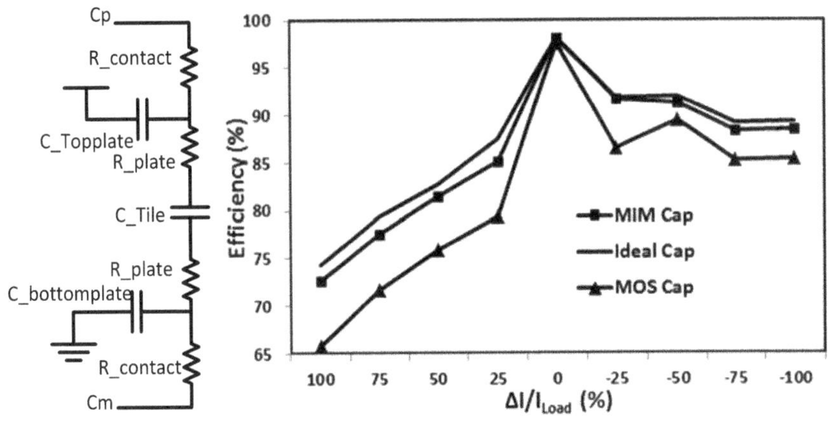

(a) (b)

4. Conclusions and Future Work

With aggressive voltage and technology scaling, power delivery is expected to be a major challenge for the semiconductor industry. With increasing power densities, reduction in power pin count and quadratic increase in off-chip power loss, traditional power delivery methods may not be efficient enough. In this work, we exploited the concept of charge recycling through voltage stacking to convert high off-chip voltage to low on-chip voltage at higher efficiency. While voltage stacking eases off-chip power delivery issues, this topology presents a unique challenge of within die noise due to interlayer current mismatch. In our work, we used a push-pull switched capacitor circuit to recycle and redistribute the charge imbalance among the stacked layers, thus reducing voltage noise. We presented a design methodology for optimizing the power loss of the switched capacitor circuit based on global sensitivity analysis. Analytically derived optimum switch width was used in simulating the switched capacitor, both for conventional and stacked load. The high efficiency of the stacked DC-DC conversion justifies the integration of voltage stacking in the 2D IC power delivery network. A novel dual boundary hysteresis-based feedback scheme was developed to regulate the SC converter. However we claim that the self-regulated stacked domains will work better as open-loop regulatora, especially for low power loads. Stacked SC regulator down-conversion (2 V–1 V) was shown at a high efficiency of 93% with current traces generated from real benchmark being used as the load. Even by incorporating the parasitic of the MIM capacitor in simulation, efficiency still ranges from 72%–91% for a load imbalance of 0–200 mA and maximum load of 400 mA. Thus with reduction of the DC-DC converter energy overhead, just-as-needed V_{dd} and near threshold computing can become even more energy efficient.

References

1. Dreslinski, R.G.; Wieckowski, M.; Blaauw, D.; Sylvester, D.; Mudge, T. Near-threshold computing: reclaiming moore's law through energy efficient integrated circuits. *Proc. IEEE* **2010**, *98*, 253–266.
2. Yu, P.; Xin, Z.; Huang, J.; Muramatsu, A.; Nomura, M.; Hirairi, K.; Takata, H.; Sakurabayashi, T.; Miyano, S.; Takamiya, M.; *et al.* Misleading Energy and Performance Claims in Sub/Near Threshold Digital Systems. In Proceedings of 2010 IEEE/ACM International Conference on Computer-Aided Design (ICCAD), San Jose, CA, USA, 7–11 November 2010; pp. 625–631.
3. Patounakis, G.; Li, Y.W.; Shepard, K.L. A fully integrated on-chip DC-DC conversion and power management system. *IEEE J. Solid-State Circuits* **2004**, *39*, 443–451.
4. Hanh-Phuc, L.; Sanders, S.R.; Alon, E. Design Techniques for fully integrated switched-capacitor DC-DC converters. *IEEE J. Solid-State Circuits* **2011**, *46*, 2120–2131.
5. Mazumdar, K.; Stan, M. Breaking the Power Delivery Wall Using Voltage Stacking. In Proceedings of the Great Lakes Symposium on VLSI, Salt Lake City, UT, USA, 3–4 May 2012; pp. 51–54.
6. Lee, S.K.; Brooks, D.; Wei, G.-Y. Evaluation of Voltage Stacking For Near-Threshold Multicore Computing. In Proceedings of the 2012 ACM/IEEE International Symposium on Low Power Electronics and Design, Redondo Beach, California, USA, 30 July–01 August 2012; pp. 373–378.
7. Rajapandian, S.; Zheng, X.; Shepard, K.L. Implicit DC-DC downconversion through charge-recycling. *IEEE J. Solid-State Circuits* **2005**, *40*, 846–852.
8. Kanev, S. Motivating Software-Driven Current Balancing in Flexible Voltage-Stacked Multicore Processors. Bachelor's Thesis, Harvard University, Cambridge, MA, USA, May 2012.
9. Mazumdar, K.; Stan, M.R. Charge Recycling On-Chip DC-DC Conversion for Near-Threshold Operation. In Proceedings of the Subthreshold Microelectronics Conference (SubVT), Waltham, MA, UK, 9–10 October 2012; pp. 1–3.
10. Brodersen, R.W.; Horowitz, M.A.; Markovic, D.; Nikolic, B.; Stojanovic, V. Methods for True Power Minimization. In Proceedings of the 2002 IEEE/ACM International Conference on Computer-Aided Design (ICCAD), New York, NY, USA, 10–14 November 2002; pp. 35–42.
11. Zhenyu, Q.; Ziegler, M.; Kosonocky, S.V.; Rabaey, J.M.; Stan, M.R. Multi-dimensional Circuit and Micro-architecture Level Optimization. In Proceedings of the 8th International Symposium on Quality Electronic Design(ISQED), San Jose, CA, USA, 26–28 March 2007; pp. 275–280.
12. Mazumdar, K.; Stan, M.R. Breaking the 3-D IC Power Delivery Wall. In Proceedings of the 2012 Asilomar Conference on Signals, Systems and Computers, Asilomar, CA, USA, 4–7 Novmber 2012.
13. Seeman, M.D. A Design Methodology for Switched-Capacitor DC-DC Converters. Ph.D. Thesis, University of California, Oakland, CA , USA, 2009.
14. Ng, V.W.S. Switched Capacitor DC-DC Converter: Superior Where the Buck Converter has Dominated. Ph.D. Thesis, University of California, Oakland, CA, USA, 2011.
15. Van Breussegem, T.M.; Steyaert, M.S.J. Monolithic capacitive DC-DC converter with single boundary-multiphase control and voltage domain stacking in 90 nm CMOS. *IEEE J. Solid-State Circuits* **2011**, *46*, 1715–1727.

16. Jain, R.; Sanders, S. A 200 mA switched capacitor voltage regulator on 32 nm CMOS and regulation schemes to enable DVFS, In Proceedings of the 14th European Conference. Birmingham, UK, 30 August–1 September 2011; pp. 1–10.
17. Fick, D.; Dreslinski, R.G.; Giridhar, B.; Gyouho, K.; Sangwon, S.; Fojtik, M.; Satpathy, S.; Yoonmyung, L.; Daeyeon, K.; Liu, N.; *et al.* Centip3De: A 3930 DMIPS/W Configurable Near-Threshold 3D Stacked System with 64 ARM Cortex-M3 Cores. In Proceedings of the 2012 IEEE International Solid-State Circuits Conference Digest of Technical Papers (ISSCC), San Francisco, CA, USA, 19–23 Feberary 2012; pp. 190–192.
18. Binkert, N.L.; Dreslinski, R.G.; Hsu, L.R.; Lim, K.T.; Saidi, A.G.; Reinhardt, S.K. The M5 simulator: Modeling networked systems. *IEEE Micro* **2006**, *26*, 52–60.

Multi-Threshold Dual-Spacer Dual-Rail Delay-Insensitive Logic (MTD^{3}L): A Low Overhead Secure IC Design Methodology

Michael Linder [1], Jia Di [1] and Scott C. Smith [2,*]

[1] Department of Computer Science and Computer Engineering, University of Arkansas; Fayetteville, AR 72701, USA; E-Mails: mlinder@uark.edu (M.L.); jdi@uark.edu (J.D.)

[2] Department of Electrical and Computer Engineering, North Dakota State University; Fargo, ND 58108, USA

* Author to whom correspondence should be addressed; E-Mail: scott.smith.1@ndsu.edu

Abstract: As portable devices become more ubiquitous, data security in these devices is becoming increasingly important. Traditional circuit design techniques leave otherwise secure systems vulnerable due to the characteristics of the hardware implementation, rather than weaknesses in the security algorithms. These characteristics, called side-channels, are exploitable because they can be measured and correlated with processed data, potentially giving an attacker insight into the device's secret data. Alternative design techniques such as dual-rail asynchronous designs are capable of minimizing these potential side-channels by decoupling them from the data being processed. However, these techniques are either expensive to implement compared to standard designs or leave exploitable imbalances in the dual-rail implementation itself. Multi-Threshold Dual-Spacer Dual-Rail Delay-Insensitive Logic (MTD^{3}L) offers security by balancing side-channels both in general and between the dual-rail signals themselves, as well as reduction in circuit overhead compared to previous secure design techniques. Results show that the Advanced Encryption Standard (AES) cores designed using MTD^{3}L exhibit similar security to previous secure techniques with substantially less area and energy overhead.

Keywords: secure hardware; side-channel; asynchronous; Advanced Encryption Standard (AES)

1. Introduction

As technology advances, more and more electronic devices store secret information such as bank accounts, identification numbers, passwords, and other private data that need to be secured from unauthorized access. Although originally considered safe and secure, hardware, just as software, is prone to attacks that force the targeted system to reveal sensitive data. Cryptographic algorithms are commonly used to protect such data. However, despite the mathematical robustness of these algorithms, their physical implementations are known to be susceptible to attacks. Non-invasive attacks on such devices take advantage of side-channel information leaked from the system, instead of trying to reverse engineer it. Such side-channel information can be power, timing, electromagnetism, and any other information that might be measured from the device during computation.

Most electronic devices running cryptographic algorithms are implemented in CMOS technology, where transistors act as voltage-controlled switches. While a circuit node is switching, electrons flow across the corresponding transistors to charge/discharge its load capacitance, thereby consuming power. Due to the fact that different transistors will be turned on/off while processing different data, causing different power consumption, power-based side-channel attacks can be implemented using the IC's transient power data. These types of power-based attacks include Differential Power Analysis (DPA) [1], and Correlation Power Analysis (CPA) [2] (which uses the Pearson product-moment correlation coefficient to guess a key). In general, these attacks acquire transient power data while the target IC performs encryption/decryption on different texts, and then use statistical algorithms to derive the key. Power-based attacks are the most powerful and prevalently implemented side-channel attacks [3], and have been successfully implemented to crack almost all cryptographic algorithms on different platforms [4–12]. A number of methods have been proposed for mitigating power-based attacks by decoupling transient power consumption from the data being processed. Techniques based on balancing power fluctuation include new CMOS logic gates, which go through a full charge/discharge cycle for each data processed [13–18]. Other power balancing methods include modifying the algorithm execution [19], compensating current at the power supply node [20], and using sub-threshold operation [21]. Additionally, many techniques for randomizing power data have been proposed [22–27].

The principle of timing-based attacks [28] is very similar to power-based ones except these attacks rely on timing fluctuations of the target circuit while processing different data patterns. Depending on the load capacitance and driving strength, the charge/discharge process during the switching activities at an internal circuit node will take different amounts of time to finish, which in turn causes different timing delays. Existing countermeasures include inserting dummy operations [29], using redundant representation [30], and unifying the multiplication operands [31].

Asynchronous circuits, especially dual-rail asynchronous circuits, possess unique characteristics that could help mitigate such attacks [32–37]. Dual-rail asynchronous circuits, such as NULL Convention Logic (NCL) [38], use two wires to represent one signal. The DATA-spacer alternation protocol ensures the number of switching of each signal to be independent from the input; instead, it is only determined by the number of data processed [39], making power variation significantly smaller than synchronous designs [40]. Nonetheless, switching activity remains unbalanced between the two rails of each signal, which most likely drive different capacitive loads; thus, DPA, High-Order DPA, or CPA can still succeed. Moreover, such dual-rail logic circuits are even more vulnerable to timing-based attacks due to their strong data-timing dependency.

To alleviate these issues, Dual-Spacer Dual-Rail Delay-Insensitive Logic (D^3L) was developed in [41] as an extension of NCL, which was capable of mitigating both power- and timing-based side-channel attacks. However, this increased security was gained at the cost of more than four times the area of the base synchronous circuit and almost twice the area of the NCL version. This paper details the design, implementation, and analysis of MTD^3L, developed by combining the D^3L paradigm with Multi-Threshold NULL Convention Logic (MTNCL) [42,43], resulting in a circuit design methodology that is just as secure as D^3L, while requiring slightly more than half the area and energy compared to D^3L.

The remainder of this paper is organized as follows. Section 2 presents an overview of the pertinent previous work, including NCL, MTNCL, and D^3L. Section 3 details the MTD^3L circuit design method. Section 4 presents AES core implementations using MTD^3L, as well as versions implemented using D^3L, NCL, and the standard synchronous technique. Section 5 details the simulation setups for determining resistance against side-channel attacks, and presents and analyzes the simulation results. Section 6 draws conclusions and provides direction for future work.

2. Previous Work

2.1. NULL Convention Logic (NCL)

NCL [38] is a commonly used delay-insensitive (DI) asynchronous logic design style, which utilizes a multi-rail one-hot scheme, such as dual-rail, to encode data. A dual-rail signal, D, consists of two wires or rails, D^0 and D^1, which may assume any value from the set {DATA0, DATA1, NULL}, as depicted in Table 1. The DATA0 state corresponds to a Boolean logic 0, the DATA1 state corresponds to a Boolean logic 1, and the NULL state corresponds to the empty set (meaning that the value of D is not yet available). The two rails are mutually exclusive, such that both rails can never be asserted simultaneously; this state is defined as an illegal state.

Table 1. Dual-rail signal.

State	D^0	D^1
NULL	0	0
DATA0	1	0
DATA1	0	1
Illegal	1	1

NCL circuits are comprised of 27 fundamental gates, which constitute the set of all functions consisting of four or fewer variables [43]. Since each rail of an NCL signal is considered a separate variable, a four variable function is not the same as a function of four literals, which would consist of eight variables for dual-rail logic (e.g., a literal includes both a variable and its complement, F and F', whereas NCL rails are never complemented, such that a dual-rail NCL signal, F, consists of two variables, F^1 and F^0, where F^0 is equivalent to F'). The primary type of threshold gate, shown in Figure 1, is the THmn gate, where $1 \leq m \leq n$. THmn gates have n inputs. At least m of the n inputs must be asserted before the output will become asserted. In a THmn gate, each of the n inputs is connected to the rounded portion of the gate; the output emanates from the pointed end of the gate; and the gate's threshold value, m, is written inside of the gate.

Figure 1. THmn threshold gate.

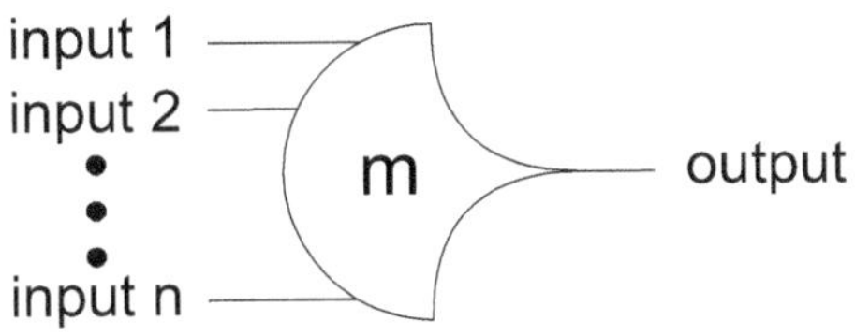

Another type of threshold gate is referred to as a weighted threshold gate, denoted as THmn w, w_1, w_2, …, w_R. Weighted threshold gates have an integer value, $m \geq w_R > 1$, applied to input R. Here, $1 \leq R < n$; where n is the number of inputs; m is the gate's threshold; and w_1, w_2, …, w_R, each >1, are the integer weights of input1, input2, …, input$R$, respectively. For example, consider the TH34W2 gate, whose $n = 4$ inputs are labeled A, B, C, and D, shown in Figure 2. The weight of input A, $W(A)$, is therefore 2. Since the gate's threshold, m, is 3, this implies that in order for the output to be asserted, either inputs B, C, and D must all be asserted, or input A must be asserted along with any other input, B, C, or D.

Figure 2. TH34w2 threshold gate: $Z = AB + AC + AD + BCD$.

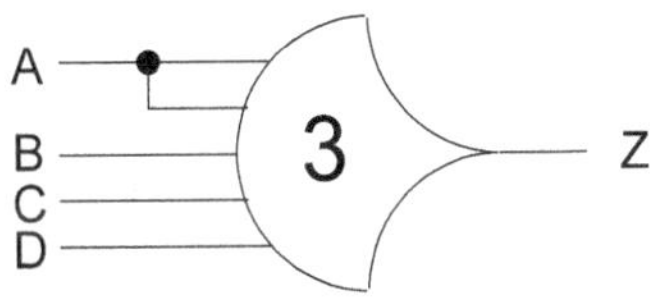

NCL threshold gates are designed with hysteresis state-holding capability, such that after the output is asserted, all inputs must be deasserted before the output will be deasserted. Hysteresis ensures a complete transition of inputs back to NULL before asserting the output associated with the next wavefront of input data. Therefore, a THnn gate is equivalent to an n-input C-element (*i.e.*, when all inputs are asserted the output is asserted; the output then remains asserted until all inputs are deasserted, at which time the output becomes deasserted) [44]; and a TH1n gate is equivalent to an n-input OR gate. NCL threshold gates may also include a reset input to initialize the output. Circuit diagrams designate resettable gates by either a d or an n appearing inside the gate, along with the gate's threshold. d denotes the gate as being reset to logic 1; n, to logic 0. These resettable gates are used in the design of DI registers.

To achieve hysteresis state-holding capability, NCL gates have both *set* and *hold* equations, where the *set* equation determines when the gate will become asserted and the *hold* equation determines when the gate will remain asserted once it has been asserted. The *set* equation determines the gate's functionality as one of the 27 NCL gates [43], whereas the *hold*1 equation is simply all inputs ORed together. The general equation for an NCL gate with output Z is: $Z = set + (Z^{-} \times hold1)$, where Z^{-} is the previous output value and Z is the new value.

To implement an NCL gate using CMOS technology, an equation for the complement of Z is also required, which in general form is: $Z' = reset + (Z'^{-} \times hold0)$, where *reset* is the complement of *hold*1 (*i.e.*, the complement of each input, ANDed together) and *hold*0 is the complement of *set*, such that the gate output is deasserted when all inputs are deasserted, and then remains deasserted while the gate's *set* condition is false. The new output value, Z, depends on the previous output value, Z^{-}, which requires internal gate feedback, as shown in Figure 3, for a static implementation [43].

Figure 3. (**a**) Static NULL Convention Logic (NCL) gate structure; (**b**) Static NCL TH23 implementation.

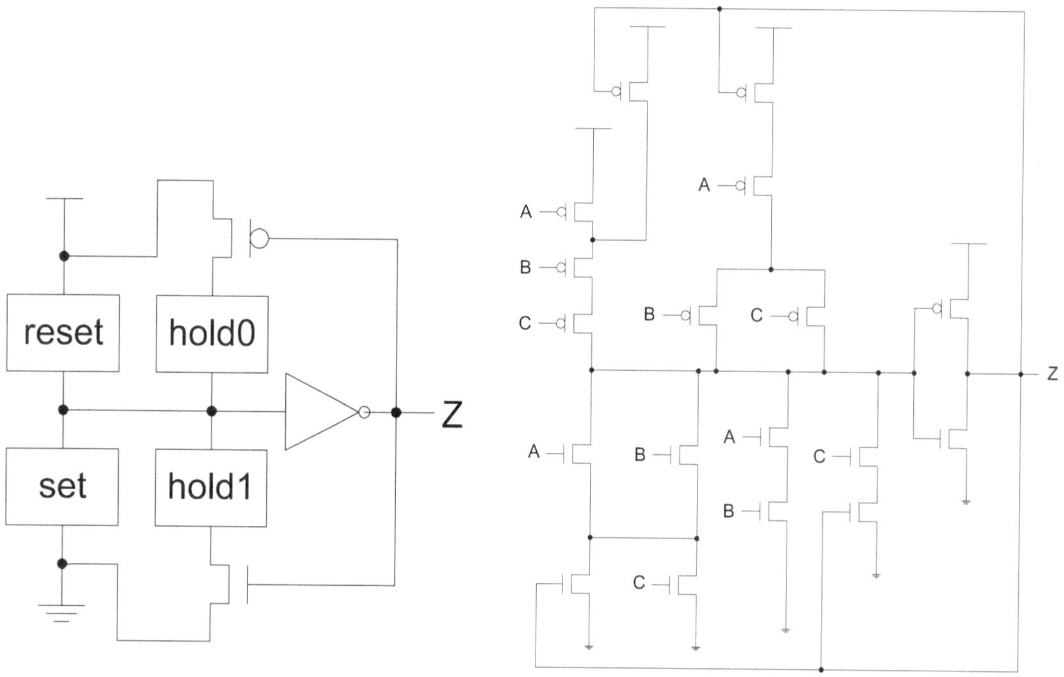

(**a**) (**b**)

The framework for NCL systems consist of DI combinational logic sandwiched between DI registers, as shown in Figure 4, which is very similar to synchronous systems. NCL systems contain at least two DI registers, one at both the input and at the output. Two adjacent register stages interact through their request and acknowledge signals, K_i and K_o, respectively, to prevent the current DATA wavefront from overwriting the previous DATA wavefront, by ensuring that the two DATA wavefronts are always separated by a NULL wavefront. The acknowledge signals are combined in the Completion Detection circuitry [43] to produce the request signal(s) to the previous register stage. NCL registration is realized through cascaded arrangements of single-bit dual-rail registers, shown in Figure 5, which pass a DATA value at the input only when K_i is request for data (rfd) (*i.e.*, logic 1) and likewise pass NULL only when K_i is request for null (rfn) (*i.e.*, logic 0). They also contain a NOR gate to generate K_o, which is rfn when the register output is DATA and rfd when the register output is NULL.

Figure 4. NCL system framework: input wavefronts are controlled by local handshaking signals and Completion Detection instead of by a global clock signal. Feedback requires at least three DI registers in the feedback loop to prevent deadlock.

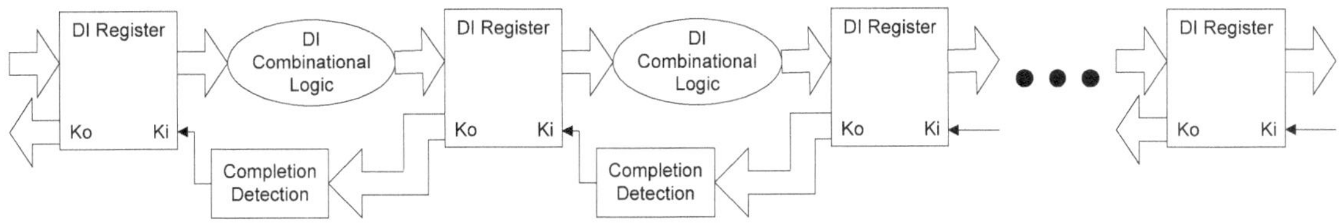

Figure 5. Single-bit dual-rail register.

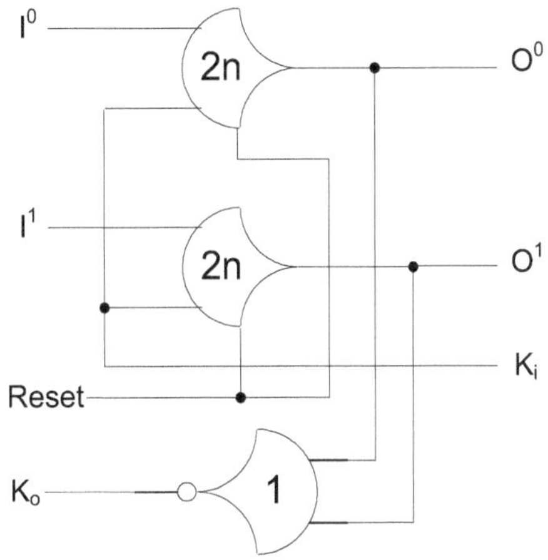

The NCL dual-rail encoding style decouples the relationship between the data being processed and the side-channel characteristics of the circuit, particularly power consumption, giving NCL circuits increased resistance to power related side-channel attacks over synchronous circuits. However, while the overall switching activity of a data bit may be balanced regardless of data, the individual encoding rails are not balanced, as illustrated in Figure 6. Furthermore, if the two rails are driving different capacitance, which is always the case in a large circuit, then attackers may be able to distinguish between the rails, allowing attackers to determine a relationship between the processed data and the circuit's side-channels.

Figure 6. NCL switching activity for passing consecutive DATA1s.

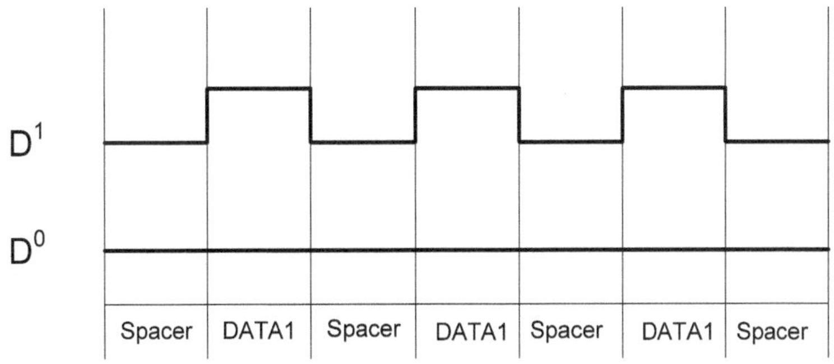

2.2. *Multi-Threshold NULL Convention Logic (MTNCL)*

MTNCL circuits utilize a sleep signal to simultaneously force all circuit elements to NULL instead of propagating a NULL input through the circuit, as described in [42]. This allows for the MTNCL gates to no longer require state-holding hysteresis logic and for the MTNCL combinational logic circuits to no longer need to be input-complete and observable, both of which significantly reduce area and power/energy, while increasing speed. The Static MTNCL (SMTNCL) gate and the Slept Early Completion and Registration Input-Incomplete (SECRII) architecture are shown in Figures 7 and 8, respectively, and are both detailed in [42].

Figure 7. (**a**) SMTNCL gate structure; (**b**) SMTNCL TH23 implementation.

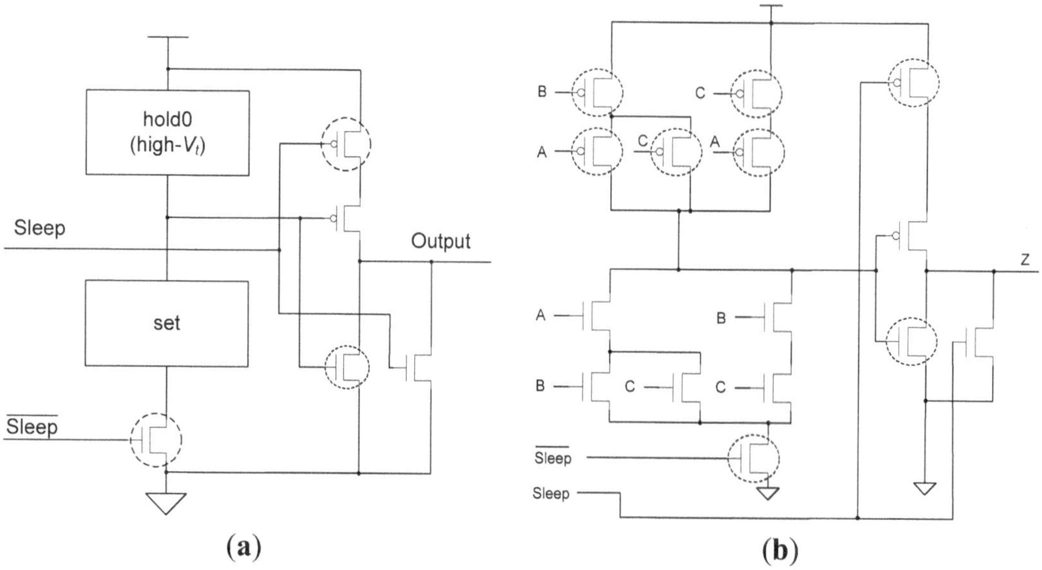

Figure 8. SECRII architecture with completion logic and registration slept.

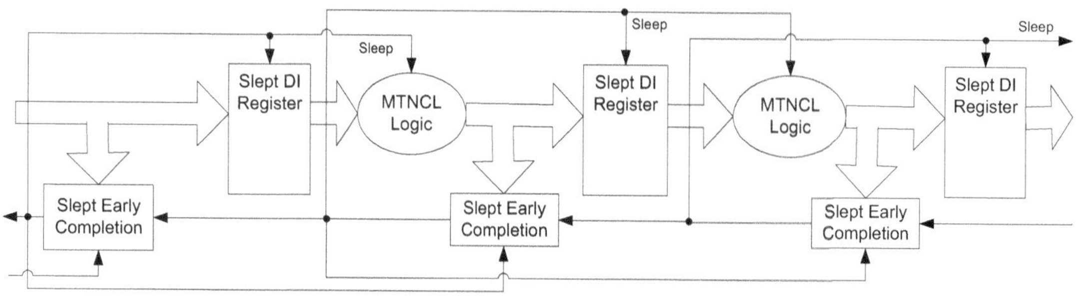

2.3. *Dual-Spacer Dual-Rail Delay-Insensitive Logic (D^3L)*

D^3L is an extension of the NCL logic style that utilizes a dual-spacer protocol, as opposed to NCL's single spacer protocol [41]. The motivation for this is the elimination of imbalanced switching activity on the two encoding wires of a data bit. By balancing this switching activity, data is further decoupled from the power consumption of the circuit, providing increased robustness against power analysis attacks.

Table 2 shows the D^{3}L encoding scheme. Like NCL, the DATA and NULL states remain the same. However, the NULL state is now called the All-Zero Spacer (AZS); and the former illegal state, where both rails are asserted, is now the All-One Spacer (AOS). The AZS and AOS are alternated between spacer cycles, implementing a dual-spacer protocol. As a result, the switching activity over a complete set of data/spacer cycles is balanced on both rails, as shown in Figure 9.

Table 2. D^{3}L encoding scheme.

State	D^0	D^1
All-Zero Spacer (AZS)	0	0
Data0	1	0
Data1	0	1
All-One Spacer (AOS)	1	1

Figure 9. D^{3}L switching activity for passing consecutive DATA1s.

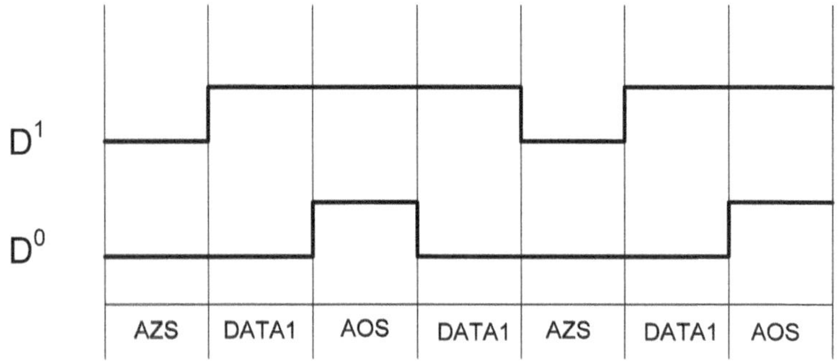

The D^{3}L threshold gates are modified versions of the NCL threshold gates. As such, the complete set of 27 NCL functions is implemented in D^{3}L. While NCL gates use hysteresis, D^{3}L gates are unable to do so to accommodate the dual-spacer protocol. As such, D^{3}L threshold gates are smaller than NCL threshold gates due to the omission of the hysteresis transistors. The removal of hysteresis, however, means that D^{3}L gates are unable to guarantee input completeness. Instead, the NCL_X technique [45] is used to provide input-completeness. This technique adds additional logic to D^{3}L functions that check the inputs and outputs of each function to create a completion signal. Since the spacer cycles in D^{3}L occur when data rails are the same, an XNOR gate can be used to detect them. The outputs from these XNOR gates are input to the completion component, along with the usual handshaking signals, to ensure that the logic is ready for the next wavefront. A downside to this technique, however, is the overhead incurred by adding large amounts of XNOR and threshold gates to the design to ensure input-completeness. An input-complete D^{3}L AND function is shown in Figure 10.

Figure 10. D[3]L input-complete AND function.

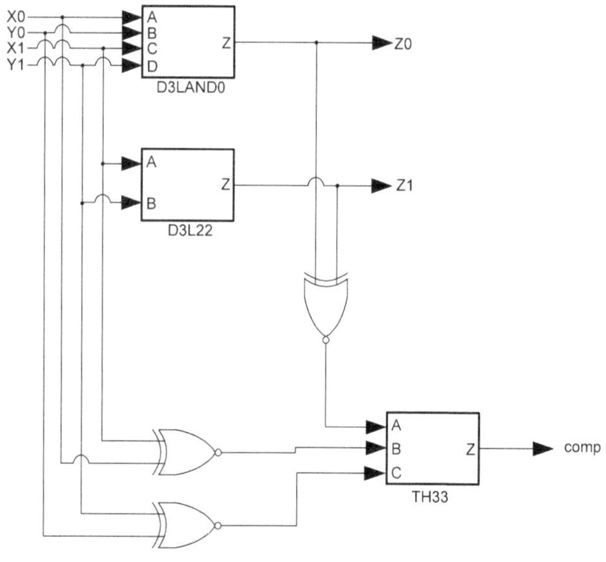

The basic D^3L Register, shown in Figure 11, is a modified NCL register. It consists of two TH22 gates which are resettable to the desired value. An XNOR gate facilitates K_o generation by checking the relative values of the register's outputs. As mentioned previously, the XNOR gate is required to detect both AZS and AOS.

Figure 11. D[3]L register.

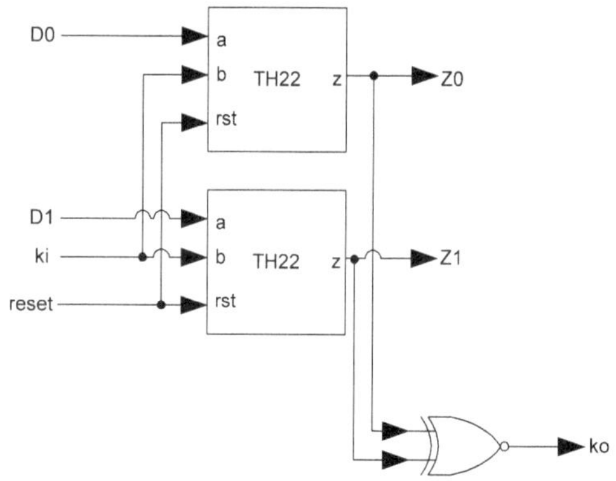

Additional logic is required to facilitate the dual-spacer protocol. NCL registers require a NULL input before they are able to accept new data. They will not recognize an all-one spacer. To fix this, extra logic capable of recognizing the all-one spacer is used to control the register's *Ki* input. This Ki Generator has four inputs: the early completion *Ki*, *ps* (previous spacer), and the two dual-rail outputs of the register. The value of *ps* is generated by a resettable TH22 gate. This value is logic 0 for an AZS and logic 1 for an AOS. The ps gate and the register must be reset to the same value. If the register is reset to DATA then the ps gate is reset to logic 0. The Ki Generator's output follows the Boolean equation $\text{Ki\_gen} = \overline{\text{Ki}}\,\overline{\text{ps}}\,(\text{Z0} + \text{Z1}) + \text{Ki ps}\,(\overline{\text{Z0}} + \overline{\text{Z1}}) + \overline{\text{Z0}}\,\overline{\text{Z1}}\,\text{Ki} + \text{Z0 Z1}\,\overline{\text{Ki}}$ resulting from the truth table shown in Table 3, yielding the transistor level implementation shown in Figure 12. If an AOS is needed then the value of *Ki_gen* will be changed to logic 1 allowing the register to latch it. Once the next data value arrives, *Ki_gen* will switch to logic 0. As a result, one of the register's TH22 gates will have two low inputs which will change its output to logic 0, latching the data. A complete D^3L register is shown in Figure 13.

Table 3. Ki Generator (KiGen) truth table.

Row	Z0	Z1	Ki	ps	Ki_gen
1	0	0	0	0	0
2	0	0	0	1	0
3	0	0	1	0	1
4	0	0	1	1	1
5	0	1	0	0	1
6	0	1	0	1	0
7	0	1	1	0	0
8	0	1	1	1	1
9	1	0	0	0	1
10	1	0	0	1	0
11	1	0	1	0	0
12	1	0	1	1	1
13	1	1	0	0	1
14	1	1	0	1	1
15	1	1	1	0	0
16	1	1	1	1	0

Figure 12. KiGen transistor diagram.

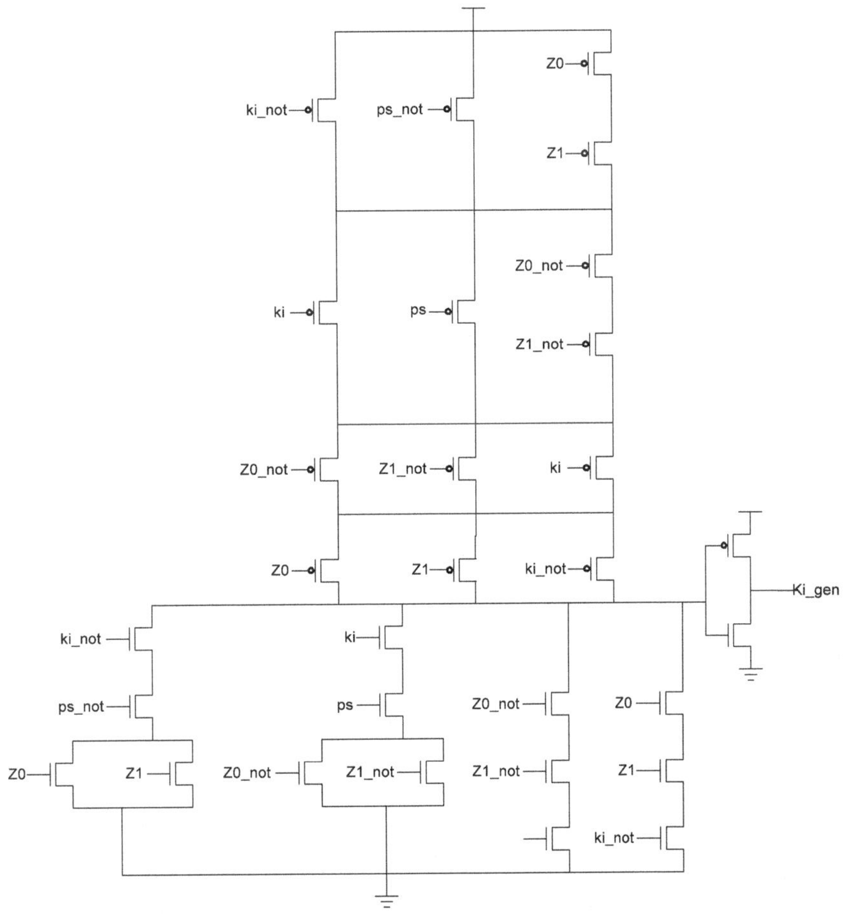

Figure 13. Complete D^{3}L register.

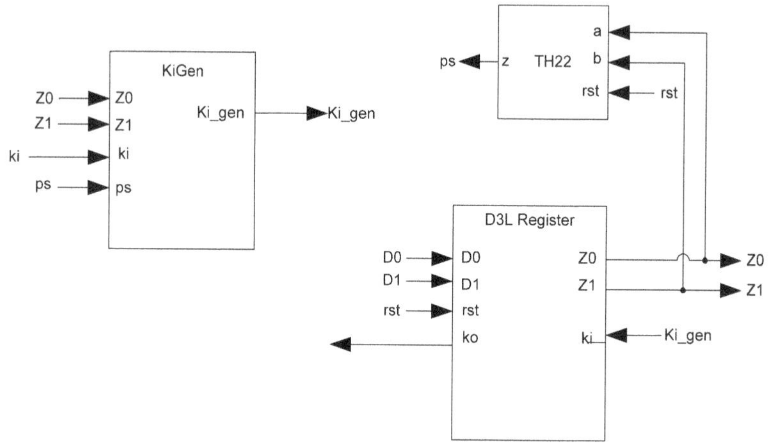

While the D^{3}L register is capable of handling the dual-spacer protocol, it is insufficient to implement ring register configurations. This is because a basic D^{3}L register is incapable of generating alternating spacers. Instead, the same spacer would pass through the ring twice causing deadlock. To solve this, a D^{3}L filter register is required for generating alternating spacers, which consists of a basic D^{3}L register with a spacer filter operating on the register's inputs. The spacer filter monitors the dual-rail input, the previous spacer, and the K_o from the register to ensure that spacers are alternated as they pass through. In a typical ring register configuration, the first two registers would be normal D^{3}L registers reset to NULL and a filter register reset to DATA0 or DATA1. When the filter register receives an AZS or AOS it outputs the alternate spacer. This ensures that the same spacer does not pass through the ring twice. Figure 14 shows the D^{3}L filter register diagram, while Figure 15 shows the transistor level schematic of the spacer filter, whose outputs are based on the following equations:

$$\mathrm{D0\_filter} = \mathrm{D0}\,\overline{\mathrm{D1}} + \overline{\mathrm{Ko}}\,\overline{\mathrm{ps}}\,\mathrm{D0} + \mathrm{Ko}\,\mathrm{ps}\,\mathrm{D0} + \mathrm{Ko}\,\mathrm{ps}\,\overline{\mathrm{D1}} + \overline{\mathrm{Ko}}\,\overline{\mathrm{ps}}\,\overline{\mathrm{D1}}$$

$$\mathrm{D1\_filter} = \overline{\mathrm{D0}}\,\mathrm{D1} + \overline{\mathrm{Ko}}\,\overline{\mathrm{ps}}\,\mathrm{D1} + \mathrm{Ko}\,\mathrm{ps}\,\mathrm{D1} + \mathrm{Ko}\,\mathrm{ps}\,\overline{\mathrm{D0}} + \overline{\mathrm{Ko}}\,\overline{\mathrm{ps}}\,\overline{\mathrm{D0}}$$

Figure 14. D^{3}L filter register.

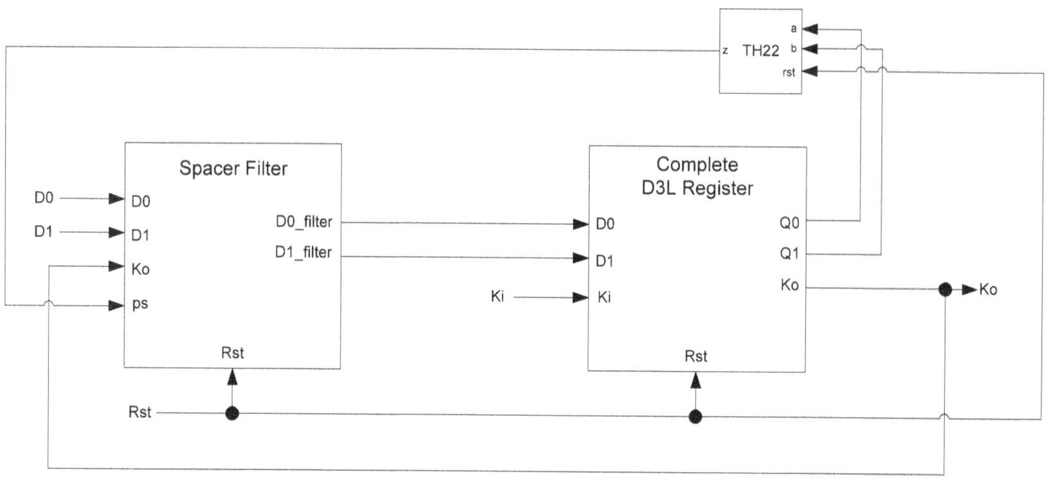

Figure 15. Spacer filter transistor level diagram.

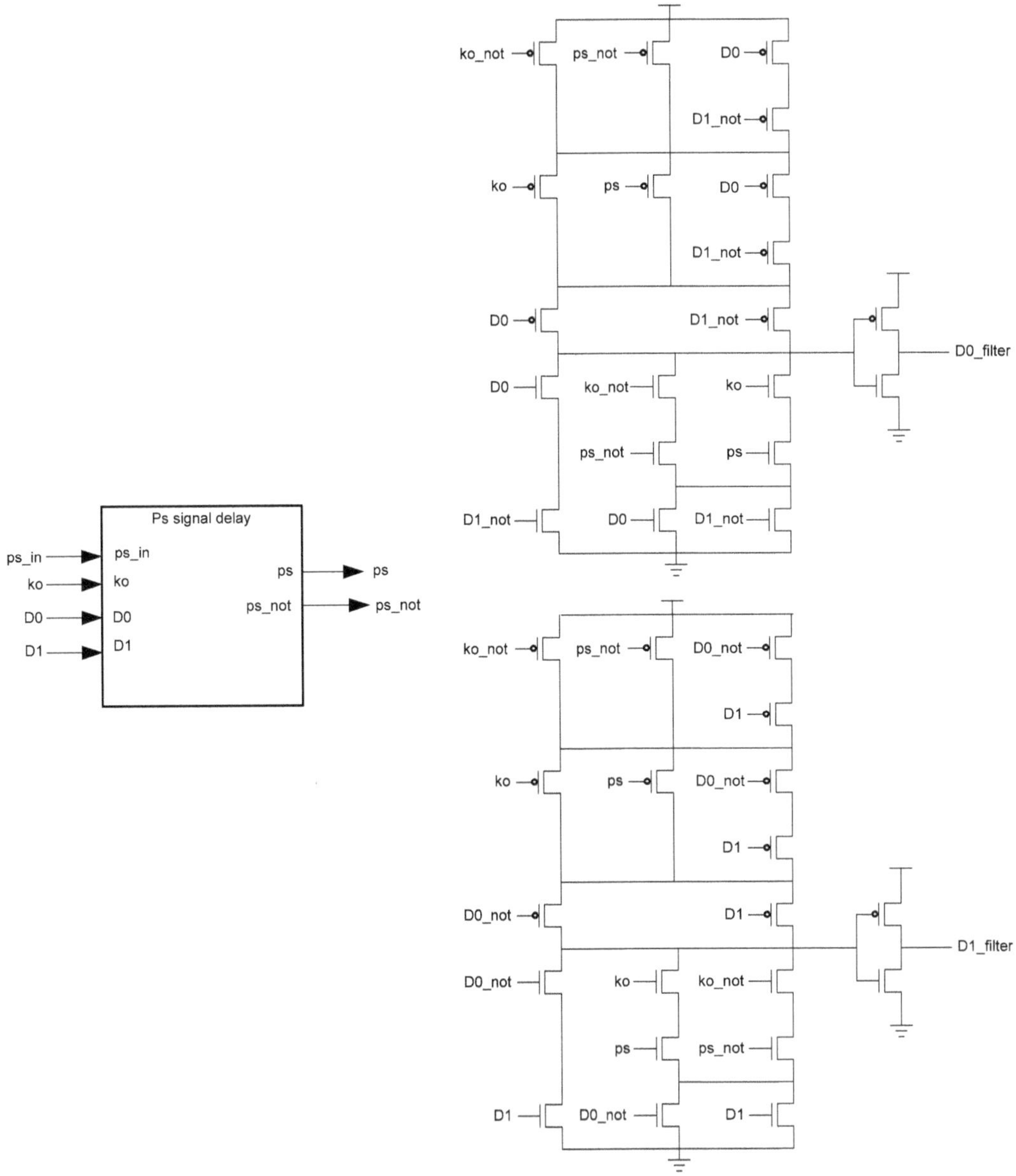

The PS signal delay component used in the spacer filter, shown in Figure 16, prevents *ps* from changing unless the register's K_o is logic 1, *i.e.*, requesting DATA. This ensures that the value of *ps* is only changed once the register receives the spacer.

Figure 16. PS signal delay component.

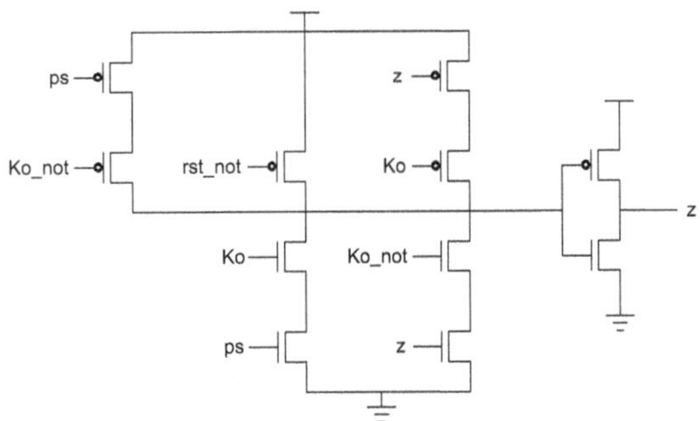

In situations where a component needs many cycles to output data but does not have input provided for each cycle, the component will not be able to receive the spacers it needs as input. Instead, a spacer generator register is used to generate these spacers for the component. A spacer generator register is a basic D^3L register with a spacer generator sitting between it and its inputs. The spacer generator keeps track of the previous spacer and generates the alternate spacer when requested regardless of the dual-rail input it receives. For example, if the previous spacer was an AZS and the register requests a spacer, the spacer generator will generate an AOS. The next time a spacer is requested, it generates an AZS. Figure 17 shows the D^3L spacer generator register; and Figure 18 shows the circuit diagram for the spacer generator component. The outputs of the spacer generator are given by the following equations:

$$\mathrm{D0\_gen} = \overline{\mathrm{Ko}}\,\overline{\mathrm{ps}}\,(\mathrm{D0}+\mathrm{D1}) + \mathrm{Ko}\,\mathrm{ps}\,(\mathrm{D0}+\overline{\mathrm{D1}}) + \mathrm{Ko}\,\mathrm{D0}\,\overline{\mathrm{D1}}$$

$$\mathrm{D1\_gen} = \overline{\mathrm{Ko}}\,\overline{\mathrm{ps}}\,(\mathrm{D0}+\mathrm{D1}) + \mathrm{Ko}\,\mathrm{ps}\,(\overline{\mathrm{D0}}+\mathrm{D1}) + \mathrm{Ko}\,\overline{\mathrm{D0}}\,\mathrm{D1}$$

Figure 17. D^3L spacer generator register.

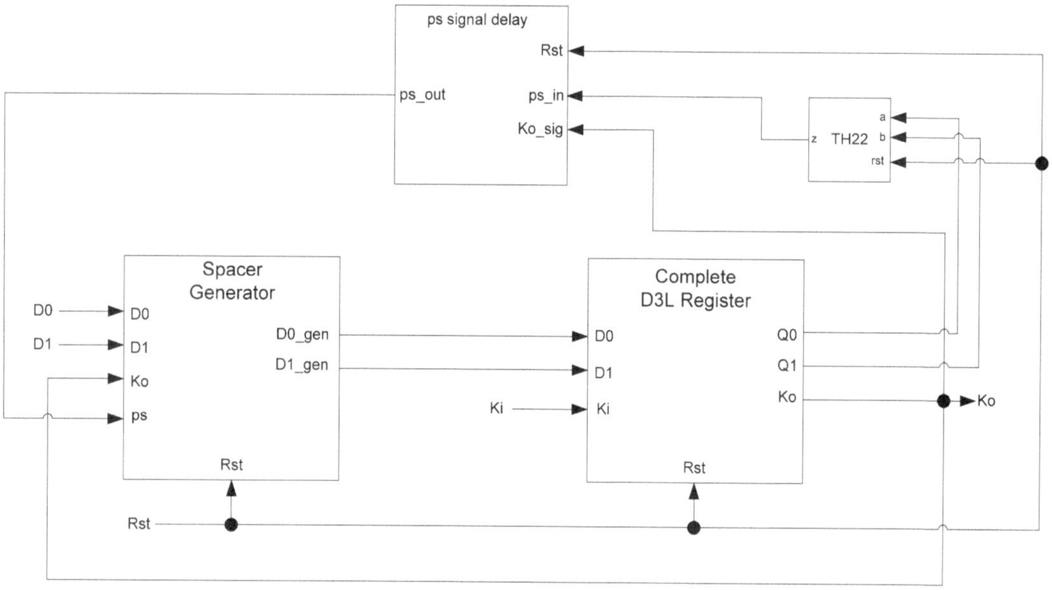

Figure 18. Spacer generator circuit diagram.

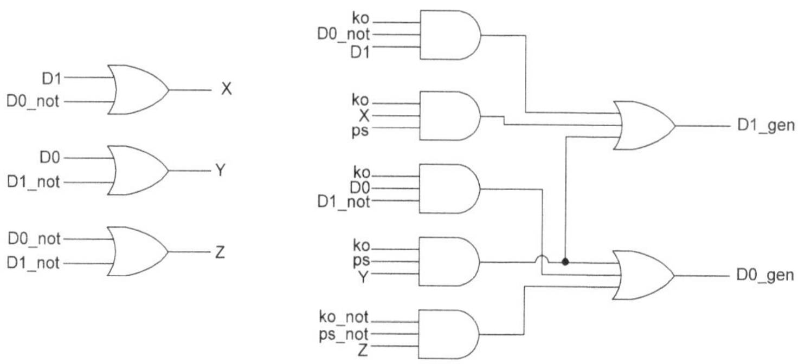

3. Multi-Threshold Dual-Spacer Dual-Rail Delay-Insensitive Logic (MTD^{3}L)

3.1. Motivation

Although the D^{3}L scheme successfully implements the dual-spacer protocol, it suffers from high overhead compared to equivalent NCL designs. This overhead comes from two sources. The first is the required NCL-X style completion logic in the form of several XNOR gates attached to each logic function. The second is the more complex registration. To eliminate the first source of overhead, the MTNCL technique can be combined with D^{3}L, to form the new secure hardware paradigm presented in this paper, MTD^{3}L.

Because D^{3}L gates do not use hysteresis, an external source is required for input completion detection. Rather than using XNOR and threshold gates, the early completion technique used in MTNCL can be utilized to ensure that requests for a spacer will only be generated when all circuit inputs are that spacer and the following stage is requesting a spacer. At this point, the combinational logic can be slept to the proper value, ensuring input-completeness. Thus, the need for extra completion checking logic is eliminated.

3.2. MTD^{3}L Gates

The addition of sleep logic to a D^{3}L gate is simple. No modification to the D^{3}L logic is required because it already matches the form of the modified NCL gates used in the MTNCL technique—a *hold*0 block and a set block. The only modification required is the addition of the sleep transistors. The sleep-to-0 transistors can be used in the same way as in SMTNCL. These transistors are responsible for the AZS transition. A similar set of transistors can be used for the AOS transition.

The sleep transistors are controlled by a pair of sleep signals, sleep-to-0 ($s0$) and sleep-to-1 ($s1$), and their complements, as shown in Table 4. These signals should not be asserted at the same time. Instead, if either of the inputs is asserted, the circuit will be slept to the appropriate value. Figure 19 shows the MTD^{3}L gate design. When $s0$ is asserted, the circuit is slept to the all-zero state. In this case, the NMOS transistor parallel to the output inverter is turned on, the NMOS transistor gating the main circuit to ground is turned off, and the PMOS transistor gating the output circuit to V_{DD} is turned off. Additionally, since $s1$ is off, the NMOS transistor controlled by $ns1$ is turned on, completing the path from the output to ground, forcing the output to logic 0. The PMOS transistor controlled by $s1$ is also turned on, allowing the main circuit to pass an output of 1 to the output inverter, preventing glitches from occurring when $s0$ is later asserted. When this happens, the output inverter will have logic 1 on its input, so it will continue to output logic 0 until new data has arrived. Similarly, when $s1$ is asserted, the circuit is slept to the AOS. The path to V_{DD} for the main circuit is turned off while the path to ground remains on, allowing a 0 to eventually reach the output inverter. The output inverter's path to ground is cut off and a direct path to V_{DD} is formed, forcing the output to be logic 1. When the sleep-to-1 state ends, the output inverter will have logic 0 on its input so the output will remain at logic 1, preventing a glitch. If neither sleep signal is asserted, the circuit operates as it would normally. All four power- and ground-gating transistors are turned on, allowing normal access to power and ground for the circuit and output inverter. The two parallel output transistors are turned off, so the output is only controlled by the output inverter. If both sleep signals happen to be asserted at once, the four power- and ground-gating transistors will be turned off, leaving the circuit in a floating state; however, this will never occur in a properly operating circuit.

Table 4. MTD³L sleep signals.

*s*0	*s*1	**Output**
0	0	Normal
0	1	All-One Spacer
1	0	All-Zero Spacer
1	1	Invalid

Figure 19. (**a**) MTD³L version 1 gate structure; (**b**) MTD³L23 version 1 implementation.

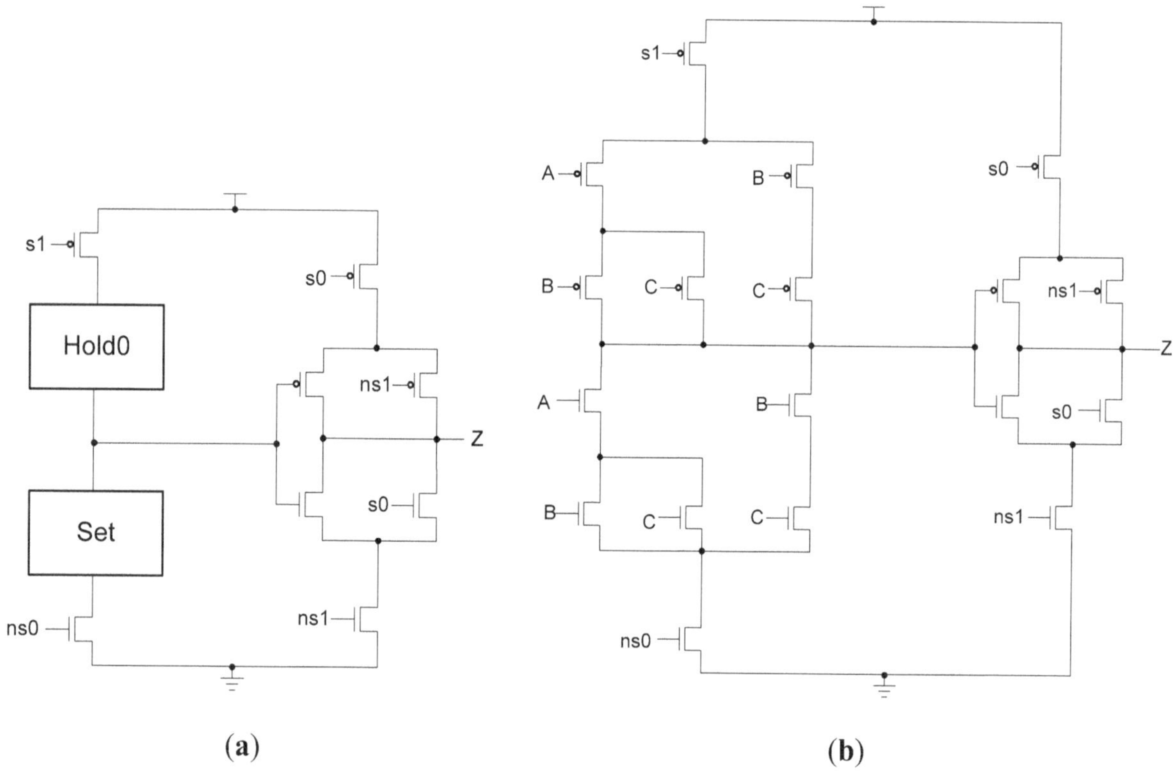

One of the drawbacks of the version 1 design is the potential for very large fanouts on the sleep signals. If the design is coarsely pipelined or the combinational logic happens to be very large, a single set of sleep signals may have to drive thousands of gates, requiring these signals to be heavily buffered. Not only must *s*0 and *s*1 be buffered but their complements will require buffering as well. To mitigate this issue and reduce the number of inputs to these gates in general, a modified design may be used to eliminate the need for the complemented sleep signals, as shown in Figure 20. This design removes the power- and ground-gating transistors from the main circuit, leaving only the four transistors on the output inverter. These four transistors are controlled by *s*0 and the complement of *s*1, allowing for the removal of *s*0's complement and *s*1 itself. Thus, only two signals must be buffered instead of four. The drawback to this technique is that the main circuit is directly exposed to power and ground, eliminating the ability to gate the circuit with high-V_t transistors.

Figure 20. (**a**) MTD^{3}L version 2 gate structure; (**b**) MTD^{3}L23 version 2 implementation.

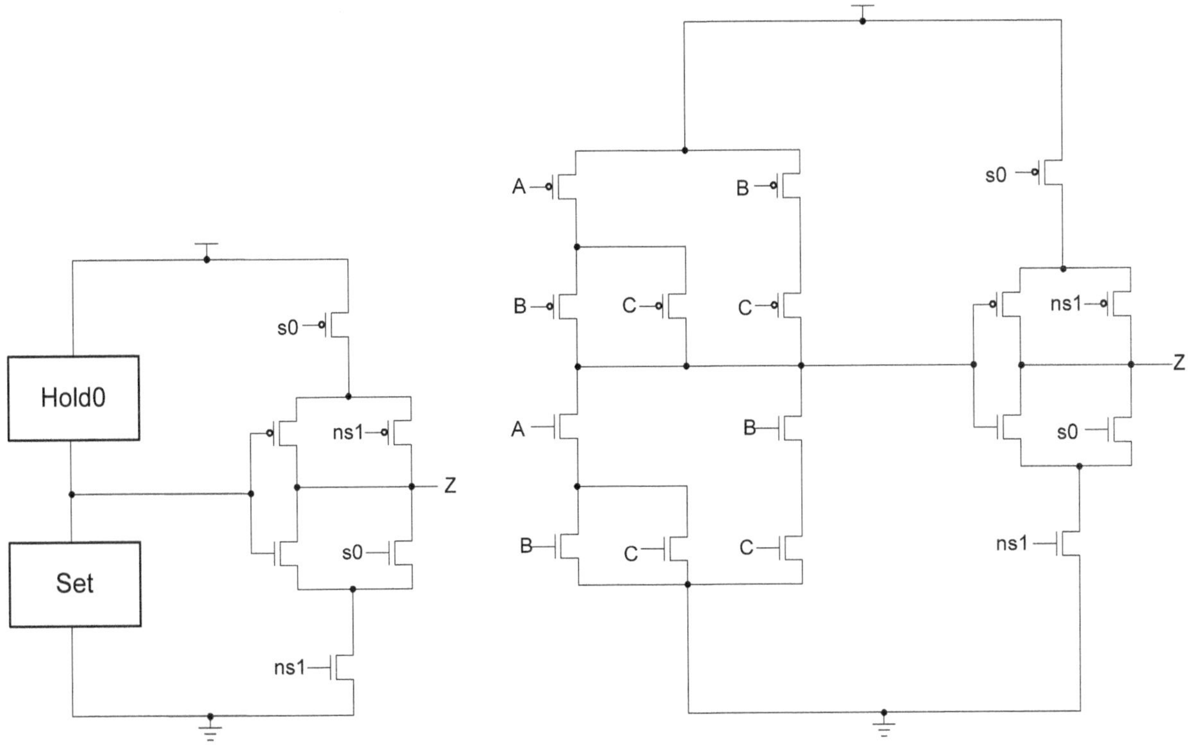

(**a**) (**b**)

3.3. *MTD^{3}L Registration*

The basic MTD^{3}L register, shown in Figure 21, is a modified NCL register. It consists of two TH22 gates which are resettable to the desired value. An XNOR gate facilitates early completion by checking the relative values of the register's inputs. If both input rails have the same value then the register has received a spacer and will request data. If the values are different, then DATA has been received so the register will request the next spacer.

Figure 21. MTD^{3}L register.

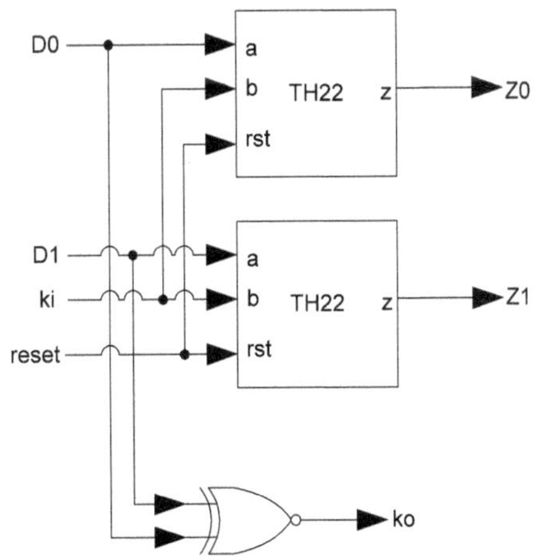

Additional logic is required to facilitate the dual-spacer protocol and early completion checking. The early completion component consists of resettable TH22 gates whose inputs are the register's K_o

and the next stage's inverted K_o. The reset state of the early completion component is logic 1 if the register's reset state is NULL and logic 0 if the register's reset state is DATA. In order to ensure input-completeness, the early completion K_o is inverted before being passed back as the register's K_i input. This prevents a partial spacer wavefront from passing through the register by ensuring that all of the register's inputs are an AOS or AZS before the spacer wavefront is allowed to pass through the register. In order to facilitate the dual-spacer protocol, the same *Ki* Generator used in D^3L registration, shown in Figure 12, is used for MTD^3L.

If the register needs to supply sleep signals then sleep signal logic is used here as well, as shown in Figure 22. This logic generates two sleep signals, *s*0 and *s*1. The values of the sleep signals are shown in Table 5. If the register's K_i is 0 then a spacer is being requested. To determine which spacer is being requested, *Ki_gen*'s value is used. If *Ki_gen* is logic 0 then an AZS is being requested; if it is logic 1 then an AOS is being requested. To avoid incorrect sleep states, a buffer is used as a delay element to ensure that the change in *Ki_gen*'s value is evaluated first. For example, if the desired change was from the no sleep state of row 2 to the sleep-to-1 state of row 3, then both *Ki_gen* and K_i will switch. If K_i switches first, a sleep-to-0 will be issued erroneously. However, if *Ki_gen* switches first then the no sleep state will be maintained until K_i changes as well, resulting in the correct sleep-to-1 state.

Figure 22. Complete MTD^3L register.

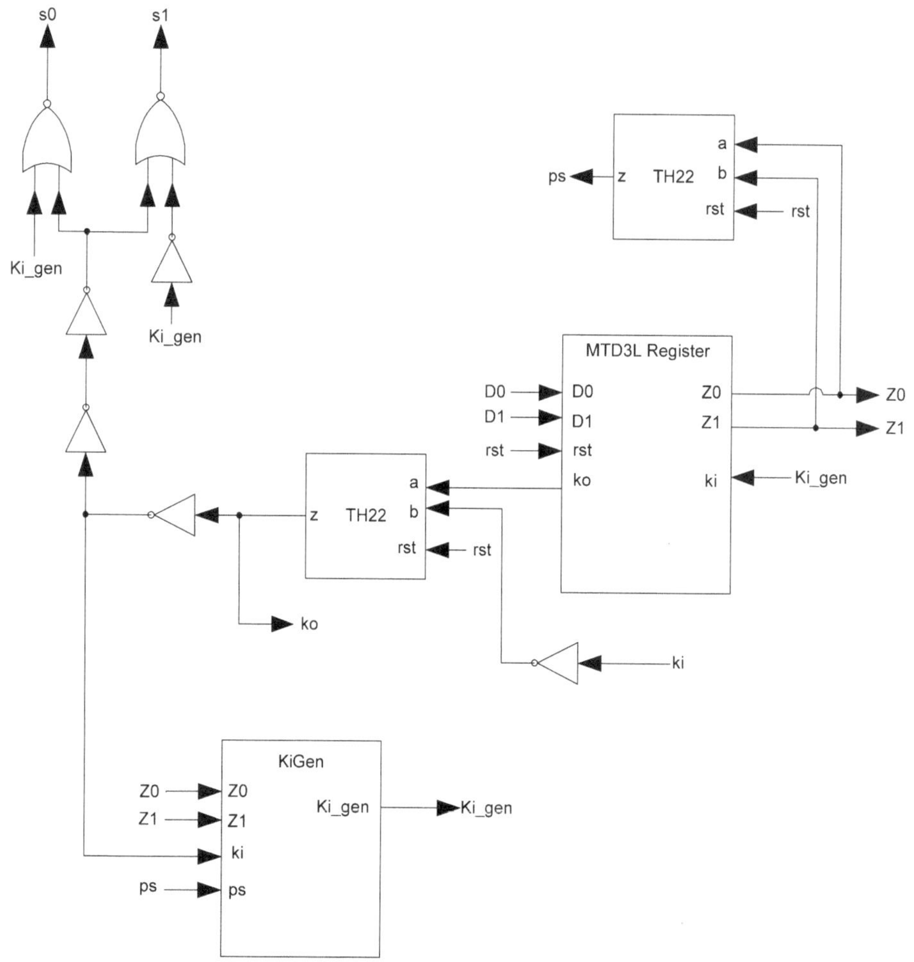

Table 5. MTD^{3}L sleep signal switching sequence.

Ki_gen	K_i	s0	s1
0	0	1	0
0	1	0	0
1	0	0	1
1	1	0	0

MTD^{3}L spacer filter and spacer generator registers, shown in Figures 23 and 24, respectively, are used in the same manner as they are used in D^{3}L circuits. The filter register is used as the final register in a register ring, shown in Figure 25. It is reset to DATA and filters the spacer that passes through the ring, alternating it so that the dual-spacer protocol is enforced. The spacer generator generates the appropriate spacer as needed regardless of the values of its inputs. Typically, these registers are the ones that generate sleep signals as they are usually the registers that are facing combinational logic as shown in Figure 25. The actual spacer filter and spacer generator components are unmodified from their D^{3}L counterparts, shown in Figures 15 and 18, respectively.

Figure 23. MTD^{3}L spacer filter register.

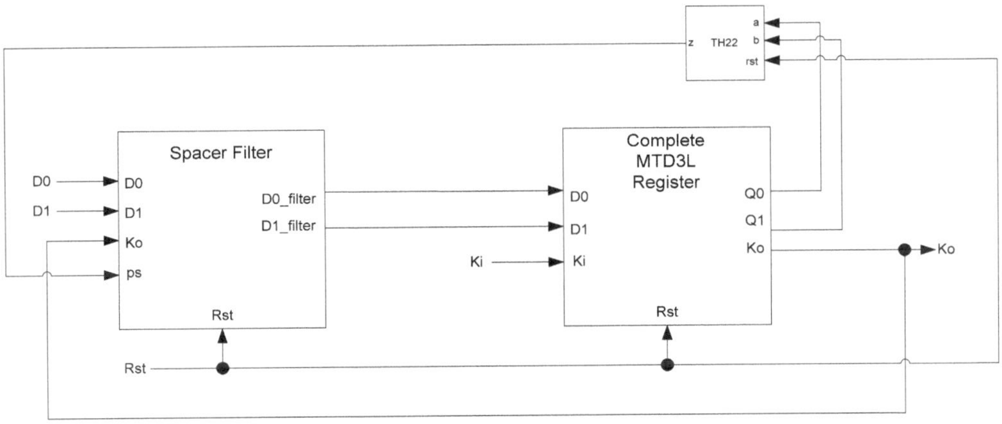

Figure 24. MTD^{3}L spacer generator register.

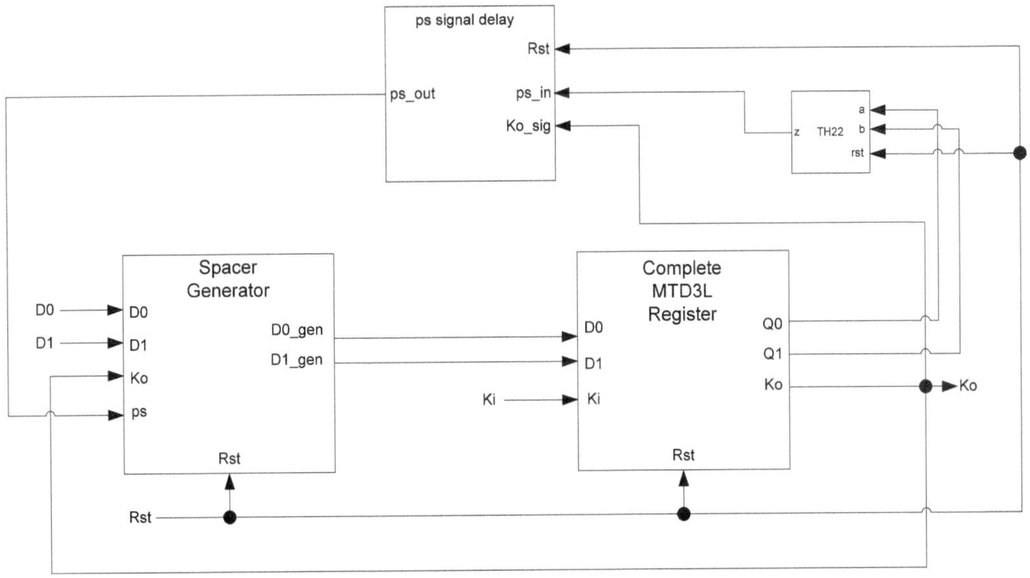

Figure 25. MTD^{3}L ring register.

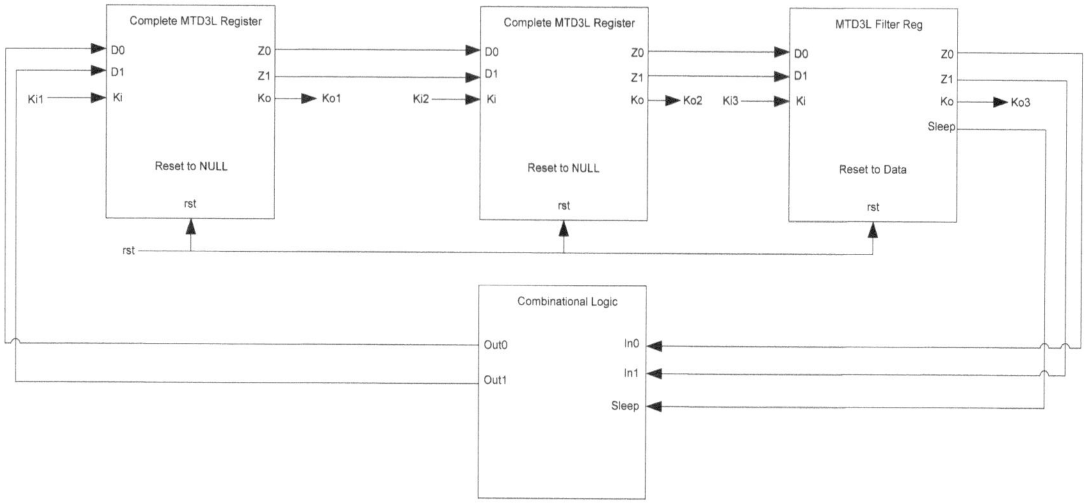

4. AES Core Implementations

4.1. MTD^{3}L Implementation

The implementation of the 128-bit AES core [46] in each design follows the same architecture. The AES transform and key expansion functions are computed in parallel. A Control block synchronizes the two functions and ensures that the correct sub-key is sent to the Transform block at the correct time. From outside, this circuit behaves as a register in terms of handshaking, so it can be easily placed into an asynchronous system. The AES core accepts an input and produces an output with one external DATA/spacer cycle. However, the AES circuit actually produces several internal DATA/spacer cycles for processing each plaintext, which are hidden from the outside. Each round of the AES algorithm consists of one of these internal cycles.

As shown in Figure 26, each design consists of five blocks: the FirstRound block, the Control block, the AESTransform block, the KeyExpansion block, and the LastRound block. The FirstRound block is a set of input registers that latch in new data and provide it to the AESTransform and KeyExpansion blocks. The AESTransform block performs the ciphertext calculation for each round of the algorithm. The KeyExpansion block calculates the subkey used in the AESTransform block. The Control block creates the control signals as well as generating the *RCon* constant which is used in the KeyExpansion block. The LastRound block performs the final round of calculations and also has a set of output registers to hold the final ciphertext.

Figure 26. Advanced Encryption Standard (AES) top level diagram.

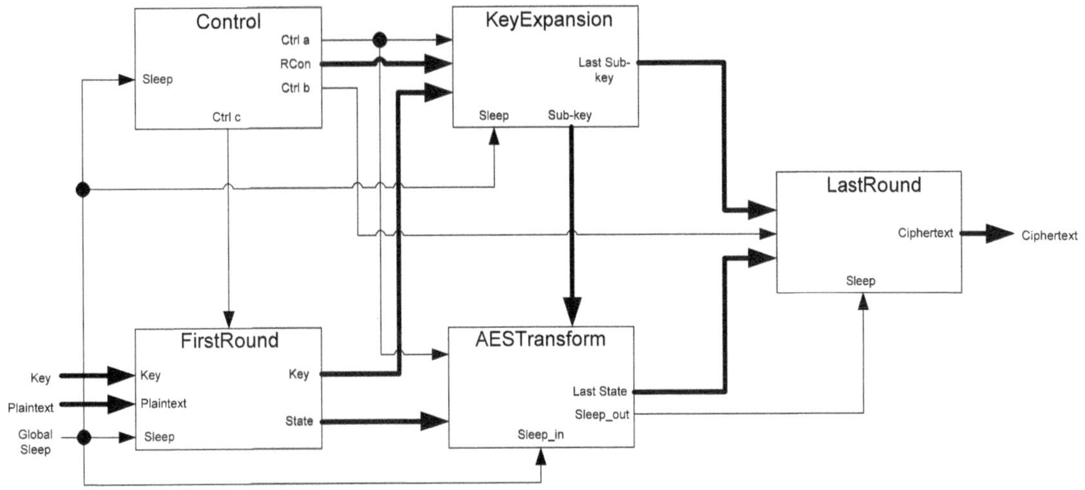

The communication between blocks, shown in Figure 27, consists of three handshaking signals: *Ki*1, *Ki*2, and *Ki*3. These signals are generated by a set of threshold gates in the top level circuit. The inputs of these threshold gates are the K_O values from registers throughout the design. The AESTransform, KeyExpansion, and Control blocks all have a three-ring register to save processed data. Each ring register's K_o is used as inputs to the top-level handshaking. Additionally, an extra K_O from the registers in the FirstRound and LastRound blocks is used as well. The resulting *Ki*1, *Ki*2, and *Ki*3 signals are used as the K_i inputs to the ring-registers mentioned previously.

Figure 27. Top level handshaking diagram.

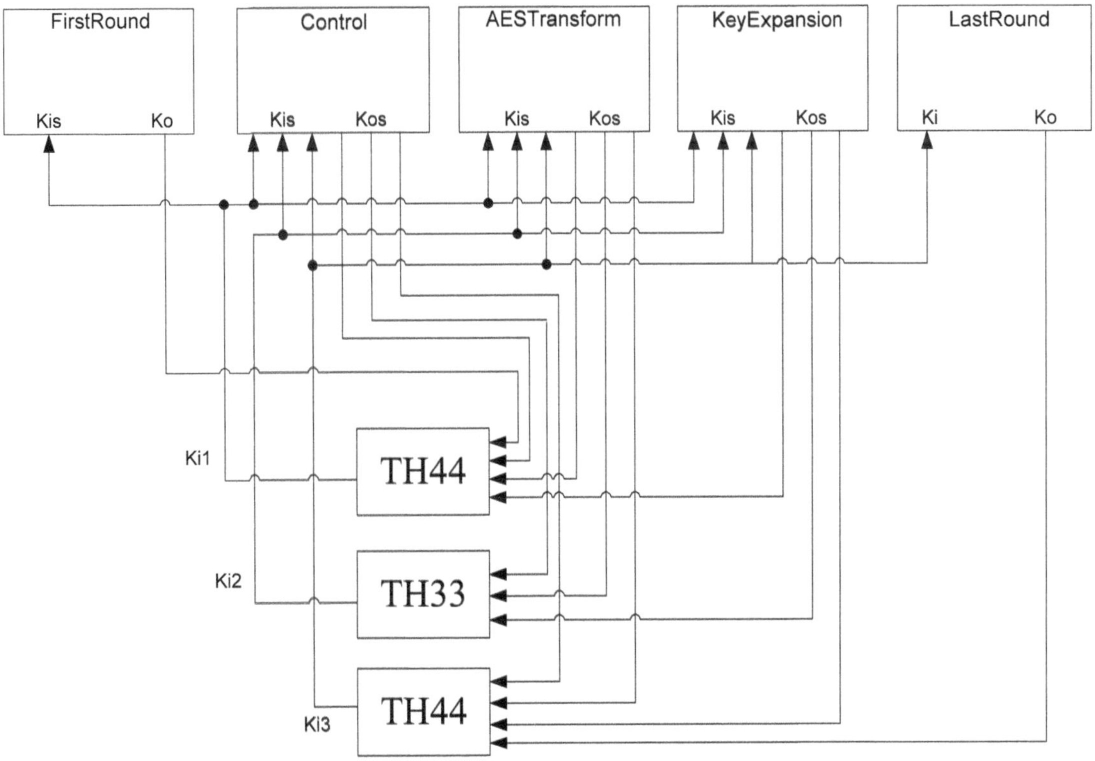

The sleep signal generation mechanism consists of two types of sleep signals: a global sleep and local sleeps. The global sleep, which is a circuit input, is meant to sleep the entire circuit between encryption stages, ensuring that everything is reset properly. This sleep is only asserted after the ciphertext is latched out and external handshaking is requesting a spacer cycle. The internal sleep signals are generated locally within each block by the ring-registers. These sleeps are asserted between calculation stages. The LastRound block, lacking registration between it and the previous AESTransform block, uses the local sleep signals generated by the AESTransform block.

The FirstRound block, shown in Figure 28, consists of a set of 128-bit registers for both the input plaintext and input keys. These are regular, non-slept generating registers. The K_i to these registers is the master K_i input for the circuit. The K_os are used to generate a reset signal for the rest of the blocks, preparing them for a fresh encryption process. This reset signal also acts as the circuit's K_o output. The data outputs of these registers go to a set of spacer-generator registers. These registers use the $Ki1$ signal as their K_i, synchronizing these registers with the first stage of the ring-registers throughout the design. Their K_os are not used. Next, an AddRound block calculates the initial addition of the plaintext and key, passing this data to the AESTransform block. The initial key is also passed to the AESTransform and KeyExpansion blocks. The sleep signals for the AddRound block are generated by the data processing spacer-generator register.

Figure 28. FirstRound block diagram.

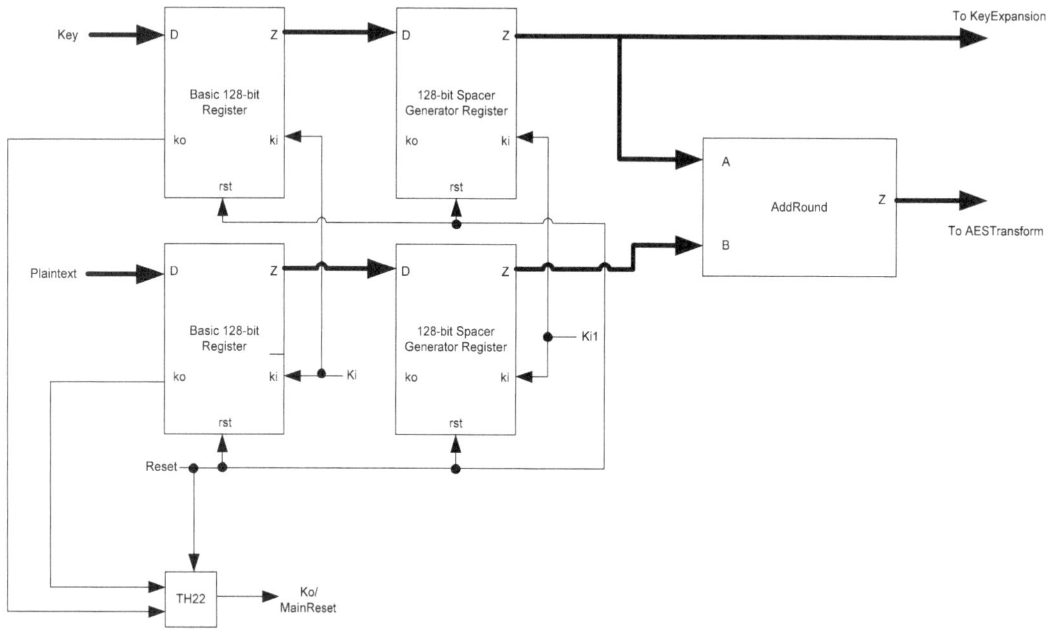

The AES Transform block, shown in Figure 29, handles the primary calculation of the ciphertext. It consists of the main AES components: AddRound, SubByte, MixColumns, and Shiftrows. A three-ring register is used to save data between each round of calculation. To control whether the input data is taken from the FirstRound or from the circuit feedback loop, a 2-to-1 multiplexer is used. The control signal of the multiplexer is supplied by the Control block. The output of the multiplexer serves as both the output of the circuit and the feedback loop to the first register in the ring. The first and second registers are normal registers while the third is a filter register. This filter register generates the sleep signals to be used by this block as well as the subsequent LastRound block.

Figure 29. AES transform block diagram.

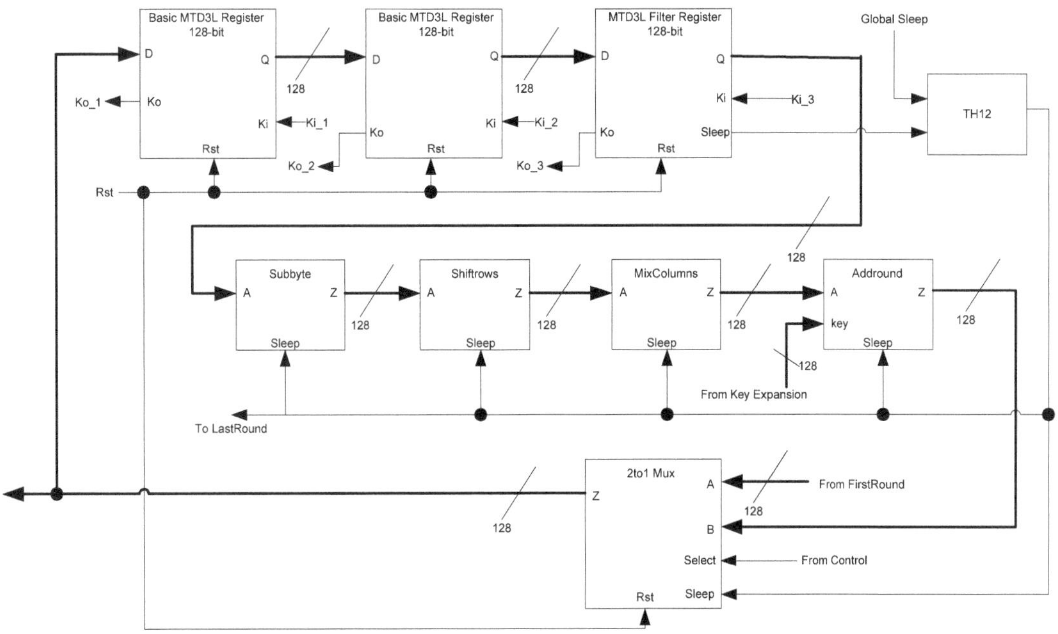

The KeyExpansion block, shown in Figure 30, is very similar to the AES Transform block. It consists of the KeyExpansion logic, a three-ring register, and a 2-to-1 multiplexer. The register and multiplexer act in the same manner as they do in the AES Transform block. The sleep signals for the KeyExpansion logic are generated by the third register in the ring, which is a filter register. The KeyExpansion logic takes the 128-bit key and transforms it into a new subkey, which is sent to the AES Transform and LastRound blocks.

Figure 30. KeyExpansion block diagram.

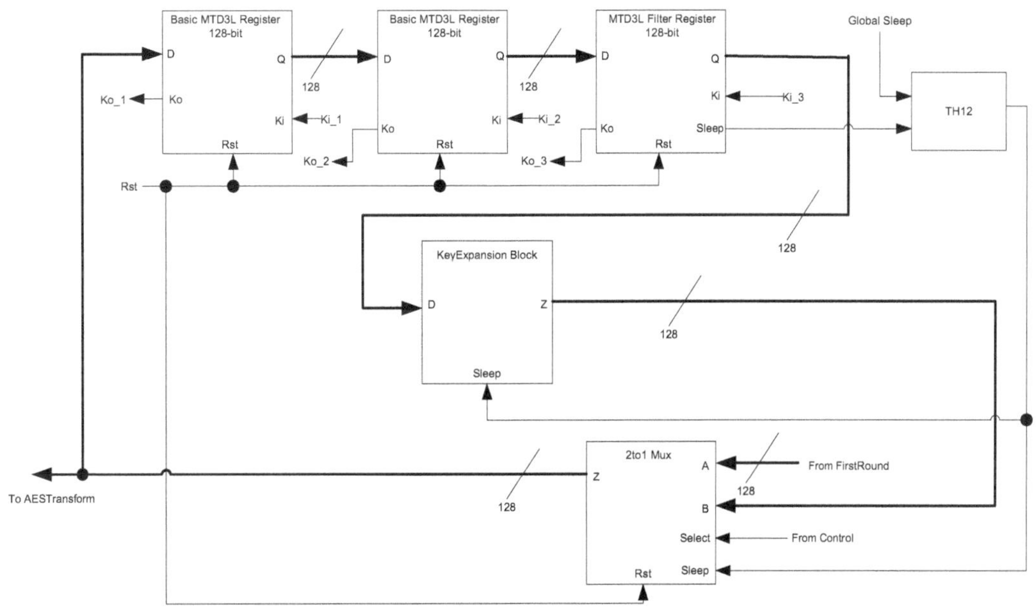

The Control block, shown in Figure 31, synchronizes all other blocks to ensure proper operation. It generates the control signals for the multiplexers throughout the design, the *RCon* constant used in the

KeyExpansion calculations, and the control signal that activates the LastRound block. The LastRound block will only latch data to the output register when this signal is received. The input to the Control block is the reset signal generated by the K_o of the FirstRound block. This signal resets the Control block's state, preparing it for a new encryption set.

The Control block is a state machine consisting of four components: a 5-bit three-ring register, the control signal logic, the increment logic, and the LastRound logic. The ring-register saves the state of the Control block. The filter register in this ring also generates the sleep signals for the block. The control signal logic generates the *RCon* signal and the multiplexer control signals. The increment logic is a counter, keeping count of the number of encryption rounds executed. When the counter reaches the round limit, the stop signal is activated.

The initial state of the Control block needs to be an AZS. However, the filter register in the ring-register is reset to data. Thus, extra logic is required to generate the initial AZS. A set of TH22n gates is used to do this. Each TH22n gate is tied to one of the signals, which is connected to both inputs of the gate. At reset, each of these gates will output a zero, giving the required spacer. Once the reset is finished, these gates will act as C-elements. Since each gate has the same signal on both inputs, its output will follow the value of the input.

Figure 31. Control block diagram.

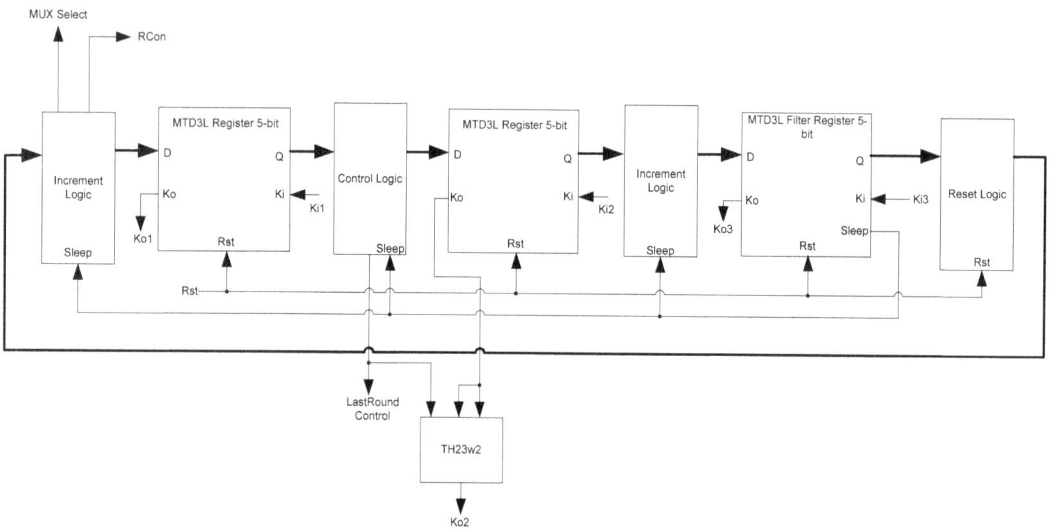

The LastRound block, shown in Figure 32, consists of the final SubByte, Shiftrows, and AddRound operations on the ciphertext as well as the final output register. The MixColumns block is not used in the final round. The inputs to this block are the last transformed ciphertext from the AESTransform block and the final subkey from the KeyExpansion block. The sleep signals are the same ones generated in the AESTransform block. The register will only latch data once the stop signal is received from the Control block, which acts as the K_i to the register. This final register is a filter register, ensuring that the output from the AES circuit is capable of generating the correct spacer and sleep signals for the following circuit block.

Figure 32. LastRound block diagram.

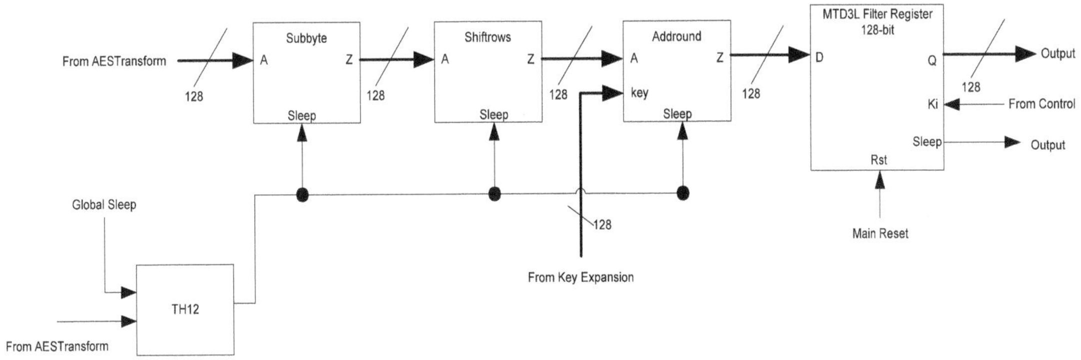

4.2. AES Implementation using D^3L

The D^3L implementation is very similar to the MTD^3L implementation. The same five blocks are used and their configurations are essentially the same. There are two primary differences between the two designs. First, since the D^3L design lacks sleep signals, a global reset is used to reset the spacer-generator registers in the FirstRound between encryptions. This reset is required for the circuit to reset properly. The second major difference is the usage of completion signals. Each combinational block has a completion signal which is used to ensure input-completeness. The completion signals (ccheck) of the FirstRound and LastRound blocks are added to the $Ki1$ calculation. In the AES Transform block, the $Ko1$ output consists of the combination of the K_o from the first register as well as the ccheck signals from the multiplexer and AES logic. This is also done in the KeyExpansion and Control blocks. The ccheck signals themselves are generated with completion checking logic inside each combinational logic block.

4.3. AES Implementation using NCL

Being extensions of the NCL style, the D^3L and MTD^3L designs are very similar to the NCL design as well. Additionally, the NCL design served as the base of the D^3L design, which was created by taking the original NCL design and replacing the gates and signals. Because of this, the structure of the blocks themselves and the combinational logic are very similar. The primary differences in the designs are found in registration, spacer generation, and input-completeness logic. Figure 33 shows the NCL AES Transform block.

NCL threshold gate logic supports input-completeness by design, so there is no need for extra input-completeness logic. Additionally, only a single register type is needed, which can be reset to the desired values. Finally, the FirstRound block, shown in Figure 34, only needs to generate NULL spacers when needed, so the spacer-generator registers are replaced with NULL generators.

Figure 33. NCL AES Transform block diagram.

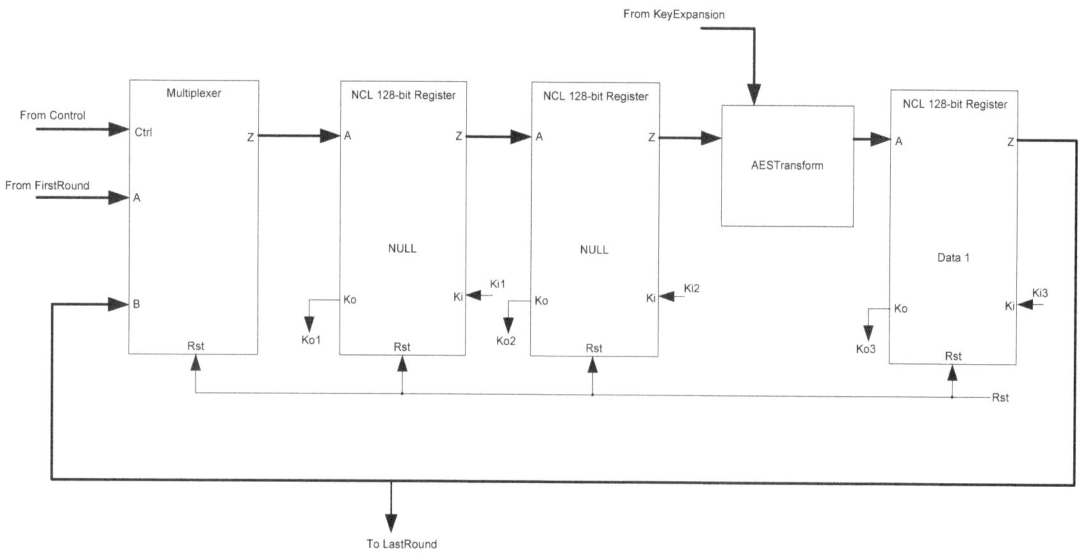

Figure 34. NCL FirstRound block diagram.

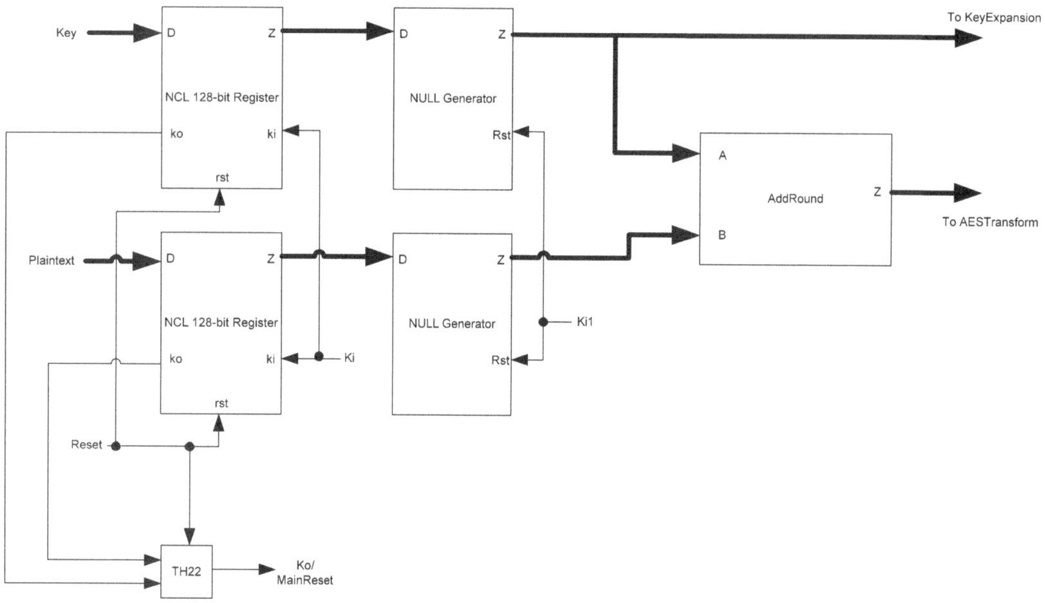

4.4. AES Implementation using Synchronous Logic

The asynchronous designs are all based off the synchronous design, which was created first. Because of this, there are several similarities between the synchronous and asynchronous designs. The synchronous design, like the other designs, consists of five logic blocks. The FirstRound block is replaced with a single AddRound block. The initial register stage is instead found in the KeyExpansion and AES Transform blocks. Registers are found in the AESTransform, KeyExpansion, and Control blocks, much like the other designs. The LastRound block contains the output register which is activated with a stop signal generated by the Control block. Two designs can be used, a single clock design and a double clock design. The single clock design was used in the power and timing simulations since it is faster. The double clock design isolates the SubByte block of the AES

Transform circuit in facilitating the implementation of the CPA attack. Figure 35 shows the synchronous AES Transform block.

Figure 35. Synchronous AES Transform block diagram.

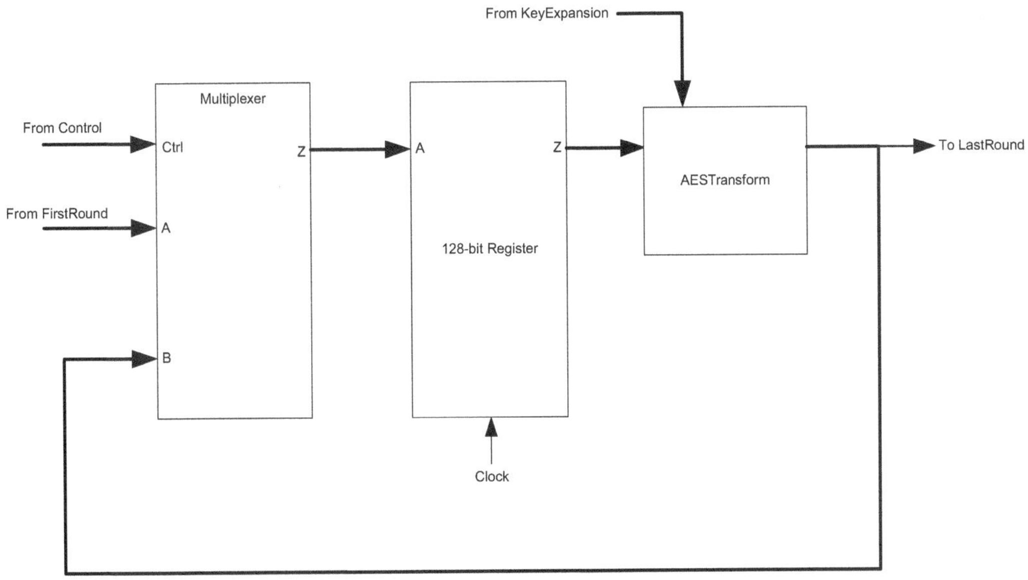

4.5. Delay Element Design

In the D^{3}L and MTD^{3}L designs, a delay element was designed to help mask the relationship between the circuit delay and the data being processed. The designed delay element is a buffer chain appended to the output of the targeted Sbox that can either operate at a high speed or a low speed. The inverters that make up this chain are n-volt controlled. A second NMOS transistor is added to the pull-down network of the inverter. An example buffer is shown in Figure 36. This transistor is controlled by an external voltage. To operate at high speed, the voltage given is the same as the designated supply voltage. To operate at a slow speed, the voltage is reduced as much as possible without deforming the inverter's output.

Figure 36. N-Volt controlled buffer.

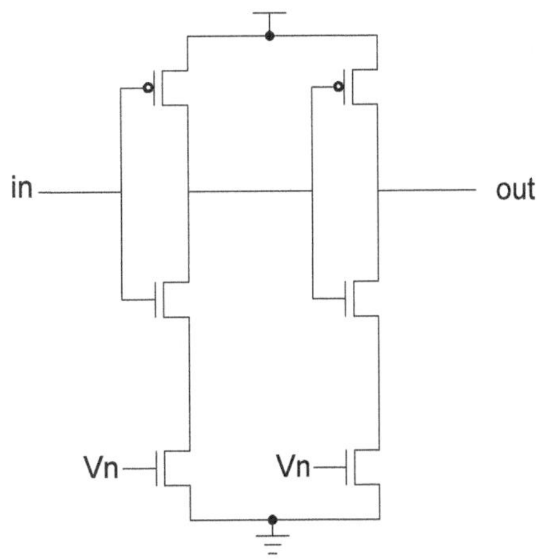

To apply different voltages levels to these inverters, a control signal is generated externally. The method of generating this control signal is not important so long as the result is sufficiently random. This control signal is used to control a pair of PMOS transistors. The source of these transistors is either a high voltage source or a low voltage source. Depending on the value of the control signal, only one of these transistors will be turned on at a time, which will supply its voltage to the delay inverters. Figure 37 shows the control logic.

Figure 37. Buffer chain with voltage control.

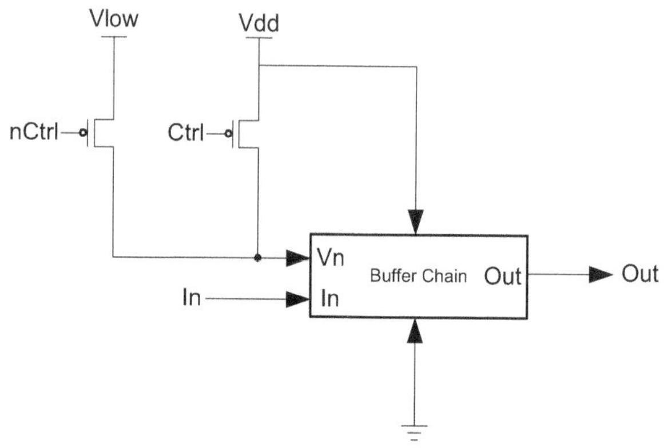

5. Results and Analysis

5.1. Simulation Environment and Data Collection

Each AES design presented in Section 4 was implemented at the transistor level using Cadence and the IBM 8RF-DM 130 nm process to evaluate and compare energy consumption, speed, area, and side-channel attack resistance. The full AES designs were used for the collection of energy, speed, and area data. All simulations were done using the Cadence UltraSim simulator. Each design was simulated using the input key 0x2b7e151628aed2a6abf7158809cf4f3c.

Because a complete evaluation of the D^3L and MTD^3L designs requires two spacer cycles, each simulation covered two complete encryptions. For the synchronous design, the simulation begins with the circuit in its reset state. Next, the key and plaintext are given and the circuit operation continues until the ciphertext is received. On the next clock cycle, a second plaintext is entered and the second encryption cycle completes. The energy and speed of the design is calculated from the reset state until the time of completion for the second encryption. The synchronous design is controlled using vector files. The NCL, D^3L, and MTD^3L designs, being asynchronous, are more difficult to simulate using vector files, due to the difficulties in anticipating when the handshaking signals should be changed. Thus, the asynchronous designs are simulated using controllers defined with VerilogA, which monitors the outputs of the design and makes adjustments to the design's inputs accordingly. The NCL simulation begins in a NULL state. The first plaintext is passed followed by another NULL state. Once this cycle completes, a second plaintext is given followed by the third NULL state. The energy and speed data is calculated from the initial state through the end of the second DATA-NULL pair. The D^3L and MTD^3L simulations are similar, following the pattern of AZS-DATA-AOS-DATA-AZS.

The circuit area for each design was determined by laying out each design and measuring its dimensions. First, the cell layouts were completed using the Virtuoso tool in Cadence. The full designs were placed and routed according to their respective Verilog netlists using Synopsys Astro. From there, the designs' area could be measured. The speed of each design was simply the time it took for two encryptions to be completed and any extra time needed until a third encryption is ready to begin. The synchronous design was tested to determine its maximum clock speed. The energy consumption of each design was determined by measuring the charge of each design of the duration of the simulation period and multiplying this by the supply voltage, which was the same for each design.

5.2. Simulation Results

As shown in Table 6, the synchronous design is the fastest. The NCL design is the slowest; and the MTD^{3}L and D^{3}L designs are in the middle. The D^{3}L design uses the most energy followed by the MTD^{3}L design and the NCL design. As explained in Section 2, the D^{3}L design suffers from significant overhead problems, which can be seen in these results, particularly with respect to energy consumption. The purpose of the MTD^{3}L design was to reduce this overhead to more reasonable levels. In this respect, the MTD^{3}L design has a 36% reduction in energy consumption over the D^{3}L design. However, the MTD^{3}L design still uses nearly 300X more energy than the synchronous design, so while some overhead reduction was achieved, more efforts are needed in order to bring the MTD^{3}L design down to desired overhead levels.

Table 6. Speed and energy comparison.

Design	Delay (ns)	Delay overhead *vs.* Synchronous design	Energy (nJ)	Energy overhead *vs.* Synchronous design
Synchronous	153	0X	1.356	0X
NCL	462	302X	2.208	163X
D^{3}L	325	212X	6.012	443X
MTD^{3}L	330	216X	3.84	283X

Table 7 presents the area of each design after cell placement in Synopsys Astro. The MTD^{3}L design sees significant overhead reduction compared to the D^{3}L design. This can be attributed to the removal of the NCL_X style completion logic. With this overhead reduction, the MTD^{3}L design is comparable to the NCL design in area.

Table 7. Area comparison.

Design	Width (μm)	Height (μm)	Total Area (mm^2)
Synchronous	1227	1223	1.50
NCL	1812	1809	3.28
D^{3}L	2503	2503	6.27
MTD^{3}L	1835	1838	3.37

The overhead of the D^{3}L design came from two sources. The first was the NCL_X completion checking. The second was from increased registration overhead caused by the need of handling two spacers. The MTD^{3}L design solves the first problem but does not solve the second. The reduction in

MTD^3L energy consumption can be attributed to the reduced completion checking overhead due to the use of early completion instead of NCL_X style checking. However, the MTD^3L design still uses the same registration scheme that the D^3L design uses, so this overhead remains.

Table 8 shows a comparison of the energy consumption of different components of the NCL and MTD^3L designs. The energy cost of the MTD^3L registration is roughly 700% greater than that of the NCL design. The combinational logic and buffering of both designs is about the same, showing that the overhead cost comes almost exclusively from the registration. Table 9 shows the differences in transistor count and average leakage of the registration for both designs, showing that the MTD^3L version requires many more transistors. This is due to the logic required to implement the dual-spacer protocol, such as the *Ki* Generator and the Spacer Filter. These components contribute significantly to the design overhead.

Table 8. Energy comparison.

Component	NCL design energy (nJ)	MTD^3L design energy (nJ)
Registration	0.252	1.836
Logic	1.512	1.26
Buffering	0.444	0.744
Total	2.208	3.84

Table 9. Registration comparison.

Design	Transistor count	Average leakage (μA)
NCL	49140	2
MTD^3L	242364	180

5.3. Side-Channel Attack Implementation

Because of the long simulation times required for the full designs, data collection for the power and timing attacks were performed with sub-circuits of each design. This is because each of these attacks requires many different simulation samples (256 samples in this case) to be successful. This number of simulations with the full designs would be impractical. The sub-circuits consist of the initial Addround and Subbyte stage of each design. This is because the Subbyte operation is the major vulnerable point to side-channel attacks. The attacks themselves focus on only one S-box of the Subbyte block, brute forcing all 256 plaintext input combinations of that S-box and attempting to extract one byte of the cipher key. It is assumed that if one byte of the key can be extracted then the other 15 bytes can be obtained as well.

While UltraSim in Cadence was used for the full simulations, it was found that Synopsys Nanosim could perform simulations in less time. Because so many simulation samples were required for the side-channel attacks, Nanosim was used to collect this information rather than Ultrasim. The simulation setup is similar but different for each design. For example, the synchronous design requires a clock and registers while the asynchronous designs do not. Test vectors for each design were generated. These vectors all use the same key as shown in Section 5.1. The plaintexts cover every combination of inputs for the first byte. All following bytes were kept at 0x00. Thus, the plaintexts followed the form 0xff000000000000000000000000000000. This setup helps reduce noise by attempting to isolate the attacked S-box. To help Nanosim keep track of a particular node, a small 1 ohm resistor was placed in series between the attacked S-box and its power supply.

The synchronous design consists of the Addround block and the Subbyte block with the Subbyte block being isolated by a pair of registers, as shown in Figure 38. Initially, the registers are reset to 0. Next, the key and plaintext are input to the Addround block. On the first clock edge, Addround's output is clocked into the Subbyte block via the first register. Once the Subbyte block computes the output, the result is latched onto the second register. The second clock edge is the critical point. At this moment, the switching activity of the attacked S-box may be observed according to the changes in the second register.

Figure 38. Synchronous side-channel attack implementation.

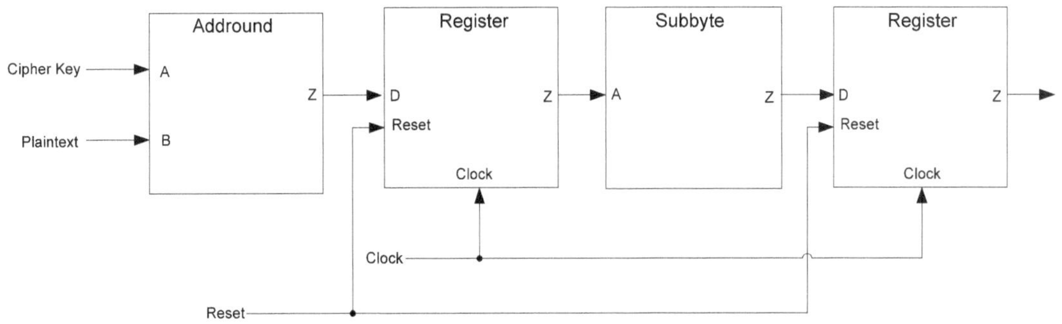

As shown in Figure 39, the NCL design consists of only the Addround and Subbyte blocks. It begins the simulation in a NULL state. Next, the key and plaintext are given as DATA. After the computation is complete, a NULL wave front is input. Thus, the NCL simulation follows a NULL-DATA-NULL pattern. For the timing attack, the simulation output is divided into two sets. The first set consists of the NULL to DATA transition. The second consists of the DATA to NULL transition. Both sets have any leading and trailing zeros removed to help align and simply the data for the timing attack.

The D^3L and MTD^3L designs also consist only of Addround and Subbyte blocks. Like the NCL simulation, these designs begin at the AZS. The plaintext and key are input as DATA. Once the circuit is finished, an AOS is given next. The same plaintext and key are input as DATA a second time. Once the second computation is complete, another AZS is given. Thus, these designs follow a complete dual-spacer pattern of AZS-DATA-AOS-DATA-AZS. The same data is given twice to help simplify the power and timing attacks. These data sets are divided into four groups for the timing attack. The first is the transition from the AZS to DATA. The second is the transition from DATA to the AOS. The third is from the AOS to DATA. The last is from DATA to the AZS. Each set has leading and trailing zeros removed much like the NCL data.

Figure 39. NCL, D^{3}L, and MTD^{3}L side-channel attack implementation.

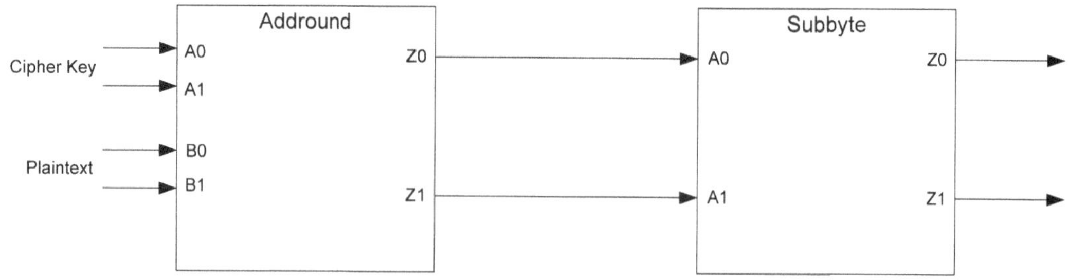

The D^{3}L and MTD^{3}L designs were also tested with the random delay elements, presented in Section 4, inserted into the targeted S-box, as shown in Figure 40. These delay elements are attached to each of the outputs of the target S-box. They are controlled by a signal generated by the bitwise XOR of the outputs of the Addround Z1 rail. The simulation setup is otherwise the same as the previous D^{3}L and MTD^{3}L simulations.

Figure 40. D^{3}L and MTD^{3}L delayed side-channel attack implementation.

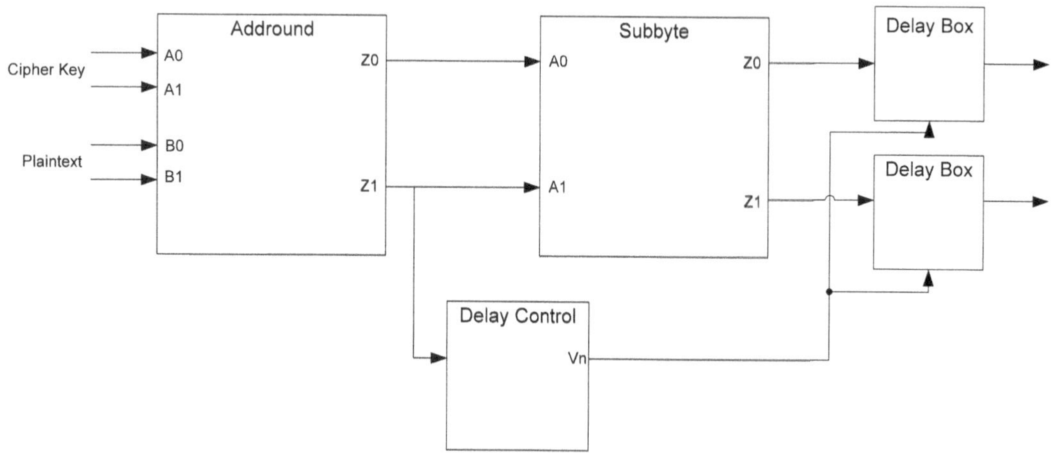

The power and timing attacks were carried out with a Java program. The program takes the simulation data and a statistical model of the design as input. The format of the simulation data depends on the type of attack being done. For the synchronous design, the only attack was a power analysis attack. The simulation input was in the form of a set of values of the current drawn by the power supply at a given time. The statistical model was the calculated Hamming distance given the key guess and the given plaintext. The NCL design was attacked using an energy based model. The simulation data began as a set of current values, similar to the synchronous data. This data was summed over the duration of the simulation period resulting in a single value representing the energy used in the simulation. The D^{3}L and MTD^{3}L designs were also attacked with an energy based model. In addition to this attack, the D^{3}L and MTD^{3}L designs were also attacked with a timing model. Similar to the energy attack, the power data was converted to a single time value by measuring the number of time steps where the circuit was drawing power. It is assumed that the circuit only draws significant current while it is calculating the ciphertext. The total number of discrete time steps above this threshold represent the time it took to complete the operation. Because the asynchronous designs are designed to balance switching activity independently of the data given, the Hamming distance model used for the synchronous attack is not sufficient for the asynchronous attacks. Instead, the expected energy and timing of these designs was estimated by counting the expected number of transistor switches for a given input combination. It is assumed that the energy and time delay of the circuit for a given set of inputs is directly proportional to the number of transistors that switch as a result of those inputs. The energy attacks were carried out over the complete simulation data. The timing attacks were carried out over the partitioned data.

5.4. Side-Channel Attack Results

Table 10 shows the results of the power- and energy-based attacks. The highest correlation out of the set of key guesses is shown for each design as if the highest correlation guess was generated by the correct key value. For the timing attacks against the asynchronous designs, which were each partitioned into several parts, only the part that resulted in the highest correlation is given. For example, the MTD^{3}L timing attack had the highest result for the first data to AOS transition, so only that result is given. All other MTD^{3}L transitions resulted in lower correlations. The synchronous attack was successful while the NCL, D^{3}L and MTD^{3}L attacks were not. The synchronous design, having no defense against power analysis, resulted in the highest correlation coefficient. This means that the key guess for this design has the most confidence. The NCL, D^{3}L, and MTD^{3}L designs each resulted in a lower coefficient and failure to guess the correct key. The D^{3}L and MTD^{3}L coefficients were very similar. This is expected because the changes from the D^{3}L design to the MTD^{3}L design should not have impacted the MTD^{3}L design's side-channel defenses. The NCL coefficient was unexpectedly the lowest, in disagreement with the results from [41], where the key was correctly guessed for the NCL circuit implemented using the IBM 0.5 μm process. This could be due to a number of reasons, such as a weakness in the statistical model, lack of resolution in the simulation data, or the unbalanced load effect in NCL being somewhat amplified in the large non-digital 0.5 μm process.

Table 10. Power/energy attack results.

Design	Attack type	Correlation coefficient	Correct key guess success/failure
Synchronous	Power	0.872	Success
NCL	Energy	0.207	Failure
D^{3}L	Energy	0.376	Failure
MTD^{3}L	Energy	0.353	Failure

Table 11 shows the results of the time based attacks. The synchronous design was not attacked in this way, since it is already known to be susceptible, while the NCL design, also known to be susceptible, was attacked to show that the attack algorithm being used for the asynchronous designs could correctly guess the key. Since the energy attack on the NCL circuit in [41] was successful, adding random delay elements to protect it from timing attacks was moot, so this was not done. For the D^{3}L and MTD^{3}L circuits, only a minimal number of random delay elements were added to reduce the correlation coefficient, since reducing area without compromising susceptibility is the main objective. None of the attacks against the D^{3}L or MTD^{3}L designs were successful; the correlations for these designs were very similar and quite low, showing weak correlation between the statistical models and the actual behavior of the designs. Adding random delay elements is not necessary to further protect the D^3L and MTD^3L circuits against timing attacks, since this had minimal impact on reducing their correlation coefficients. Again, the MTD^3L design performed very similarly to the D^3L design. These results as well as the results given in Section 5.2 show that the MTD^3L design offers similar security to the D^3L design while having much less overhead with regard to area and power consumption.

Table 11. Timing attack results.

Design	Correlation coefficient	Correct key guess success/failure
NCL	0.400	Success
D^3L	0.337	Failure
MTD^3L	0.366	Failure
Delayed D^3L	0.326	Failure
Delayed MTD^3L	0.367	Failure

6. Conclusions and Future Work

This paper develops MTD^3L, an enhancement to the D^3L technique presented in [41], by combining D^3L with MTNCL. Because the D^3L technique suffered significant overhead with respect to area and power consumption, a method of reducing overhead to make D^3L more reasonable was necessary. The resulting MTD^3L method shows significant reductions in power consumption and area overhead, and demonstrated the same strength as the D^3L technique against power and timing analysis attacks. The end result is that the MTD^3L technique is a substantial improvement over the original D^3L technique, and is applicable for designs where security is much more important than power consumption or area, since MTD^3L still has significant overhead compared to the original synchronous design. Hence, more work is required to reduce the overhead even further.

The majority of the overhead is from the increased complexity of the registration stage. This complexity was also present in the original D^3L technique. The overhead required to implement the dual-spacer protocol, particularly in the filter registers, makes D^3L and MTD^3L registers much larger and consume much more power than NCL or synchronous registers. Therefore, course-grained pipelines will have much less overhead than fine-grained ones. Future work includes investigations of possible ways to reduce this registration overhead, bringing the MTD^3L implementation cost closer to that of the NCL design.

Note that although complete layouts of the circuits were done, including place and route, all simulations were performed on the transistor-level circuits. This was because post-parasitic extraction simulations consumed all of the server memory and caused the system to crash. Since both power and timing attacks are strongly correlated to circuit parasitics, the authors have been actively applying for funding to fabricate these circuits so that hardware measurements can be made.

Conflicts of Interest

The authors declare no conflict of interest.

References

1. Jaffe, J.; Kocher, P.; Jun, B. Differential Power Analysis. In Proceedings of 19th International Advances in Cryptology Conference, Santa Barbara, CA, USA, 16-20 August 2009; pp. 388–397.
2. Brier, E.; Clavier, C.; Olivier, F. Correlation Power Analysis with a Leakage Model. In Proceedings of Cryptographic Hardware and Embedded Systems (CHES) 2004, Cambridge, MA, USA, 11-13 August 2004; pp. 16–29.
3. Blake, I.; Seroussi, G.; Smart, N.; Cassels, J.W.S. *Advances in Elliptic Curve Cryptography*; Cambridge University Press: New York, NY, USA, 2005.
4. Messerges, T.; Dabbish, E.; Sloan, R. Investigations of Power Analysis Attacks on Smartcards. In Proceedings of Workshop on Smartcard Technology, McCormick Place South Chicago, IL, USA, 10-11 May 1999; p. 17.
5. Coron, J. Resistance against Differential Power Analysis for Elliptic Curve Cryptosystems. In Proceedings of 1st International Workshop on Cryptographic Hardware and Embedded Systems, Worcester, MA, USA, 12-13 August 1999; pp. 292–302.
6. Boer, B.; Lemke, K.; Wicke, G. A DPA Attack against the Modular Reduction within a CRT Implementation of RSA. In Proceedings of 4th International Workshop on Cryptographic Hardware and Embedded Systems, Redwood Shores, CA, USA, 13-15 August 2002; pp. 228–243.
7. Serner, S.; Colin, W. More Detail for a Combined Timing and Power Attack against Implementations of RSA. In Proceedings of the Institute of Mathematics and its Applications (IMA) International Conference, Cirencester, UK, 16-18 December 2003; pp. 245–263.
8. Ors, S.; Gurkaynak, F.; Oswald, E.; Preneel, B. Power-Analysis Attack on an ASIC AES implementation. In Proceedings of International Conference on Information Technology: Coding and Computing, The Orleans, NV, USA, 5-7 April 2004; pp. 546–552.
9. Boracchi, G. *A Study on the Efficiency of Differential Power Analysis on AES S-Box*; Technical Report, Elettronica e Informazione, Politecnico di Milano: Milano, Italy, 2007.
10. Chari, S.; Jutla, C.; Rao, J.; Rohatgi, P. A Cautionary Note Regarding Evaluation of AES Candidates on Smart Cards. In Proceedings of 2nd Advanced Encryption Standard Candidate Conference, Rome, Italy, 22-23 March 1999; pp. 133–147.
11. Berna, O.; Elisabeth, O.; Bart, P. Power-Analysis Attacks on an FPGA—First Experimental Results. In Proceedings of 5th International Workshop on Cryptographic Hardware and Embedded Systems (CHES), Cologne, Germany, 8-10 September 2003; pp. 35–50.
12. Ors, S.; Gurkaynak, F.; Oswald, E.; Preneel, B. Power-Analysis Attack on an ASIC AES Implementation. In Proceedings of International Conference on Information Technology: Coding and Computing, Las Vegas, Nevada, USA, 5-7 April 2004; pp. 546–552.
13. Mace, F.; Standaert, F.; Quisquater, J.; Legat, J. A Design Methodology for Secured ICs Using Dynamic Current Mode Logic. In Proceedings of 15th International Workshop on Integrated Circuit and System Design, Power and Timing Modeling, Optimization and Simulation (PATMOS), Leuven, Belgium, 21-23 September 2005; pp. 550–560.
14. Verbauwhede, I.; Tiri, K.; Hwang, D.; Schaumont, P. Circuits and Design Techniques for Secure ICs Resistant to Sidechannel Attacks. In Proceedings of IEEE International Conference on Integrated Circuit Design and Technology (ICICDT), Padova, Italy, 24-26 May 2006.

15. Aigner, M.; Mangard, S.; Menicocci, R.; Olivieri, M.; Scotti, G.; Trifiletti, A. A Novel CMOS Logic Style with Data Independent Power Consumption. In Proceedings of International Symposium on Circuits and Systems, Kobe, Japan, 23-26 May 2005; pp. 1066–1069.
16. Lin, K.; Fan, S.; Yang, S.; Lo, C. Overcoming Glitches and Dissipation Timing Skews in Design of DPA Resistant Cryptographic Hardware. In Proceedings of Design, Automation & Test in Europe Conference & Exhibition, Nice, France, 16-20 April 2007; pp. 1265–1270.
17. Sundaresan, V.; Rammohan, S.; Vemuri, R. Power Invariant Secure IC Design Methodology Using Reduced Complementary Dynamic and Differential Logic. In Proceedings of IFIP International Conference on Very Large Scale Integration and System-on-Chip (VLSI-SoC), Atlanta, USA, 15–17 October 2007; pp. 1–6.
18. Kulikowski, K.; Venkataraman, V.; Wang, Z.; Taubin, A. Power Balanced Gates Insensitive to Routing Capacitance Mismatch. In Proceedings of Design, Automation & Test in Europe Conference & Exhibition, Munich, Germany, 10-14 March 2008; pp. 1280–1285.
19. Wang, Y.; Leiwo, J.; Srikanthan, T.; Jianwen, L. An Efficient Algorithm for DPA-resistant RSA. In Proceedings of IEEE Asia Pacific Conference on Circuits and Systems (APCCAS), Singapore, 4–7 December 2006; pp. 1659–1662.
20. Mesquita, D.; Techer, J.; Torres, L.; Sassatelli, G.; Cambon, G.; Robert, M.; Moraes, F. Current Mask Generation—A Transistor Level Security Against DPA Attacks. In Proceedings of 18th Symposium on Integrated Circuits and Systems Design, Florianolpolis, Brazil, 4–7 September 2005; pp. 115–120.
21. Haider, S.; Nazhandali, L. Utilizing Sub-threshold Technology for the Creation of Secure Circuits. In Proceedings of International Symposium on Circuits and Systems (ISCAS), Seattle, WA, USA, 18–21 May 2008; pp. 3182–3185.
22. Hasan, M. Power analysis attacks and algorithmic approaches to their countermeasures for koblitz curve cryptosystems. *IEEE Trans. Comput.* **2001**, *50*, 1071–1083.
23. Corsonello, P.; Perri, S.; Margala, M. An Integrated Countermeasure against Differential Power Analysis for Secure Smart-Cards. In Proceedings of International Symposium on Circuits and Systems (ISCAS), Island of Kos, Greece, 21–24 May 2006; pp. 5611–5614.
24. Yang, S.; Wolf, W.; Vijaykrishnan, N.; Serpanos, D.; Xie, Y. Power Attack Resistant Cryptosystem Design—A Dynamic Voltage and Frequency Switching Approach. In Proceedings of Design, Automation & Test in Europe Conference & Exhibition, Munich, Germany, 7–11 March 2005; pp. 64–69.
25. Baddam, K.; Zwolinski, M. Evaluation of Dynamic Voltage and Frequency Scaling as a Differential Power Analysis Countermeasure. In Proceedings of 20th International Conference on VLSI Design held jointly with 6th International Conference on Embedded Systems, Bangalore, India, 6–10 January 2007; pp. 854–862.
26. Ambrose, J.; Ragel, R.; Parameswaran, S. RIJID—Random Code Injection to Mask Power Analysis based Side Channel Attacks. In Proceedings of Design Automation Conference (DAC), Yokohama, Japan, 23–26 January 2007; pp. 489–492.
27. Rivain, M.; Dottax, E.; Prouff, E. Block Ciphers Implementations Provably Secure Against Second Order Side Channel Analysis. In Proceedings of Fast Software Encryption, Lausanne, Switzerland, 10-13 February 2008.

28. Kocher, P.C. Timing Attacks on Implementations of Diffie-Hellman, RSA, DSS, and Other Systems. In Proceedings of 16th International Advances in Cryptology Conference, Santa Barbara, CA, USA, 20-24 August 2006; pp. 388–397.
29. Chevallier-Mames, B.; Ciet, M.; Joye, M. Low-cost solutions for preventing simple side-channel analysis—Side-channel atomicity. *IEEE Trans. Comput.* **2004**, *53*, 760–768.
30. Page, D.; Smart, N. Parallel cryptographic arithmetic using a redundant montgomery representation. *IEEE Trans. Comput.* **2004**, *53*, 1474–1482.
31. Hodjat, A.; Hwang, D.; Verbauwhede, I. A Scalable and High Performance Elliptic Curve Processor with Resistance to Timing Attacks. In Proceedings of International Conference on Information Technology: Coding and Computing, Las Vegas, USA, 4–6 April 2005; pp. 538–543.
32. Sokolov, D.; Murphy, J.; Bystrov, A.; Yakovlev, A. Design and analysis of dual-rail circuits of security applications. *IEEE Trans. Comput.* **2005**, *54*, 449–460.
33. Bouesse, G.; Renaudin, M.; Dumont, S.; Germain, F. DPA on Quasi Delay Insensitive Asynchronous Circuits— Formalization and Improvement. In Proceedings of Design, Automation & Test in Europe Conference & Exhibition, Munich, Germany, 7–11 March 2005; pp. 424–429.
34. Verbauwhede, I.; Tiri, K.; Hwang, D.; Schaumont, P. Circuits and Design Techniques for Secure ICs Resistant to Sidechannel Attacks. In Proceedings of IEEE International Conference on Integrated Circuit Design and Technology (ICICDT), Padova, Italy, 24-26 May 2006.
35. Shang, D.; Burns, F.; Bystrov, A.; Koelmans, A.; Sokolov, D.; Yakovlev, A. High-security asynchronous circuit implementation of AES. *IEE Proc. Comput. Digit. Tech.* **2006**, *153*, 71–77.
36. Kulikowski, K.; Venkataraman, V.; Wang, Z.; Taubin, A.; Karpovsky, M. Asynchronous Balanced Gates Tolerant to Interconnect Variability. In Proceedings of International Symposium on Circuits and Systems (ISCAS), Seattle, WA, USA, 18–21 May 2008; pp. 3190–3193.
37. Baddam, K.; Zwolinski, M. Path switching: A technique to tolerate dual rail routing imbalances. *Des. Autom. Embed. Syst.* **2008**, *12*, 207-220.
38. Fant, K.; Brandt, S. NULL Convention Logic™: A Complete and Consistent Logic for Asynchronous Digital Circuit Synthesis. In Proceedings of Application Specific Systems, Architectures and Processors, Chicago, IL, USA, 19-23 August 1996; pp. 261–273.
39. Di, J.; Yang, F. D^3L—A Framework on Fighting Against Non-Invasive Attacks to Integrated Circuits for Security Applications. In Proceedings of 3rd IASTED International Conference Circuits, Signals, and Systems, Marina del Rey, CA, USA, 24–26 October 2005; pp. 73–78.
40. Wu, J.; Kim, Y.; Choi, M. Low-Power Side-Channel Attack-Resistant Asynchronous S-Box Design for AES Cryptosystems. In Proceedings of Great Lakes Symposium on VLSI (GLSVLSI), Providence, RI, USA, 16–18 May 2010; pp. 459–464.
41. Cilio, W.; Di, J.; Smith, S.C.; Thompson, D.R. Mitigating Power- and Timing-Based Side-Channel Attacks Using Dual-Spacer Dual-Rail Delay-Insensitive Asynchronous Logic. *Microelectron. J.* **2013**; 44, 258–269.
42. Di, J.; Smith, S.C. Ultra-Low Power Multi-Threshold Asynchronous Circuit Design. *U.S. Patent: 7,977,972 B2*, 12 July 2011.
43. Smith, S.C.; Di, J. *Designing Asynchronous Circuits Using NULL Convention Logic* (*NCL*); Synthesis Lectures on Digital Circuits and Systems, Morgan & Claypool: Oak View Drive San Rafael, CA, USA, July 2009.

44. Muller, D.E. Asynchronous Logics and Application to Information Processing. In *Switching Theory in Space Technology*; Stanford University Press: St. Redwood, CA, USA, 1963; pp. 289–297.
45. Kondratyev, A.; Lwin, K. Design of asynchronous circuits using synchronous CAD tools. *J. IEEE Des. Test* **2002**, *19*, 107–117.
46. National Institute of Standards and Technology, Federal Information Processing Standard 197, the Advanced Encryption Standard (AES). Available online: http://csrc.nist.gov/publications/fips/fips197/fips-197.pdf (accessed on 20 October 2013).

6

Compact Modeling Solutions for Oxide-Based Resistive Switching Memories (OxRAM)

Marc Bocquet [1,*], Hassen Aziza [1], Weisheng Zhao [2], Yue Zhang [2], Santhosh Onkaraiah [1,3], Christophe Muller [1], Marina Reyboz [3], Damien Deleruyelle [1], Fabien Clermidy [3] and Jean-Michel Portal [1]

[1] Institut Matériaux Microélectronique Nanosciences de Provence(IM2NP), Aix-Marseille Université, Centre national de la recherche scientifique (CNRS), UMR 7334, 13284 Marseille, France; E-Mails: hassen.aziza@im2np.fr (H.A.); santhosh.onkaraiah@im2np.fr (S.O.); Christophe.muller@im2np.fr (C.M.); damien.deleruyelle@im2np.fr (D.D.); jean-michel.portal@im2np.fr (J.-M.P.)

[2] Institut d'Électronique Fondamentale(IEF), University of Paris-Sud, Centre national de la recherche scientifique (CNRS), UMR 8622, F91405 Orsay, France; E-Mails: weisheng.zhao@u-psud.fr (W.Z.); yue.zhang@u-psud.fr (Y.Z.)

[3] Commissariat l'Énergie Atomique - Laboratoire dÉlectronique et de Technologie de lInformation (CEA-Léti), 38054 Grenoble, France; E-Mails: marina.reyboz@cea.fr (M.R.); fabien.clermidy@cea.fr (F.C.)

* Author to whom correspondence should be addressed; E-Mail: marc.bocquet@im2np.fr

---

Abstract: Emerging non-volatile memories based on resistive switching mechanisms attract intense R&D efforts from both academia and industry. Oxide-based Resistive Random Acces Memories (OxRAM) gather noteworthy performances, such as fast write/read speed, low power and high endurance outperforming therefore conventional Flash memories. To fully explore new design concepts such as distributed memory in logic, OxRAM compact models have to be developed and implemented into electrical simulators to assess performances at a circuit level. In this paper, we present compact models of the bipolar OxRAM memory based on physical phenomenons. This model was implemented in electrical simulators for single device up to circuit level.

Keywords: compact modeling; RRAM; OxRAM; design

1. Introduction

Memory devices based on resistive switching materials are currently pointed out as promising candidates to replace conventional non-volatile memory devices based on charge-storage beyond 2xnm-technological nodes [1–3]. Indeed, compared to conventional floating gate technologies, Resistive RAMs (so-called RRAM) gather fast write/read operations, low power consumption, CMOS voltage compatibility and high endurance. Moreover, the resistive memory element consists of a simple Metal/Insulator/Metal (MIM) stack. In this way, one of the major advantages of resistive switching memories is their capability, whatever the underlying physics is, to be integrated in the back-end-of-line enabling NVM solutions to be distributed over CMOS logic. Relying on different based on different physical mechanisms, various RRAM technologies are now categorized in the ITRS. The Redox Memory category, covered in this study, includes Conductive Bridge RAM (CBRAM) [4] and Oxide Resistive RAM (OxRAM) [5] both of which exhibit a bipolar behavior, (*i.e.*, switching relying on voltage polarity) (cf Figure 1a). Conversely, RRAM technologies referred to as Thermo-Chemical Memories (TCM) [6], or fuse-antifuse memories, are mostly based on nickel oxide (NiO) and exhibit a unipolar behavior (*i.e.*, switching relying on voltage amplitude) as show in Figure 1b.

Figure 1. Typical I–V characteristic of resistive memories: (**a**) Bipolar behavior; (**b**) Unipolar behavior.

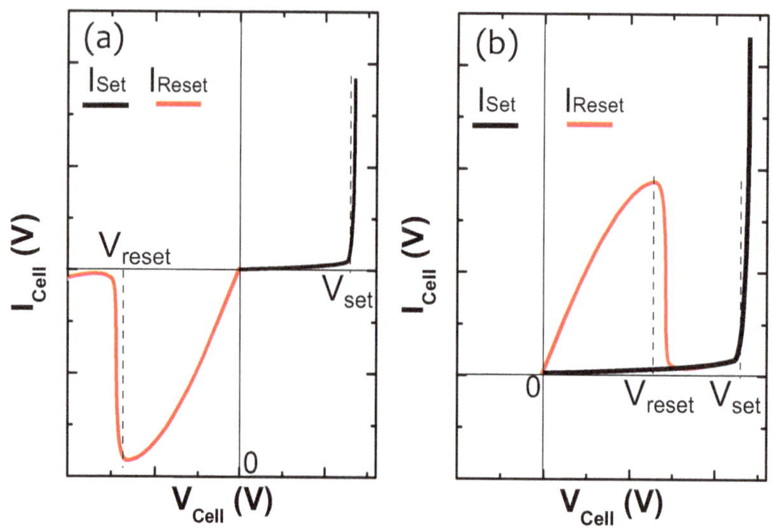

For the OxRAM memory elements addressed in this paper, the MIM structure is generally composed of metallic electrodes sandwiching an active layer, usually an oxygen-deficient oxide. A large number of resistive switching oxides, like HfO_2, Ta_2O_5, NiO, TiO_2 or Cu_2O, are reported in the literature [7–10]. The Valency Change Mechanism (VCM) occurs in specific transition metal oxides and is triggered by a migration of anions, such as oxygen vacancies.

After an initial electroforming step, the memory element may be reversibly switched between a High Resistance State (HRS) and a Low Resistance State (LRS). The electroforming stage corresponds to a voltage-induced resistance switching from an initial very high resistance state (virgin state) to a conductive state. In the case of bipolar switching, bipolar voltage sweeps are required to switch the memory element (Figure 1a). Resistive switching in an OxRAM element corresponds to an abrupt change between a HRS (R_{HRS}) and a LRS (R_{LRS}). This resistance change is achieved by applying specific voltage to the structure (*i.e.*, V_{Set} and V_{Reset}). Generally, the electroforming voltage is superior to these voltages. However, several groups have demonstrated forming-free structures by adjusting the stoichiometry of the active layer [11–13].

In this paper, compact models of the bipolar OxRAM memory is presented. This model was implemented in electrical simulators for single device up to circuit level.

2. Compact Model for OxRAM Cells

Even if OxRAM technology is still in its *infancy*, it is broadly accepted that the field-assisted motion of oxygen vacancies governs the bipolar resistance switching [14]. The proposed OxRAM modeling approach [15–17] relies on electric field-induced creation/destruction of Conductive Filament (CF) within the switching layer. The model is based on a single master equation in which both *set* and *reset* operations are accounted simultaneously and control the radius of the conduction pathway (r_{CF}). Figure 2 depicts the proposed model for the switchable MIM structure.

Figure 2. Formed and dissolved conductive filament resulting from *set* and *reset* operations respectively in Metal/Insulator/Metal (MIM) structure.

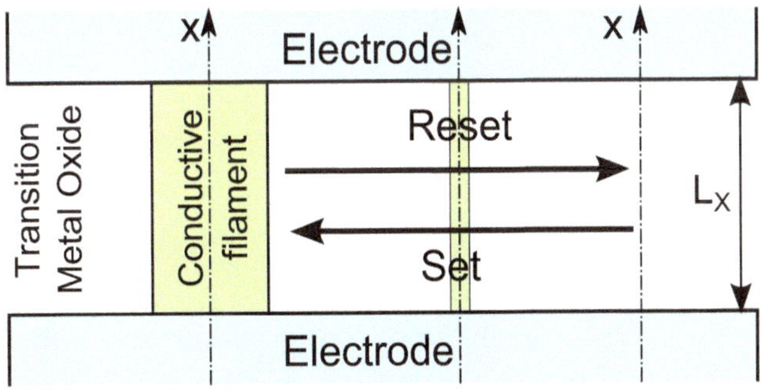

The *set* (*resp. reset*) process can be described by an electrochemical kinetic equation relying on the Butler-Volmer equation [18]. In the LRS, where conduction is controlled by the CF, charge transport is assumed to be ohmic according to previous reports in the literature [19,20]. However, HRS is actually dominated by leakage current within the oxide layer. To take into account a lot of trap assisted current [Poole-Frenkel, Schottky emission, Space Charge Limited Current (SCLC)], a power low between the cell current and the applied bias has been considered. The model assumes an uniform CF radius and electric field in the cell where temperature elevation (triggered by Joule effect) may accelerate redox reaction rates. In this way the local temperature of the filament is given by [15]:

$$T = T_{amb} + \frac{V_{Cell}^2}{8 \cdot kth} \cdot \left(\frac{r_{CF}^2}{r_{CFmax}^2} \cdot (\sigma_{CF} - \sigma_{OX}) + \sigma_{OX} \right) \qquad (1)$$

where T_{amb} is the ambient temperature; V_{Cell} is the voltage applied between the top and the bottom electrodes; k_{th} is the thermal conductivity and σ_{CF} (*resp.* σ_{OX}) is the electrical conductivity of the conductive filament (*resp.* oxide).

The *Set* operation is modeled based on the Butler-Volmer equation through the electrochemical reduction rate (τ_{Red}):

$$\frac{1}{\tau_{Red}} = A_{RedOx} \cdot e^{-\dfrac{E_a - q \cdot \alpha_{Red} \cdot V_{Cell}}{k_b \cdot T}} \tag{2}$$

where k_b is the Boltzmann constant.

Similarly, *reset* concerns the local dissolution of the CF and accounted by the oxidation rate (τ_{Ox}):

$$\frac{1}{\tau_{Ox}} = A_{RedOx} \cdot e^{-\dfrac{E_a + q \cdot \alpha_{Ox} \cdot V_{Cell}}{k_b \cdot T}} \tag{3}$$

where E_a is the activation energy; α_{Red} and α_{Ox} are the transfer coefficient (ranging between 0 and 1); A_{RedOx} is the nominal redox rate. Hence, the growth/dissolution of the filament results from the inter-play between both redox reaction velocities through the following master equation:

$$\frac{dr_{CF}}{dt} = \frac{r_{CF_{max}} - r_{CF}}{\tau_{Red}} - \frac{r_{CF}}{\tau_{Ox}} \tag{4}$$

where the local CF radius (r_{CF}) is comprised between zero and a maximal value ($r_{CF_{max}}$). To allow implementation into electrical simulation tools, a discrete writing is required. If the time step is sufficiently small, τ_{Red} et τ_{Ox} may be assumed as constant. The discrete form of Equation (4) is then given by Equation (5). Solving the differential Equation (4) step by step allows a better convergence of simulation tools.

$$r_{CF_{i+1}} = \left(r_{CF_i} - \frac{\tau_{eq}}{\tau_{Red}}\right) \cdot e^{\dfrac{-\Delta t}{\tau_{eq}}} + \frac{\tau_{eq}}{\tau_{Red}} \tag{5}$$

where $\tau_{eq} = \dfrac{\tau_{Red} \cdot \tau_{Ox}}{\tau_{Red} + \tau_{Ox}}$.

Finally, the total current in the OxRAM includes two components: one is related to the conductive species (I_{CF}) [15] the other concerns conduction through the oxide (I_{OX}):

$$I_{CF} = \frac{V_{Cell}}{L_x} \cdot \left(r_{CF}^2 \cdot \pi \cdot (\sigma_{CF} - \sigma_{OX}) + r_{CFmax}^2 \cdot \pi \cdot \sigma_{OX}\right) \tag{6}$$

$$I_{OX} = A_{HRS} \cdot S_{Cell} \cdot \left(\frac{V_{Cell}}{L_x}\right)^{\alpha_{HRS}} \tag{7}$$

where L_x is the oxide thickness and S_{Cell} is the total area of the device. Finally, the total current flowing through the cell is:

$$I_{Cell} = I_{OX} + I_{CF} \tag{8}$$

These equations were then implemented within an ELDO compact model following the flowchart given in Figure 3. At each call of the OxRAM instance during a transient simulation, the previous state of the filament as well as the applied voltage are provided to the model in order to take into account the memory effect. The new filament state and the current are then computed as function of these inputs and the time step.

Figure 3. Program flowchart employed for numerical simulation of Oxide-based Resistive Random Acces Memories (OxRAM) memory devices.

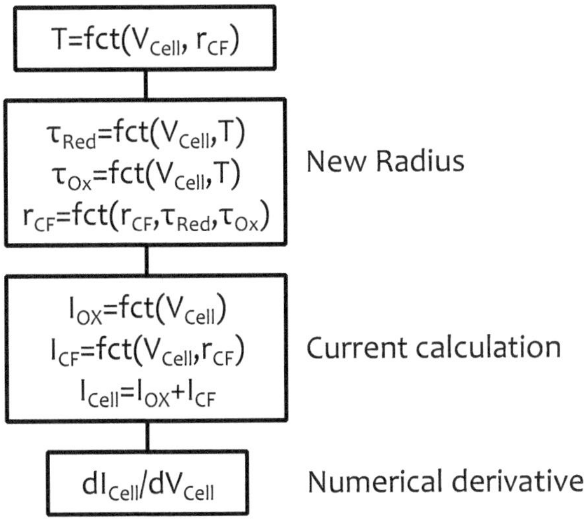

3. Model Validation

The compact model was calibrated on recent electrical data measured on HfO_2-based OxRAM devices [21]. To validate the proposed theoritical approach, the model was confronted to quasi-static and dynamic experimental data extracted from the literature. Figure 4a shows quasi-static *set* and *reset* $I(V)$ characteristics measured on HfO_2-based memory elements. In this study, the memory elements consisted in a Ti/HfO_2/TiN stack with a hafnium oxide thickness of $10\,nm$. The description of the cell manufacture is presented in [21]. Using the *set* parameters given in Table 1, the present model shows an excellent agreement with experimental data for both *set* and *reset* operations.

Table 1. Physical parameters used for Bipolar simulations.

Parameters	Values	Parameters	Values
$r_{CF_{max}}$	$20\,nm$	L_x	$10\,nm$
$Scell$	$1\mu m \times 1\mu m$	A_{RedOx}	$1 \times 10^9\,s^{-1}$
A_{RedOx}	$1 \times 10^9\,s^{-1}$	E_a	$0.95\,eV$
α_{Red}	0.85	A_{HRS}	$5 \times 10^{-9}\,A/(V^2)$
α_{HRS}	2	α_{Ox}	0.85
σ_{Ox}	$0.1\,m \cdot S$	σ_{CF_0}	$12.5 \times 10^5\,m \cdot S$
T_{amb}	$300\,K$	K_{th}	$0.8\,W/(K \cdot m)$

Figure 4. (**a**) Experimental I(V) (■); and (**b**) *set*; and (**c**) *reset* voltage as a function of the programming ramp speed measured on a HfO_2-based memory structures presented in [21] and corresponding simulation results (−).

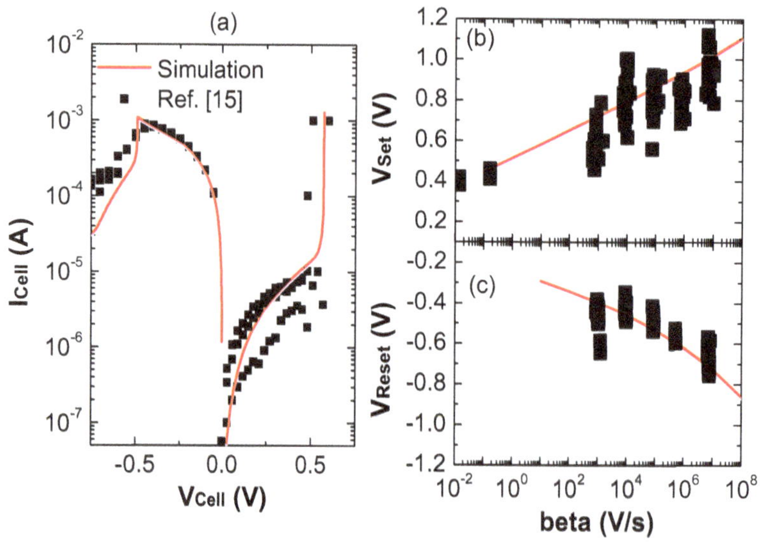

Figure 4b shows that the proposed model also satisfactorily catches the impact of the experimental increase of V_{Set} and V_{Reset} with the programming ramp speed. Moreover, the effect of the *set* current limitation on the *reset* current is also taken into account by this compact model (Figure 5). It is interesting to note that this behavior appears for unipolar and bipolar memory [11]. However, in our study, only the bipolar structures will be studied.

Figure 5. Maximum current during the *reset* operation (I_{Reset}) as a function of the maximum current during the preceding *set* operation ($I_{CompSet}$). Experimental data were extracted from Reference [5,8,10,22–24].

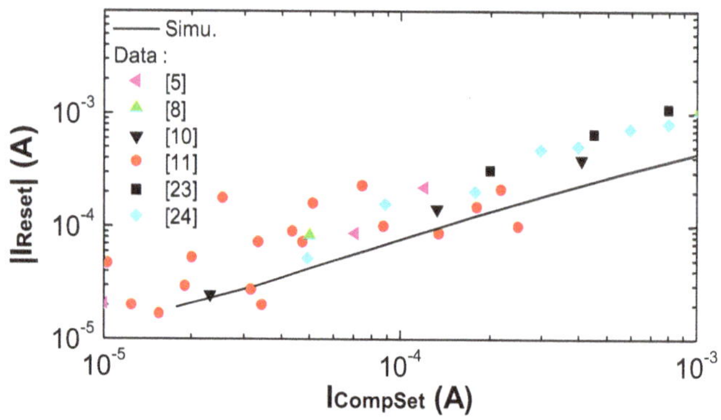

Figure 6 illustrates the transient current response of the cell when a voltage ramp is applied to the cell (Figure 4b). A significant cell voltage discontinuity is observed during the *set* operation. This behavior highlights the self acceleration of *set* mechanism. Indeed, when the applied voltage is below the *set* voltage, the resistance continuously decreases. Let us mention that our model, which already includes a thermal activation of *set* operation, should be able to take into account this effect once the parasitic capacitances originating from the measurement setup are provided.

Figure 6. Dynamic measurement of OxRAM (HfO_2-based memory) and corresponding simulation results.

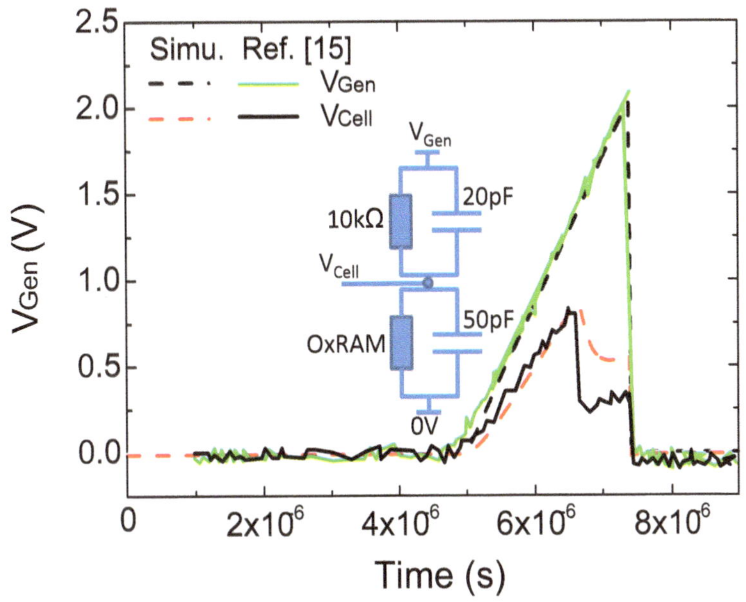

To fully validate the compact model and its integration into the electrical simulator, Figure 7 gives an example of bipolar OxRAM cells simulated at a circuit level, *i.e.*, surrounded by MOS transistors.

Figure 7. Electrical simulation of 2T-1R OxRAM structure.

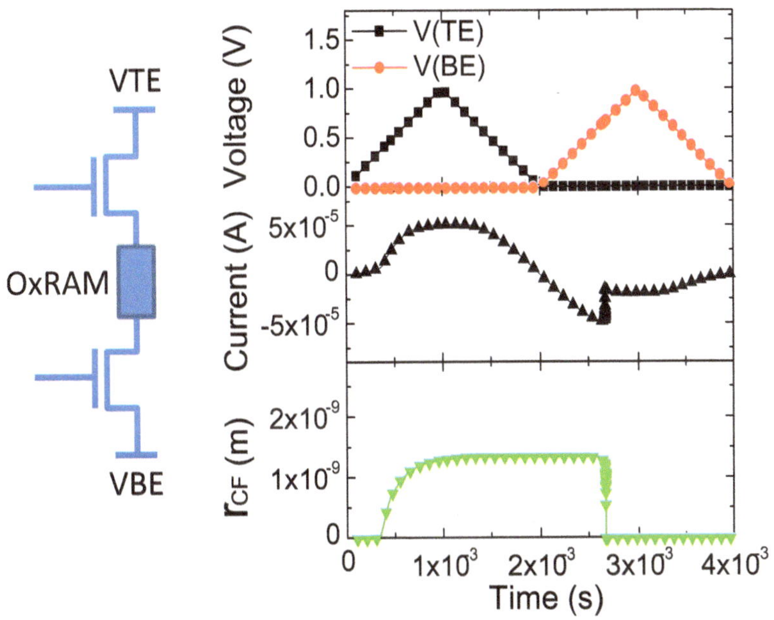

These models have been successfully used to simulate new MOS-RRAM cells like a NVM flip-flop [25], Non-Volatile SRAM [26] and OxRAM memory array [27].

4. Model Application

4.1. OxRAM Reliability Evaluation versus *OxRAM Variability* [16]

In this section, an investigation in the impact of OxRAM variability on the memory array performances is proposed [16]. Indeed, variability in advanced IC designs has emerged as a roadblock and significant efforts of process and design engineers are required to decrease its impact.

Since the cell variability is calibrated on silicon using the previous OxRAM model, only the realistically possible variations are reported in this study. A large number of Monte Carlo simulations are performed to provide the statistics needed to characterize variability. Cell variations are introduced and simulated sequentially using an electrical simulator. The goal is to track an important shift of reliability parameters.

Figure 8 presents the elementary array used for simulation: it is constituted, a row decoder, a column decoder and a sense amplifier for the read operation. Memory array cells are first placed in an erase state. Then, the memory array programming is done in two cycles. First, all memory cells are set (logical "1"), then the memory array is reset (logical "0"). Logical failures can be detected at the output of the sense amplifier during the read operation after *set/reset*.

Figure 8. *Three* × *three* OxRAM memory array.

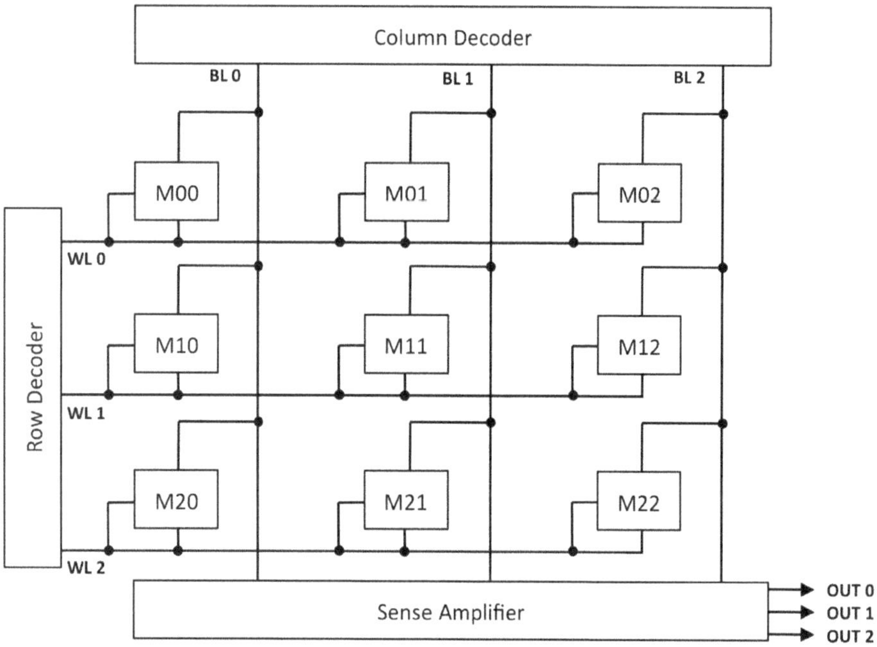

The best way to monitor the impact of variability on OxRAM electrical parameters is to plot the OxRAM hysteresis in transient mode (*i.e.*, cell current evolution versus cell voltage difference during a Write/Erase cycle). Figure 9 shows the impact of the memory array cell variability (9 cells) on the circuit hysteresis.

Figure 9. Variability impact on $I - V$ hysteresis the memory array.

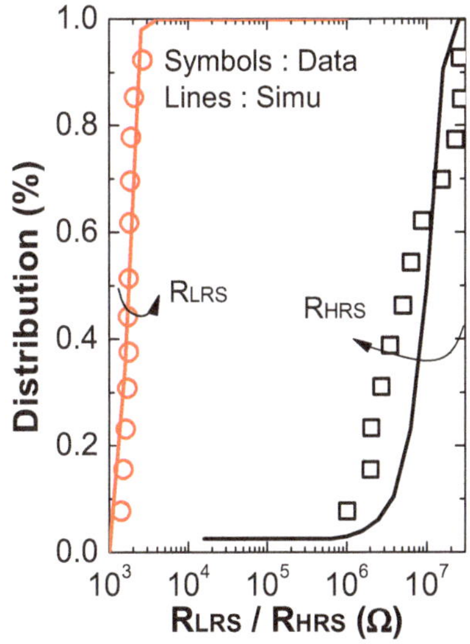

It appears clearly that R_{LRS} distribution can be severely impacted by cell variability. V_{Reset} and V_{Set} parameters can also suffer from cell variability but in a lesser extent. Notice that R_{LRS} and R_{HRS} are extracted @0.5 V (read conditions). Moreover, V_{Set} and V_{Reset} are extracted @40 μA, at a circuit level.

R_{LRS} and R_{HRS} distributions are plotted in Figure 10. Results are presented for cell variability included in the range $\pm 10\%$ of the median value of the considered card model parameters (solid bar). Results are also provided for cell variability included in the range $\pm 20\%$ (dashed bar).

Figure 10. (**a**) R_{LRS} and (**b**) R_{HRS} distributions *versus* cell variability.

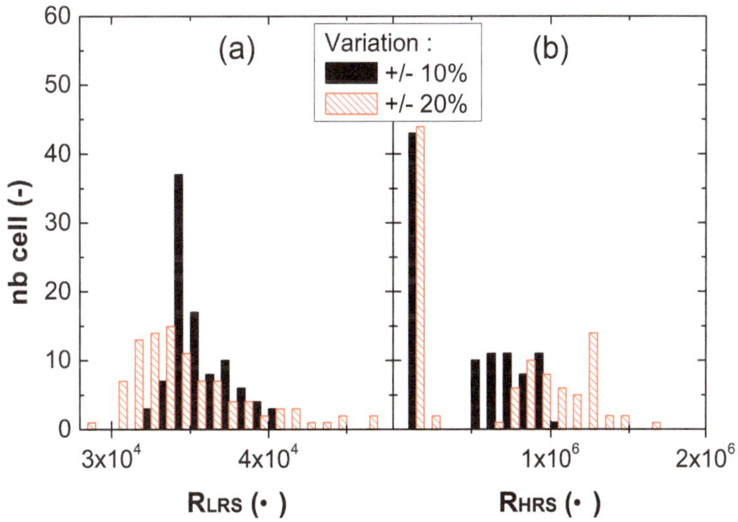

At 10%, a spreading of R_{LRS} and R_{HRS} parameters is observed. The spreading increases significantly at 20%. Although these values are related to a specific OxRAM technology, a good feedback can be provided to designers to optimize the sensing circuitry according to the level of controllability of the fabrication process.

V_{Reset} and V_{Set} distributions are plotted in Figure 11a,b. Here again, the initial spreading (solid bar) increases (dashed bar) according to the variability increase. These results are of prime importance as this study predicts an increase of V_{Set} to the value of $1.5\ V$. This means that the programming signals provided to the cell needs to reach at least $1.5\ V$ for cell to be programmed properly.

Figure 11. (**a**) V_{Set} and (**b**) V_{Reset} distributions *versus* cell variability.

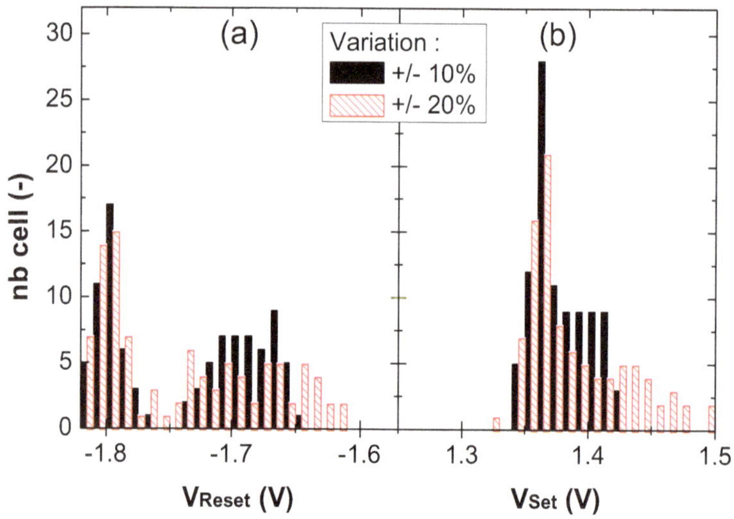

4.2. Diffrential Precharge Sense Amplifier for CRS Bitcell [26]

The read operation of data stored in cross-point resistive switching memory is currently one of the major challenges to develop this approach. Indeed, sneak path or destructive read with complementary resistive switching element are a strong limit to develop this type of architecture. Moreover, the resistance ratio (R_{HRS}/R_{LRS}) and the process variations have to be considered when designing a sense solution. A sense amplifier performing with high reliability is then required. Figure 12 shows a pre-charge based sense amplifier, which has demonstrated the best tolerance to different sources of variation, while keeping high speed and low power. In this sense amplifier, the read operation is performed in two phases:

- 1st Phase: The sense amplifier is first connected to the bit-line of the selected word with SEN set to "1" and the circuit is pre-charged with PCH equals "0";
- 2nd Phase: The data stored in the 2R cell can be evaluated to logic level at the output Q as PCH is changed to "1" and WL is pulled down to "0".

The Figure 13 validates the ability of the architecture to successfully read in parallel a full word. The model presented before has allowed us to assess the robustness of sense towards the variability of OxRAM or CMOS transistors [26]; and the validated complete crossbar architecture based on 2R complementary [27].

Figure 12. Pre-Charged Sense Amplifier for data sensing. It consists of a pre-charge sub-circuit (MPC0, MPC1), a pair of inverters (MNA0-1, MPA0-1), which act as an amplifier [26,27].

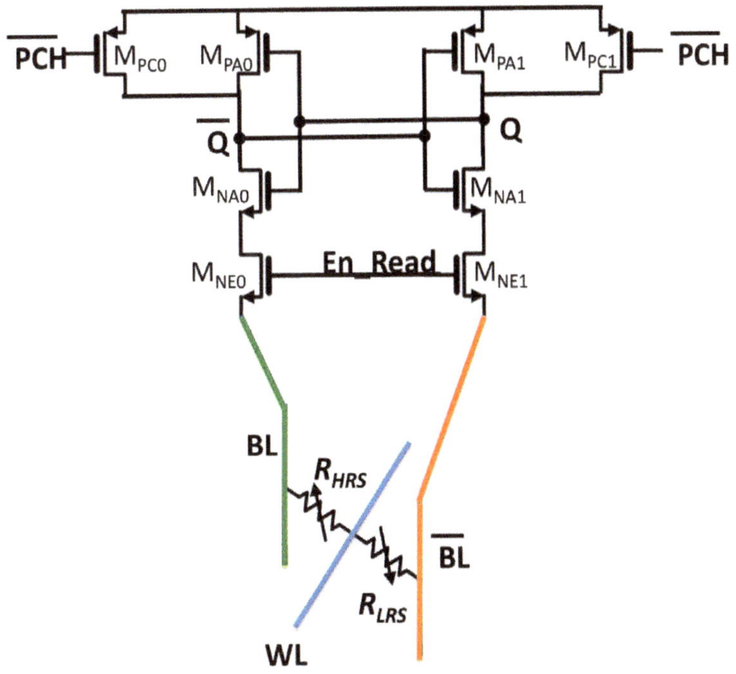

Figure 13. Simulation results with read phase of a selected cell [26,27].

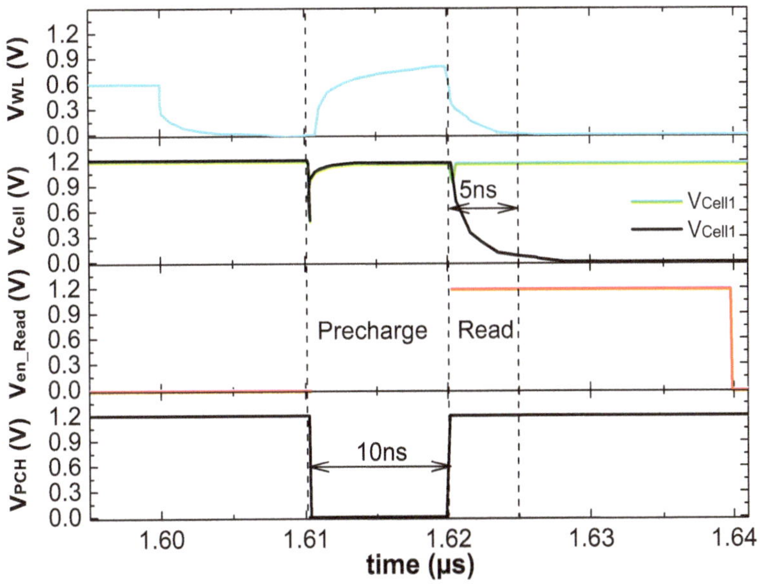

5. Conclusions

In conclusion, this paper deals with a compact model well suited to simultaneously describing *set* and *reset* operations in bipolar resistive switching memories based on HfO_2-based memory devise. By gathering local electrochemical reactions and a thermal mechanism in a single master equation, the model enables us to account for both the creation and destruction of conductive filaments. The simulation results satisfactorily match quasi-static and dynamic experimental data published in the literature on resistive switching devices. In addition, the model was implemented into circuit simulators. It has been successfully used in many circuits and has enabled the prediction of relevant trends required for designing innovative memory matrix architectures or proposing distributed memories solutions.

Acknowledgements

This work was supported in part by the ANR project DIPMEM (Design and Demonstration of Digital IP based on Emerging Non-Volatile MEMories) under grant ANR-12-NANO-0010-04. The authors would like also to thanks Amara Amara and Costin Anghel from ISEP (Paris), Thomas Cabout and Olivier Thomas from CEA-Leti for fruitful discussions on hybrid CMOS-OxRAM circuits.

Conflicts of Interest

The authors declare no conflicts of interest.

References

1. Hong, S. Memory Technology Trend and Future Challenges. In Proceedings of the IEEE International Electron Devices Meeting, San Francisco, CA, USA, 6–8 December 2010; pp. 292–295.
2. Govoreanu, B.; Kar, G.S.; Chen, Y.Y.; Paraschiv, V.; Kubicek, S.; Fantini, A.; Radu, I.P.; Goux, L.; Clima, S.; Degraeve, R.; *et al.* 10×10 nm^2 Hf/HfO$_x$ Crossbar Resistive RAM with Excellent Performance, Reliability and Low-Energy Operation. In Proceedings of the IEEE International Electron Devices Meeting, Washington, DC, USA, 5–7 December 2011; pp. 729–732.
3. Higuchi, K.; Iwasaki, T.O.; Takeuchi, K. Investigation of Verify-Programming Methods to Achieve 10 Million Cycles for 50 nm HfO_2 ReRAM. In Proceedings of the 2012 4th IEEE International Memory Workshop, Milan, Italy, 20–23 May 2012; pp. 119–122.
4. Kund, M.; Beitel, G.; Pinnow, C.U.; Rohr, T.; Schumann, J.; Symanczyk, R.; Ufert, K.D.; Muller, G. Conductive Bridging RAM (CBRAM): An Emerging Non-Volatile Memory Technology Scalable to Sub 20 nm. In Proceedings of the IEEE International Electron Devices Meeting, Washington, DC, USA, 5 December 2005; pp. 754–757.
5. Lee, H.Y.; Chen, P.S.; Wu, T.Y.; Chen, Y.S.; Wang, C.C.; Tzeng, P.J.; Lin, C.H.; Chen, F.; Lien, C.H.; Tsai, M.J. Low Power and High Speed Bipolar Switching with a Thin Reactive Ti Buffer Layer in Robust HfO_2 Based RRAM. In Proceedings of the IEEE International Electron Devices Meeting, San Francisco, CA, USA, 15–17 December 2008; pp. 1–4.
6. Baek, I.G.; Lee, M.S.J.; Seo, S.; Seo, D.H.; Suh, D.S.; Park, J.C.; Park, S.O.; Kim, H.S.; Yoo, I.K.; Chung, U.I.; *et al.* Highly Scalable Nonvolatile Resistive Memory Using Simple Binary Oxide Driven by Asymmetric Unipolar Voltage Pulses. In Proceedings of the IEEE International Electron Devices Meeting, San Francisco, CA, USA, 13–15 December 2004; pp. 587–590.
7. Waser, R.; Aono, M. Nanoionics-based resistive switching memories. *Nat. Mater.* **2007**, *6*, 833–840.

8. Seo, S.; Lee, M.J.; Seo, D.H.; Jeoung, E.J.; Suh, D.S.; Joung, Y.S.; Yoo, I.K.; Hwang, I.R.; Kim, S.H.; Byun, I.S.; *et al.*. Reproducible resistance switching in polycrystalline NiO films. *Appl. Phys. Lett.* **2004**, *85*, 5655–5657.
9. Kim, W.G.; Sung, M.G.; Kim, S.J.; Kim, J.Y.; Moon, J.W.; Yoon, S.J.; Kim, J.N.; Gyun, B.G.; Kim, T.W.; Kim, C.H.; *et al.* Dependence of the Switching Characteristics of Resistance Random Access Memory on the Type of Transition Metal Oxide. In Proceedings of the IEEE European Solid State Device Research Conference, Sevilla, Spain, 14–16 September 2010; pp. 400–403.
10. Fang, T.N.; Kaza, S.; Haddad, S.; Chen, A.; Wu, Y.C.; Lan, Z.; Avanzino, S.; Liao, D.; Gopalan, C.; Choi, S.; *et al.* Erase Mechanism for Copper Oxide Resistive Switching Memory Cells with Nickel Electrode. In Proceedings of the IEEE International Electron Devices Meeting, San Francisco, CA, USA, 11–13 December 2006; pp. 1–4.
11. Nardi, F.; Larentis, S.; Balatti, S.; Gilmer, D.C.; Ielmini, D. Resistive switching by voltage-driven ion migration in bipolar RRAM—Part I: Experimental study. *IEEE Trans. Electron Devices* **2012**, *59*, 2461–2467.
12. Fang, Z.; Yu, H.Y.; Singh, N.; Lo, G.Q.; Kwong, D.L. HfO_x/TiO_x/HfO_x/TiO_x Multilayer-based sith excellent uniformity. *IEEE Electron Device Lett.* **2011**, *32*, 566–568.
13. Kim, W.; Park, S.I.; Zhang, Z.; Yang-liauw, Y.; Sekar, D.; Wong, H.P.; Wong, S.S. Forming-Free Nitrogen-Doped AlOX RRAM with Sub-μA Programming Current. In Proceedings of the 2011 Symposium on VLSI Technology (VLSIT), Honolulu, HI, USA, 14–16 June 2011; pp. 22–23.
14. Gao, B.; Yu, S.; Xu, N.; Liu, L.F.; Sun, B.; Liu, X.Y.; Han, R.Q.; Kang, J.F.; Yu, B.; Wang, Y.Y. Oxide-Based RRAM Switching Mechanism: A New Ion-Transport-Recombination Model. In Proceedings of the IEEE International Electron Devices Meeting, San Francisco, CA, USA, 15–17 December 2008; pp. 1–4.
15. Bocquet, M.; Deleruyelle, D.; Muller, C.; Portal, J.M. Self-consistent physical modeling of set/reset operations in unipolar resistive-switching memories. *Appl. Phys. Lett.* **2011**, *98*, doi:10.1063/1.3605591.
16. Aziza, H.; Bocquet, M.; Portal, J.M.; Muller, C. Evaluation of OxRAM Cell Variability Impact on Memory Performances through Electrical Simulations. In Proceedings of the 11th Annual Non-Volatile Memory Technology Symposium (NVMTS), Shanghai, China, 7–9 November 2011; pp. 1–5.
17. Aziza, H.; Bocquet, M.; Portal, J.M.; Muller, C. Bipolar OxRRAM Memory Array Reliability Evaluation Based on Fault Injection. In Proceedings of the 2011 IEEE 6th International Design and Test Workshop (IDT), Beirut, Lebanon, 11–14 December 2011; pp. 78–81.
18. Bard, A.J.; Faulkner, L.R. *Electrochemical Methods: Fundamentals and Applications*; John Wiley & Sons, Inc.: New York, NY, USA, 2001; pp. 92–100.
19. Akinaga, H.; Shima, H. Resistive Random Access Memory (ReRAM) based on metal oxides. *Proc. IEEE* **2010**, *98*, 2237–2251.
20. Wong, H.S.P.; Lee, H.Y.; Yu, S.; Chen, Y.S.; Wu, Y.; Chen, P.S.; Lee, B.; Chen, F.T.; Tsai, M.J. Metal-oxide RRAM. *Proc. IEEE* **2012**, *100*, 1951–1970.

21. Cagli, C.; Buckley, J.; Jousseaume, V.; Cabout, T.; Salaun, A.; Grampeix, H.; Nodin, J.F.; Feldis, H.; Persico, A.; Cluzel, J.; *et al.* Experimental and Theoretical Study of Electrode Effects in HfO_2 Based RRAM. In Proceedings of the 2011 International Electron Devices Meeting (IEDM), Washington, DC, USA, 5–7 December 2011; pp. 28.7.1–28.7.4.
22. Nardi, F.; Ielmini, D.; Cagli, C.; Spiga, S.; Fanciulli, M.; Goux, L.; Wouters, D.J. Control of filament size and reduction of reset current below 10 μA in NiO resistance switching memories. *Solid-State Electron.* **2011**, *58*, 42–47.
23. Kinoshita, K.; Tsunoda, K.; Sato, Y.; Noshiro, H.; Yagaki, S.; Aoki, M.; Sugiyama, Y. Reduction in the reset current in a resistive random access memory consisting of NiO_x brought about by reducing a parasitic capacitance. *Appl. Phys. Lett.* **2008**, *93*, doi:10.1063/1.2959065.
24. Tsunoda, K.; Kinoshita, K.; Noshiro, H.; Yamazaki, Y.; Iizuka, T.; Ito, Y.; Takahashi, A.; Okano, A.; Sato, Y.; Fukano, T.; *et al.* Low Power and High Speed Switching of Ti-Doped NiO ReRAM under the Unipolar Voltage Source of Less than 3V. In Proceedings of the IEEE International Electron Devices Meeting (IEDM 2007), Washington, DC, USA, 10–12 December 2007; pp. 767–770.
25. Portal, J.M.; Bocquet, M.; Deleruyelle, D.; Muller, C. Non-volatile flip-flop based on unipolar ReRAM for power-down applications. *J. Low Power Electron.* **2012**, *8*, 1–10.
26. Aziza, H.; Makosiej, A.; Palma, G.; Portal, J.M.; Bocquet, M.; Thomas, O.; Clermidy, F.; Reyboz, M.; Onkaraiah, S.; Muller, C.; *et al.* Operation and stability analysis of bipolar OxRRAM-based non-volatile 8T2R SRAM as solution for information back-up. *Solid-State Electron.* **2013**, *90*, 99–106.
27. Zhao, W.S.; Zhang, Y.; Klein, J.O.; Querlioz, D.; Chabi, D.; Ravelosona, D.; Chappert, C.; Portal, J.M.; Bocquet, M.; Aziza, H.; *et al.* Crossbar Architecture Based on 2R Complementary Resistive Switching Memory Cell. In Proceedings of the 2012 IEEE/ACM International Symposium on Nanoscale Architectures (NANOARCH), Amsterdam, the Netherlands, 4–6 July 2012; pp. 85–92.

Design of Processors with Reconfigurable Microarchitecture

Andrey Mokhov *, Maxim Rykunov, Danil Sokolov and Alex Yakovlev

School of Electrical and Electronic Engineering, Newcastle University, Newcastle upon Tyne NE1 7RU, UK; E-Mails: maxim.rykunov@ncl.ac.uk (M.R.); danil.sokolov@ncl.ac.uk (D.S.); alex.yakovlev@ncl.ac.uk (A.Y.)

* Author to whom correspondence should be addressed; E-Mail: andrey.mokhov@ncl.ac.uk

---

Abstract: Energy becomes a dominating factor for a wide spectrum of computations: from intensive data processing in "big data" companies resulting in large electricity bills, to infrastructure monitoring with wireless sensors relying on energy harvesting. In this context it is essential for a computation system to be adaptable to the power supply and the service demand, which often vary dramatically during runtime. In this paper we present an approach to building processors with reconfigurable microarchitecture capable of changing the way they fetch and execute instructions depending on energy availability and application requirements. We show how to use Conditional Partial Order Graphs to formally specify the microarchitecture of such a processor, explore the design possibilities for its instruction set, and synthesise the instruction decoder using correct-by-construction techniques. The paper is focused on the design methodology, which is evaluated by implementing a power-proportional version of Intel 8051 microprocessor.

Keywords: microarchitecture; microprocessor; reconfigurability; power proportionality; Conditional Partial Order Graphs

---

1. Introduction

Since 2007, our society has used more energy for browsing the Internet than for air travel [1]. It is also predicted that the energy (and environmental) footprint of computation and data traffic will steadily increase in the near future: Data centres will grow and so will the network infrastructure together with the number of terminal nodes of the global information network such as computers, phones, gadgets and other connected cyber-physical devices (so called Internet of Things) [1]. Energy-efficiency of components at all levels of the computation hierarchy is thus becoming a major concern for the microelectronics industry. A serious factor impeding the progress in addressing this concern is a wide gap between the ways of how energy efficiency is approached by hardware and software engineers, and this gap is emphasised by the lack of common understanding between the two communities.

In this paper we discuss an approach to bridging this gap by developing a shared design criterion, called *power-proportionality*, on the basis of which both electronics and programming solutions can be judged. A computing system, for it to be considered power-proportional, has to keep power consumption and computation load proportional to each other. That is, an idle system would ideally consume no power, whereas given a small energy budget the system would adapt itself by reducing its computation flow and lowering the delivered Quality-of-Service (QoS), and still remain functional; see Section 2 for a detailed discussion. The state-of-the-art systems have generally poor power-proportionality; for example, the servers used in data centres typically consume 50%–60% of peak power under 10% of peak load [2]. We hypothesise that modern systems exhibit poor power proportionality because they are designed to operate in a narrow scope of conditions, typically optimised for either high performance or low power consumption. This approach is inherently flawed because all the design effort is focused on one particular operation mode disregarding others. Another issue that makes it difficult to handle unpredictable environmental conditions is the absence of a cross-layer feedback-control loop between energy supply components (in hardware) and an operating system or an active application (in software), see Figure 1: software is unable to steer the computation with the aim of maximising QoS by prioritising tasks, trading accuracy for energy, and, in general, by strategic planning of energy resource utilisation.

Figure 1. Cross-layer power proportionality.

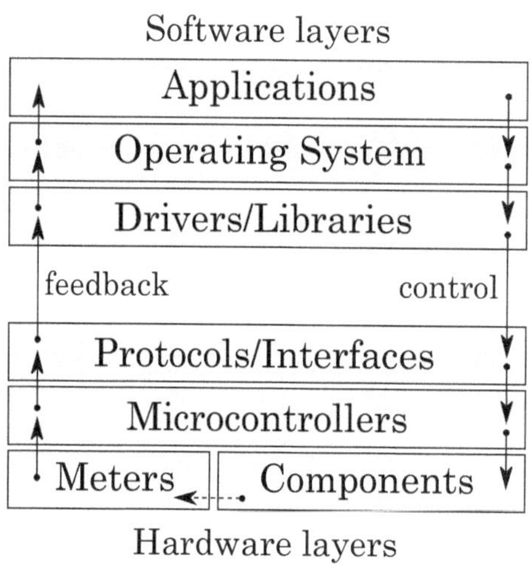

In this paper we discuss cross-layer hardware-software mechanisms for online energy monitoring and reconfiguration. The particular focus is on methodology for design and implementation of power-proportional microprocessors, capable of adapting to varying operating conditions (such as low or even unstable voltage level) and application requirements in runtime. We use a popular Intel 8051 microprocessor as our running example to demonstrate the soundness of the approach. The obtained reconfigurable implementation compares favourably to existing specialised (non-reconfigurable) Intel 8051 designs. Our approach is different from conventional power optimisation techniques, such as [3], because we extend the microarchitecture of the processor and allow its runtime reconfiguration, thereby facilitating the cross-layer feedback-control loop shown in Figure 1.

The paper is organised as follows. Section 2 discusses the challenges in design of reconfigurable processors and introduces the concept of power-proportional computing. Section 3 gives a brief background on the Conditional Partial Order Graph (CPOG) model which we use for the specification and synthesis of processor microarchitectures. We then focus specifically on design of reconfigurable microarchitectures in Section 4. This is followed by a case study on design and implementation of a reconfigurable and power-proportional version of the Intel 8051 core in Section 5.

2. Reconfigurability and Power Proportionality

How can one build a reconfigurable and adaptable processor? The first thought could be to implement it in the *field-programmable gate array* (FPGA) technology allowing static and, sometimes, dynamic reconfiguration. Such a processor would be capable of adjusting its internal structure or behaviour by rewiring interconnections between its components or even by changing functionality at the level of individual gates. This technology can provide a very fine-grain control over a system at runtime, however, the associated overheads are extremely high. In particular, in terms of energy consumption, FPGAs are typically 10–15 times more expensive than *application-specific integrated circuits* (ASICs); see a collection of recent FGPA *vs.* ASIC interviews with industrial experts in [4].

Since the fine-grain reconfigurability offered by FPGAs is overly costly in the context of energy-constrained systems, one shall consider the coarse-grain reconfigurable architectures in the ASIC realm. They significantly lower the overheads by dropping reconfigurability in datapath components and utilising their custom designed versions instead. The control logic and interconnect fabric, however, retain the capability to reconfiguration. In this setting, the key design and implementation challenge is to formally describe and synthesise the controller whose task is to coordinate *hard system resources* (the datapath components and interconnects) according to the runtime information on availability of *soft system resources* (energy and time); the latter can also include information on hardware faults in a system, thereby allowing the controller to bypass faulty components whenever possible.

Figure 2 illustrates the concept of *power-proportionality* and its duality. On the leftmost chart one can see dependency of power consumption on service demand. Ideally, the former should grow proportionally to the latter: when there is no service demand, the system should consume (near) zero power, while any increase in service demand should cause a proportional increase in power consumption until the point of saturation, when the system can no longer serve all the incoming requests. Our design goal is to reduce the interval of wasted energy as much as possible. On the right hand side of Figure 2 one can see the dual chart illustrating dependency of service provision on power supply. Indeed, an ideal computation system should provide service proportionally to the inflow of energy. From this perspective, our design goal is to build computation systems that can start delivering certain level of QoS on as low power supply as possible, thus minimising the interval of unutilised energy.

Figure 2. Power proportionality: power supply/consumption *vs.* service demand/provision.

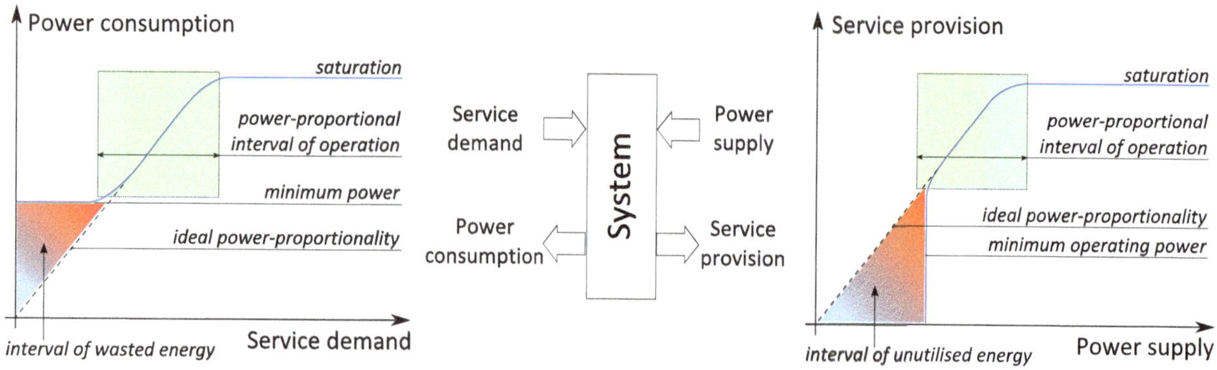

The duality of power-proportionality creates a four dimensional space of power-service interplay: a computation system can be considered a black box which has two inputs (service demand and power supply) and two outputs (service provision and power consumption) as illustrated in the centre of Figure 2. The combined design goal is therefore to achieve proportionality between the inputs and outputs of the system. In this paper we show how to design a processor microarchitecture with this goal in mind.

3. Conditional Partial Order Graphs

The conventional approach to control logic specification and synthesis is to employ *Finite State Machines* (FSMs) [5,6] or interpreted *Petri Nets* (PNs) [7] as an underlying modelling formalism. Within this approach the designer explicitly describes the controller's behaviour for each combination of available resources and operating conditions. The number of such combinations and corresponding behaviours grows extremely fast with the size and degree of adaptability of the system. This leads not only to the *state space explosion problem*, but also to explosion of the specification size [8], thus slowing down the synthesis tools, reducing productivity, and increasing the overall cost of ASIC development.

Our approach is based on the crucial observation that *the controller's behaviours are strongly related to each other* in different operating conditions. Indeed, when a system configuration is changed incrementally, e.g., a datapath component goes offline and another one is used in its place, the overall behaviour of the controller is affected in the same incremental manner, therefore one would want to have a joint behaviour specification for these two configurations.

It has been demonstrated that the FSM and PN formalisms are not well-suited to describing families of many related behaviours [8] and the design methodologies based on them have poor scalability in the context of reconfigurable systems. As an appropriate alternative, the *Conditional Partial Order Graph* model was introduced by Mokhov *et al.* [8,9]. The model was devised to allow implicit description of families of related behaviours in a compact form as will be demonstrated in Sections 3 and 4. The preliminary investigations have shown that the model provides a scalable specification and synthesis framework, capable of producing reconfigurable circuits with only 20% of overheads in terms of latency and power in comparison to their non-reconfigurable counterparts. In this paper we further develop the approach and apply it to a more sophisticated real-life example, namely, the Intel 8051 microprocessor, a popular benchmark for power optimisation methods and new microarchitectural solutions, e.g., [3,10].

Formally, a Conditional Partial Order Graph [8] (further referred to as CPOG or *graph* for short) is a quintuple $H = (V, E, X, \rho, \phi)$ where:

- V is a set of *vertices* which correspond to events (or atomic actions) in a modelled system;
- $E \subseteq V \times V$ is a set of *arcs* representing dependencies between the events;
- *Operational vector* X is a set of Boolean variables. An *opcode* is an assignment $(x_1, x_2, \ldots, x_{|X|}) \in \{0,1\}^{|X|}$ of these variables. An opcode selects a particular partial order from those contained in the graph;
- $\rho \in \mathcal{F}(X)$ is a *restriction function,* where $\mathcal{F}(X)$ is the set of all Boolean functions over variables in X. ρ defines the *operational domain* of the graph: X can be assigned only those opcodes $(x_1, x_2, \ldots, x_{|X|})$ which satisfy the restriction function, *i.e.*, $\rho(x_1, x_2, \ldots, x_{|X|}) = 1$;
- Function $\phi : (V \cup E) \to \mathcal{F}(X)$ assigns a Boolean *condition* $\phi(z) \in \mathcal{F}(X)$ to every vertex and arc $z \in V \cup E$ in the graph. Let us also define $\phi(z) \stackrel{\text{df}}{=} 0$ for $z \notin V \cup E$ for convenience.

CPOGs are represented graphically by drawing a labelled circle ○ for every vertex and drawing a labelled arrow ⟶ for every arc. The label of a vertex v consists of the vertex name, a colon and the vertex condition $\phi(v)$, while every arc e is labelled with the corresponding arc condition $\phi(e)$. The restriction function ρ is depicted in a box next to the graph; operational variables X can therefore be observed as parameters of ρ. Figure 3a shows an example of a CPOG with $|V| = 5$ vertices and $|E| = 7$ arcs. There is a single operational variable x; the restriction function is $\rho(x) = 1$, hence both opcodes $x = 0$ and $x = 1$ are allowed. Vertices $\{a, b, d\}$ have constant $\phi = 1$ conditions and are called *unconditional*, while vertices $\{c, e\}$ are *conditional* and have conditions $\phi(c) = x$ and $\phi(e) = \overline{x}$ respectively. Arcs also fall into two classes: *unconditional* (arc $c \to d$) and *conditional* (all the rest). As CPOGs tend to have many unconditional vertices and arcs we use a simplified notation in which conditions equal to 1 are not depicted in the graph, see Figure 3b.

Figure 3. Graphical representation of Conditional Partial Order Graphs (CPOGs): (**a**) Full notation; (**b**) Simplified notation.

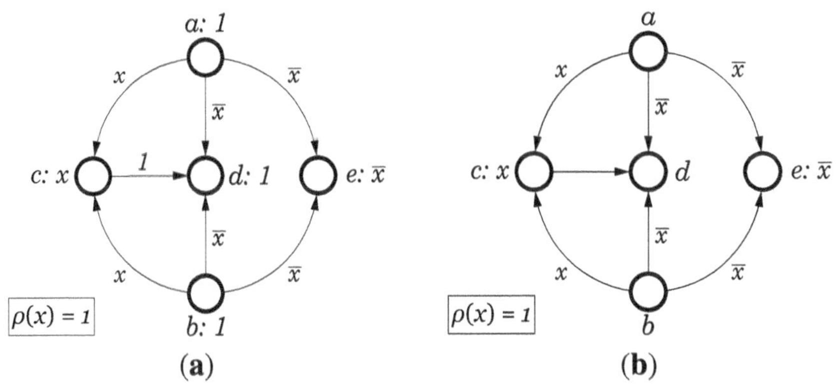

The purpose of conditions ϕ is to "switch off" some vertices and/or arcs in a CPOG according to a given opcode, thereby producing different *CPOG projections*. An example of a graph and its two projections is presented in Figure 4. The leftmost projection is obtained by keeping in the graph only those vertices and arcs whose conditions evaluate to 1 after substitution of variable x with 1 (such projections are conventionally denoted by $H|_{x=1}$). Hence, vertex e disappears (shown as a dashed circle ◌), because its condition evaluates to 0: $\phi(e) = \overline{x} = \overline{1} = 0$. Arcs $\{a \to d, a \to e, b \to d, b \to e\}$ disappear for the same reason; they are shown as dashed arrows ⇢. The rightmost projection is obtained in the same way with the only difference that variable x is set to 0; it is denoted by $H|_{x=0}$, respectively. Note that although the condition of arc $c \to d$ evaluates to 1 (in fact it is constant 1) the arc is still excluded from the resultant graph because one of the vertices it connects, viz. vertex c, is excluded and naturally an arc cannot appear in a graph without one of its vertices. Each of the obtained projections can be regarded as specification of a particular behavioural scenario of the modelled system, e.g., as specification of a processor instruction. Potentially, a CPOG $H = (V, E, X, \rho, \phi)$ can specify an exponential number of different instructions (each composed from atomic actions in V) according to one of $2^{|X|}$ different possible opcodes. We can now establish a formal correspondence between CPOGs and instruction sets.

Figure 4. CPOG projections: $H|_{x=1}$ (left) and $H|_{x=0}$ (right).

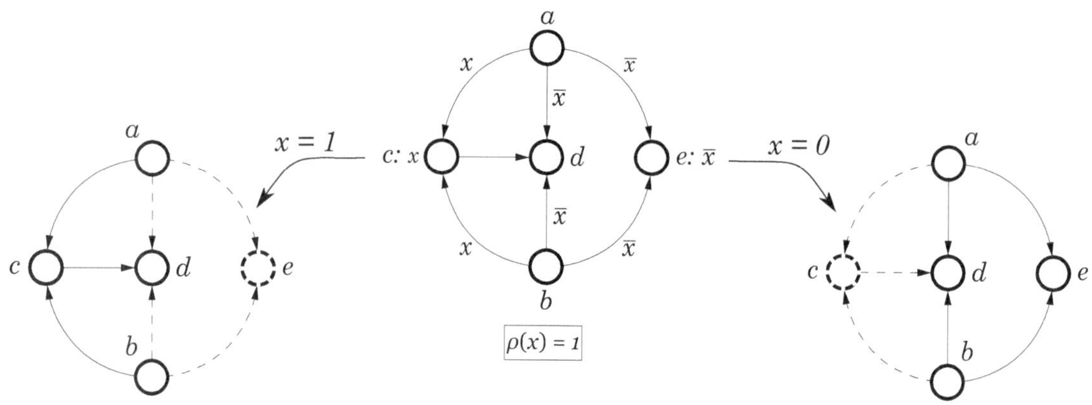

An *instruction* is a pair $\mathsf{I} = (\psi, P)$, where ψ is the *instruction opcode* (a vector of Boolean values), and P is a partial order that specifies the *instruction behaviour* or *microprogram* [11].

An *instruction set*, denoted by IS, is a set of instructions with unique opcodes, that is, for any $\mathsf{IS} = \{\mathsf{I}_1, \mathsf{I}_2, \ldots, \mathsf{I}_n\}$, such that $\mathsf{I}_k = (\psi_k, P_k)$, all opcodes ψ_k must be different.

Given a CPOG $H = (V, E, X, \rho, \phi)$ there is a natural correspondence between its projections and instructions: an opcode $\psi = (x_1, x_2, \ldots, x_{|X|})$ induces a partial order $H|_\psi$, and paired together they form an instruction $\mathsf{I}_\psi = (\psi, H|_\psi)$ according to the above definition. This leads to the following formal link between CPOGs and instruction sets. A CPOG $H = (V, E, X, \rho, \phi)$ is a *specification* of an instruction set $\mathsf{IS}(H)$ defined as a union of instructions $(\psi, H|_\psi)$ which are allowed by the restriction function ρ:

$$\mathsf{IS}(H) \stackrel{\text{df}}{=} \{(\psi, H|_\psi),\ \rho(\psi) = 1\} \tag{1}$$

Note that this definition automatically guarantees that no two instructions have the same opcode.

4. Design of Reconfigurable Microarchitectures

In this section we demonstrate how to use the CPOG model to build large instruction sets from basic building elements, how to introduce reconfigurability, and finally how to automatically derive an implementation for the specified microarchitecture.

4.1. Building Instruction Sets

CPOGs can be deconstructed by means of projections (Section 3). The opposite operation, that is constructing a CPOG out of parts, is called *composition*. This operation is essential for building large instruction sets from smaller ones, thus enabling modularisation, reuse, and incremental development.

Let instruction sets IS_1 and IS_2 be specified with graphs $H_1 = (V_1, E_1, X, \rho_1, \phi_1)$ and $H_2 = (V_2, E_2, X, \rho_2, \phi_2)$, respectively, as in Equation (1). Then their *composition* has the CPOG specification $H = (V_1 \cup V_2, E_1 \cup E_2, X, \rho_1 + \rho_2, \phi)$, where the vertex/arc conditions ϕ are defined as

$$\forall z \in V_1 \cup V_2 \cup E_1 \cup E_2,\ \phi(z) \stackrel{\text{df}}{=} \rho_1\phi_1(z) + \rho_2\phi_2(z)$$

We call H the *CPOG composition* of H_1 and H_2 and denote this operation as $H = H_1 \cup H_2$. Composition is defined only if $\rho_1 \cdot \rho_2 = 0$, as otherwise $\mathsf{IS}(H_1)$ and $\mathsf{IS}(H_2)$ contain instructions with the same opcode making the result ambiguous. It is possible to formally prove that $\cup$ is a commutative and associative operation and that $\mathsf{IS}(H) = \mathsf{IS}(H_1) \cup \mathsf{IS}(H_2)$ using the algebraic methods presented in [8], which leads to the following important result: $\mathsf{IS}(H_1 \cup H_2) = \mathsf{IS}(H_1) \cup \mathsf{IS}(H_2)$.

Once an instruction set is built, it usually undergoes a set of transformations and optimisations which include modification of the set of computational units, changing opcodes of instructions or groups of instructions, modifying the instruction pipeline, *etc.* Most of these transformations are outside the scope of this work and are defined in [9]. In the following subsections we introduce two new transformaitons.

4.2. Mode-Specific Components

There are many ways in which we can implement a particular datapath component (e.g., a multiplier) and they all differ in their performance and energy-efficiency characteristics. If the energy budget of the system varies in runtime it may be beneficial to include more than one implementation in the processor: in this way, it would be possible to choose the most appropriate one in runtime according to the availability of energy and current application requirements. The CPOG model provides a convenient way of introducing such mode-specific component implementations into the processor microarchitecture.

Let $H = (V, E, X, \rho, \phi)$ be the specification of an instruction set and $v \in V$ correspond to a datapath component whose implementation we would like to make mode-specific. This can be achieved by the CPOG transformation called *vertex splitting*. The operation transforms H in the following way:

- Vertex v is replaced by two vertices v_0 and v_1, corresponding to two mode-specific implementations of the component. In other words, vertex v is split into two new ones $\{v_0, v_1\}$;
- Arcs (u, v) are replaced with two arcs (u, v_0) and (u, v_1) and, similarly, arcs (v, u) are replaced with two arcs (v_0, u) and (v_1, u). Arc conditions are preserved;
- Set of variables X is extended with a new *mode variable* m that is used to determine which of the two implementations is active. Mode variables are changeable in runtime from software, thus facilitating the cross-layer feedback control illustrated in Figure 1;
- Function ρ is unchanged, thus instruction opcodes are not affected by this transformation and the new microarchitecture has no penalties in terms of compactness of the instruction encoding;
- The new vertices get conditions $\phi(v_0) = \overline{m}\phi(v)$ and $\phi(v_1) = m\phi(v)$, where $\phi(v)$ is the condition that was assigned to the removed vertex v.

The vertex splitting transformation is illustrated in Figure 5. This transformation increases area and leakage current, but the latter can be mitigated by power gating using the mode variable m to disconnect inactive components from the power supply. As a side benefit, vertex splitting improves the system's fault tolerance: if one of the two components fails, we can disable it and use the other component regardless of the current performance *vs.* energy-efficiency strategy. In this way the system will be able to maintain the correct functionality but the degree of power-proportionality will be decreased.

Figure 5. Vertex splitting.

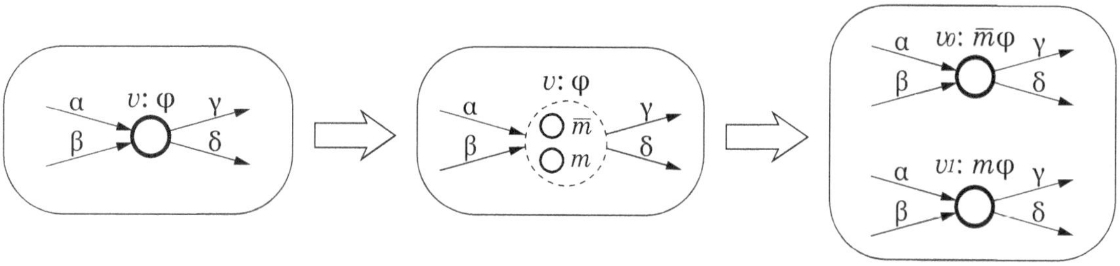

Note that this transformation can also be defined in terms of CPOG composition of two mode-specific instruction sets: $\overline{m}H_0 \cup mH_1$, where H_0 (resp., H_1) corresponds to CPOG H with vertex v replaced with vertex v_0 (respectively, v_1). The resulting CPOG specifies the same instruction set but can be derived in a simpler and more compositional manner.

4.3. Concurrency Management

It is often the case that an instruction can be executed with different degrees of concurrency. For example, we can typically perform an ALU operation either concurrently with fetching the next instruction opcode or in sequence with it.

Consider two actions a and b. If they have no mutual dependencies in terms of functionality then they can seemingly be executed concurrently. However, they both consume a certain amount of power and this fact should not be ignored when designing power-aware systems. If concurrent execution of a and b exceeds the available power budget, the supply voltage may drop significantly which can potentially lead to the failure of the whole system. On the other hand, executing a and b only in sequence may be too conservative and will result in a slower system. How can we deal with this situation?

We propose to provide support for both execution scenarios in the microarchitecture and choose the most appropriate one according to the current power budget in runtime. Using the CPOG specification model this can be achieved by making the arc between vertices a and b conditional: $\phi(a \rightarrow b) = m$. Now if the mode variable m is equal to 0 then components a and b will be executed concurrently (the arc disappears when $m = 0$); otherwise, when $m = 1$ execution of a and b will be scheduled in sequence, hence limiting the peak power consumption.

4.4. Synthesis of Microcontroller

As soon as all the intended manipulations with the instruction set are performed, we can proceed to the stage of mapping the resultant CPOG into Boolean equations to produce a physical implementation of the specified microcontroller. In order to descend from the abstract level of atomic actions to the physical level of digital circuits the following signal-level refinements are necessary.

To interface with an asynchronous datapath component a it is possible to use the standard request-acknowledgement handshake (req_a, ack_a), as shown in Figure 6. In case of a synchronous component b the request signal is used to start the computation but, as there is no completion detection, the acknowledgement signal has to be generated using a matched delay [12]. Also, there are cases when a matched delay has to be replaced with a counter connected to the *clock signal* to provide an accurate multi-cycle delay—see the interface of component c in the same figure. Note that we do not explicitly show synchronisers [13] in the diagram; it is assumed that components b and c are equipped with the necessary synchronisation mechanisms to accept asynchronous requests from the microcontroller.

Figure 6. Datapath interface architecture.

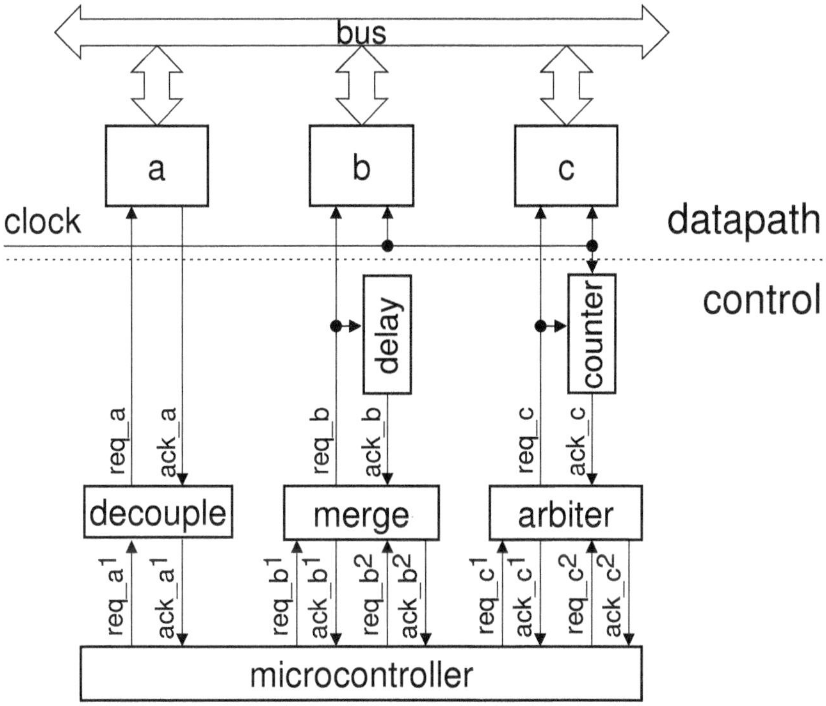

To explicitly specify handshake signals it is possible to perform a graph transformation explained in Figure 7. Every atomic action a^1 is split into a pair of events req_a^1+ and ack_a^1+ standing for rising transitions of the corresponding handshake signals. If there are two occurrences of an atomic action, e.g., b^1 and b^2, then both vertices are split (we use superscripts to distinguish different occurrences of the same event), etc. Semantically, when an atomic action a^1 is ready for execution, the controller should issue the request signal req_a^1 to component a; then the high value of the acknowledgement signal ack_a^1 will indicate completion of a.

Notice that the microcontroller does not reset handshakes until all of them are complete. This leads to a potential problem: a component cannot be released until the instruction execution is finished. To deal with this we decouple the microcontroller from the component, see box "decouple" in Figure 6 and its gate-level implementation in Figure 8a. Also, when a component b is used twice in an instruction we combine two handshakes $(req\_b^{1,2}, ack\_b^{1,2})$ into one using the merge controller, see Figure 8b. Merge controllers can only be used if the requests are mutually exclusive. If this is not the case, as e.g., for concurrent actions c^1 and c^2, then we have to set an arbiter guarding access to the component. Its implementation consists of the merge controller and the mutual exclusion (ME) element [13], see Figure 8c.

Figure 7. Signal-level refinement.

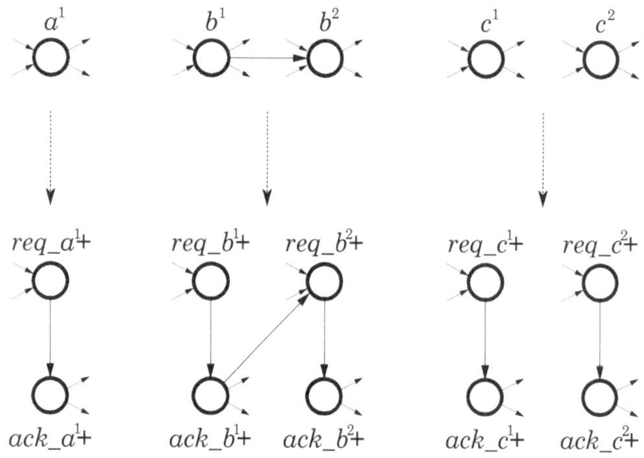

Figure 8. Handshake controllers: (**a**) Decouple; (**b**) Merge and (**c**) Arbiter.

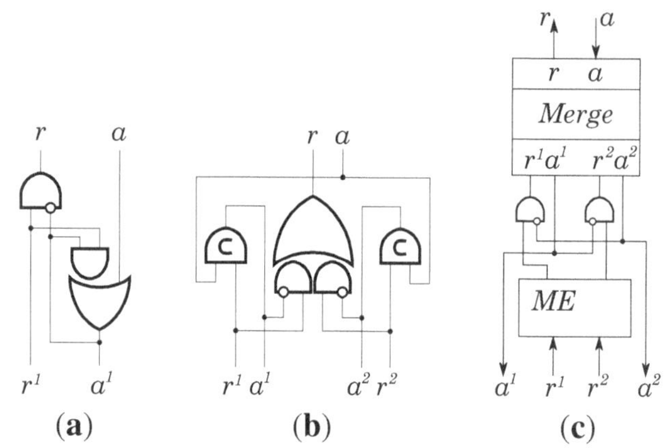

Finally, the refined graph can be mapped into Boolean equations. An event associated with vertex $v \in V$ is enabled to fire (req_v+ is excited) when all the preceding events $u \in V$ have already fired (ack_u have been received) [8]:

$$req\_v = \phi(v) \cdot \prod_{u \in V} \left(\phi(u) \cdot \phi(u \to v) \Rightarrow ack\_u\right)$$

where $a \Rightarrow b$ stands for Boolean implication indicating "b if a" relation. Mapping is a simple structural operation, however the obtained equations may be suboptimal and should undergo the conventional logic minimisation [8,11] and technology mapping [7] procedures.

It is interesting to note that the size of the microcontroller does not depend on the number of instructions directly. There are $\Theta(|V|^2)$ conditions ϕ in all the resultant equations; the average size of these conditions is difficult to estimate, but in practice we found that the overall size of the microcontroller never grows beyond $\Theta(|V|^2)$.

5. Reconfigurable 8051 Microcontroller

In this section we describe our power-proportional implementation of the Intel 8051 microprocessor with reconfigurable microarchitecture designed using the CPOG model.

The instruction set of Intel 8051 contains 255 instructions and our implementation supports all of them. We generally follow the standard 8051 architecture with two on-chip RAMs; the program is stored in a reprogrammable off-chip ROM. However several important changes were made:

- Our implementation is asynchronous, which gives certain advantages over clocked designs in case of varying operating conditions [10]. The communication between the control logic and the datapath units is arranged by means of request and acknowledgement control signals;
- The 16-bit datapath was modified. The computation units were split into pairs as described in Section 4.2 to achieve power-proportionality; each unit in a pair was optimised either for high performance or energy-efficiency. Table 1 describes the computational units. By using adjustable delay lines we achieved robust operation in a wide range of supply voltages (see Section 5.2);
- Instructions are executed according to application/environmental requirements, and taking into account the functional correctness of components, thus addressing the issues of fault tolerance.

Figure 9 shows the top-level view of the microprocessor architecture. The shaded block corresponds to the central control unit (the main design objective), the rest of the blocks correspond to datapath units. The top-level CPOG specification of the microprocessor is shown in Figure 10. Note that the main datapath block, the Arithmetic Logic Unit (ALU), is a complicated subsystem with its own control unit and multi-mode operation; its microarchitecture was designed similarly to the top-level one.

Figure 9. Architecture of the designed microprocessor (dashed lines represent a 1 bit wide control signal line, the bit-width of other connections is shown in brackets).

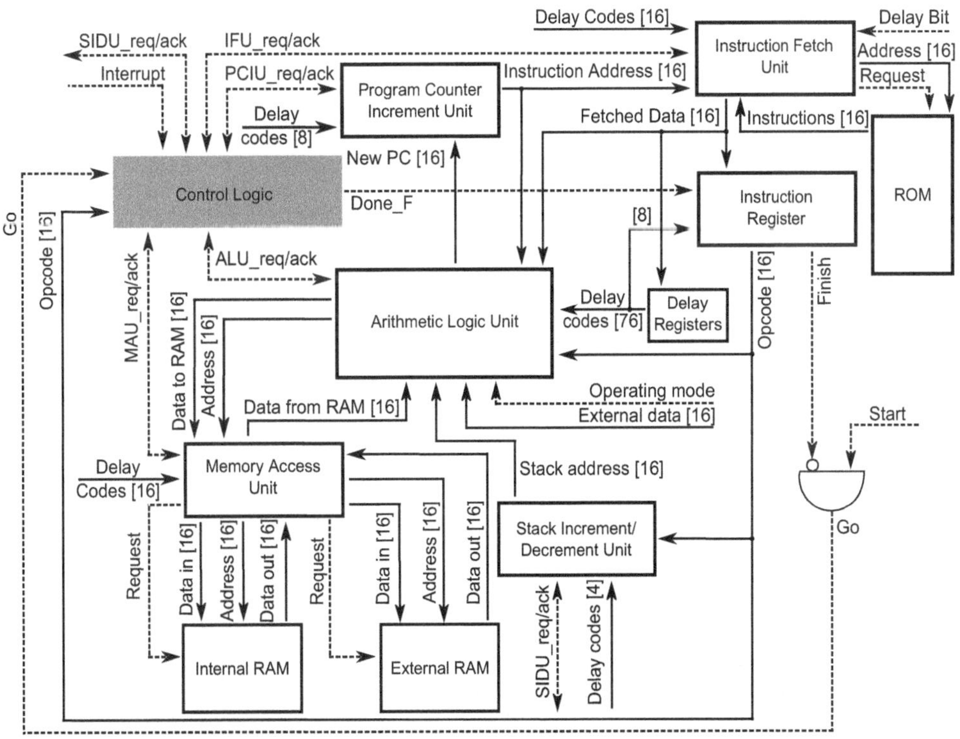

Figure 10. CPOG model of Intel 8051 control logic.

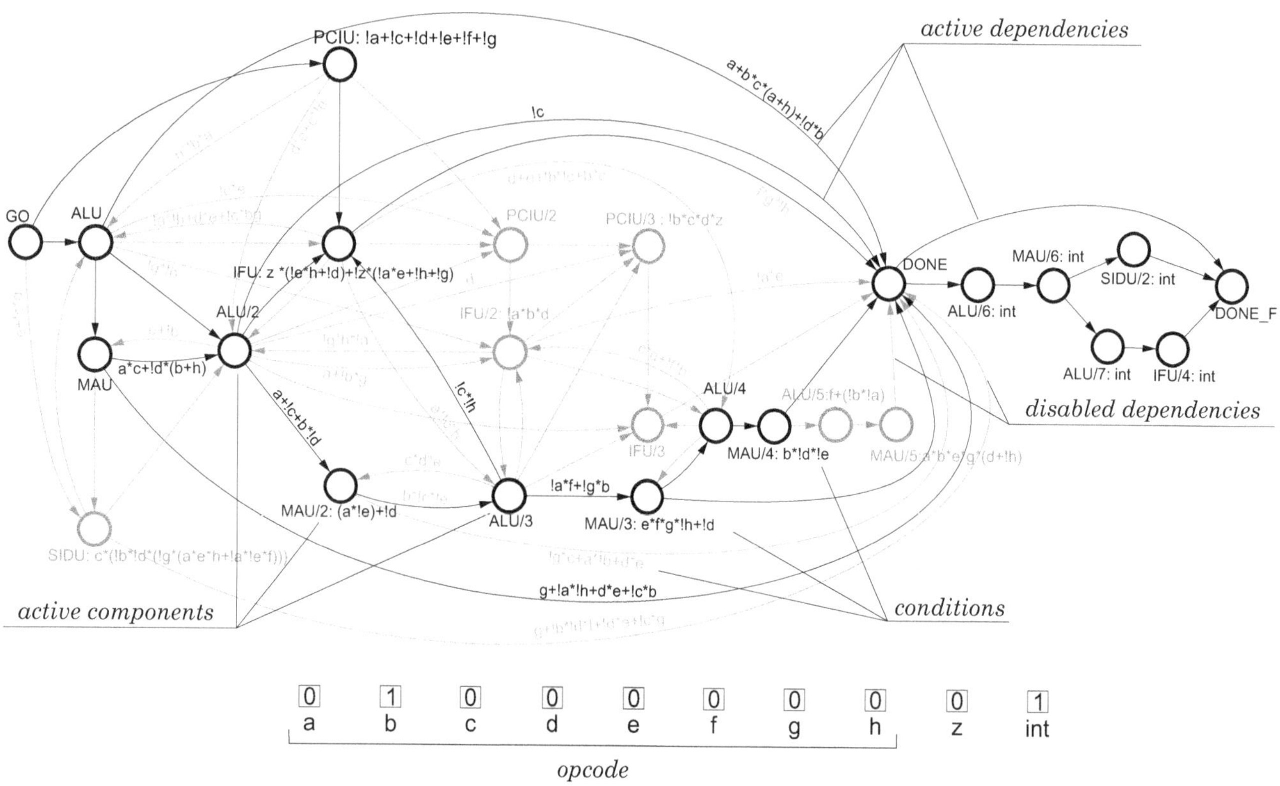

5.1. Design Flow

The first stage in microprocessor design is the specification of instructions. Each instruction corresponds to a schedule of primitive computation steps such as data transfer, arithmetic operation, memory access, *etc*., performed by datapath components. Overall, the proposed microprocessor design flow consists of the following steps:

Extraction of datapath components. We refine the architecture to the level of computational units and behavioural scenarios. There are several ways in which this can be done: (i) direct refinement from software; (ii) specification with Architecture Description Languages [14,15] and (iii) formalisation of existing instruction specifications (e.g., ISA manual). In our case we had a complete description of the microprocessor, its instruction set, and architecture details [16], therefore the last option was the most suitable. Table 1 describes five computational units extracted from the 8051 instruction set.

Partial order specification of instructions. The behaviour of each instruction is captured by a partial order (PO) in which the required computation units need to be activated. Different instructions often have the same PO representations; we grouped all 8051 instructions into 37 classes according to their POs, with the largest class containing 60 instructions.

Figure 11 shows the PO representation of one of the classes, which corresponds to a group of 17 instructions such as, for example, *ADD A, immediate* that performs addition of the accumulator and an immediate constant. We first read data from the accumulator (ALU followed by MAU) while concurrently fetching an immediate constant (PCIU followed by IFU). Then we perform addition (ALU/2) concurrently with another increment of PC, and finally write the result back into accumulator

(MAU) and fetch the next instruction (IFU/2). Many other instructions have the same behaviour: *ADDC A, immediate*; *SUBB A, immediate*; *ORL A, immediate*; *ANL A, immediate*, etc.

Table 1. List of datapath units.

Name	Description
PCIU	Program Counter Increment Unit is responsible for incrementing the Program Counter—a special register pointing to the address of the next instruction in the program memory.
IFU	Instruction Fetch Unit loads the operational code of the current instruction from the program memory using the address stored in Program Counter.
MAU	Given an address pointer and data, the Memory Access Unit can access the internal and external RAM.
ALU	Arithmetic Logic Unit is the main computation unit responsible for performing arithmetic and logic operations in the microprocessor. To make it adjustable towards a wide range of operating conditions and application requirements, we designed two sets of ALU datapath components, each optimised to work in a specific mode: one optimised for performance and the other one for energy efficiency.
SIDU	Stack pointer Increment/Decrement Unit, as the name suggests, is in charge of incrementing and decrementing the stack pointer.

Figure 11. Partial order (PO) for ADD instruction.

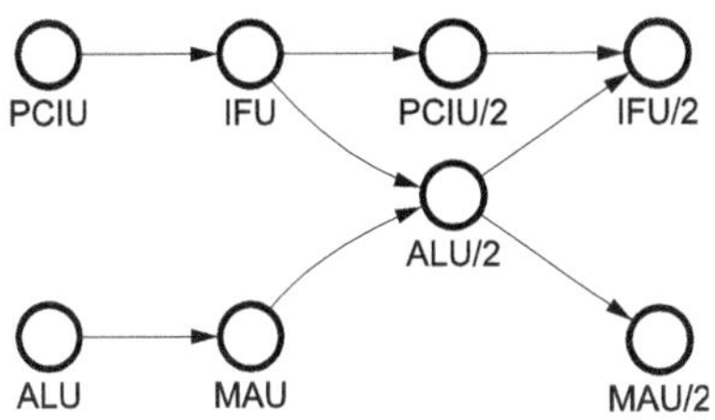

Encoding of partial orders. In order to distinguish between the POs, we need to encode them. The codes can either be assigned arbitrarily or can be provided as part of system specification. Importantly, the size and latency of the final microcontroller circuit depends significantly on the chosen encoding of the scenarios. There are several types of encoding that can be used, such as one hot, matrix, balanced, and Huffman encoding [17]. A new technique for optimal encoding was recently introduced in [18], but it was unable to process our set of POs due to high complexity of the optimal encoding algorithm, hence we used Huffman encoding in our design.

CPOG synthesis. Once all instructions are represented as POs and encoded, we can synthesise a CPOG containing all of them as described in Section 4. We used the Workcraft framework [19] which automates the specification and synthesis stages. The resulting CPOG is shown in Figure 10.

Mapping. The final stage is mapping of the CPOG representation into a set of logic gates to produce a physical implementation (a gate-level netlist) of the specified microcontroller. Finally, we can translate the obtained Boolean equations into VHDL, Verilog or other HDL language to use in standard EDA flow.

5.2. Adaptability Features

To enable power-proportionality in a wide range of power supply and service demand conditions, we designed two sets of computation units: one optimised for performance and the other one optimised for energy efficiency. Depending on whether there is a shortage of energy or on any other restrictions imposed by an active application, we can choose the most appropriate datapath unit to be used during an instruction execution. The decision can be made at different levels of system control: software level, sensors, external signals, *etc.*

Our implementation of self-timed datapath components is based on bundled-data approach, where each component is accompanied with a matching delay line to signal completion of its computation. In order to correctly function in a wide range of operating conditions (e.g., supply voltage or temperature), a component needs to adjust the latency of its completion signal. We addressed this issue by use of *adjustable delay lines*, whose latency is selectable by the *delay code* at runtime, as shown in Figure 12.

Figure 12. Adjustable delay lines.

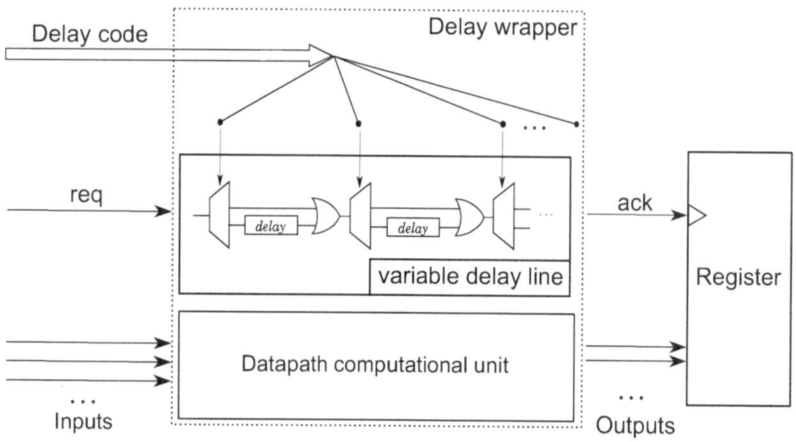

5.3. Experiments

We implemented the microprocessor in STMicroelectronics 130nm CMOS technology process and conducted experiments to evaluate the level of adaptability we can achieve using the presented approach. In this section we report results of our measurements. The nominal supply voltage of the chip is 1.2 V; we varied the supply voltage in the range from 0.22 V to 1.5 V and observed the following:

- **0.89 V to 1.5 V**: the chip works in full capability mode;
- **0.74 V to 0.89 V**: at 0.89 V the RAM starts to fail; the chip operates using internal registers only;
- **0.22 V to 0.74 V**: at 0.74 V the Program Counter unit starts to fail, so the chip gets stuck at executing the same instruction forever. The control logic synthesised using the CPOG model continues to operate correctly down to 0.22 V.

It should be noted that within the above-indicated range of 0.22–0.74 V, somewhere below 0.5 V, most of the microprocessor's hardware is subjected to work in the near-threshold and even sub-threshold regime. Our implementation was based on standard STM cells, therefore we were not surprised by the fact that some registers like the Program Counter began to fail in this range. However, the important knowledge that we gleaned from testing our processor was that that the control circuitry, even though built of standard cells, was still operational in the sub-threshold regime. This shed light on how we can build future reconfigurable and power-proportional systems that would be able to operate in parts under different supply voltage levels.

We can conclude that our chip can work in dual-memory mode in the voltage range from 0.89 V to 1.5 V. This mode is to be used for applications requiring high-performance and large memory. The single-memory mode available in the voltage range from 0.74 V to 0.89 V can be used for low-power and scratchpad memory applications.

In terms of performance, we measured the range from 67 millions of instructions per second at the nominal voltage of 1.2 V down to 2700 instructions per second at 0.25 V. Note that since the Program Counter fails below 0.74 V the processor starts executing the same instruction forever. Figure 13 shows measurements of latency and power consumption of the ADD instruction run in the voltage range from 0.88 V to 1.2 V. In our expertiments other instructions exhibited similar latency/power characteristics.

Figure 13. Measurements for ADD instruction.

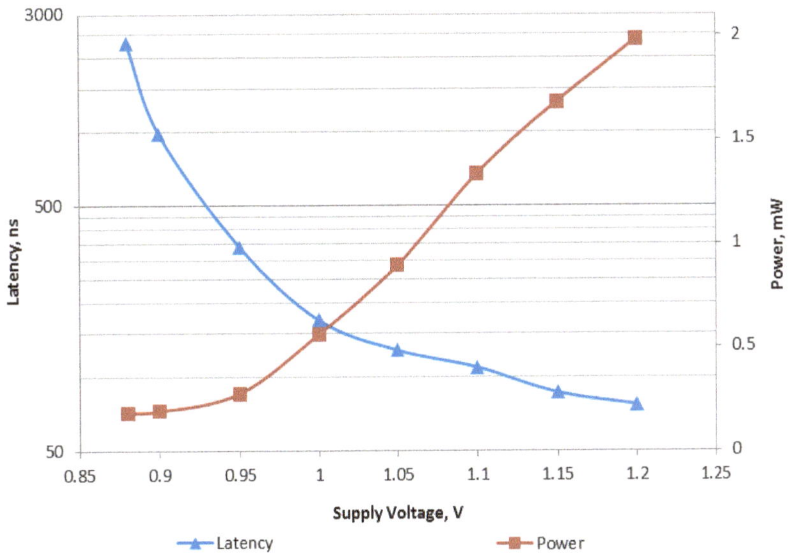

To compare our implementation with existing Intel 8051 designs we run the Dhrystone V2.1 testbenches [20], see the summary of obtained results in Table 2. Sync 80C51, H8051 and DS89C420 are synchronous designs; the first two employ non-pipeline architectures and the last one has a pipelined architecture. Async 80C51 is an asynchronous counterpart of the Sync 80C51 design. Lutonium is another asynchronous implementation which utilises a highly parallel processing with a deep pipeline architecture; it has high MIPS rate and power consumption. Nanyang A8051 is a self-timed, ultra low power (and hence low performance) implementation. After normalising the results with respect to the technology difference, our implementation is clearly placed between ultra low power design (runs twice as fast) and the highly parallel version (consumes 20 times less power).

Table 2. Performance comparison with other 8051 versions.

Processor	Technology	11 cmMIPS	Average power, mW	MIPS per W	Energy, pJ per instruction
Sync 80C51 [21]	3.3 V, 350 nm	4	40	100	10000
Async80C51 [21]	3.3 V, 350 nm	4	9	44	2250
H8051 [22]	3.3 V, 350 nm	4	44.7	89.5	11175
Lutonium [23]	1.8 V, 180 nm	200	100	1800	500
DS89C420 [24]	1.1 V, 350 nm	11	18.52	600	1684
Nanyang A8051 [25]	1.1 V, 350 nm	0.6	0.07	8000	130
Lutonium [23]	1.1 V, 180 nm	100	20.7	4830	207
Proposed 8051	**1.2 V, 130 nm**	**1.5**	**0.74**	**2027**	**493**

6. Conclusions

We presented an approach to design of processors with reconfigurable microarchitecture for power-proportional computation systems. The approach was tested by designing and implementing the Intel 8051 microprocessor, whose control core is capable of operating in the range from 0.2 V to 1.5 V of supply voltages. Unfortunately due unreliability of the Program Counter the processor is only usable starting from 0.74 V. Our future research is to push the boundaries of power-proportionality even further by using the self-timed SRAM capable of operating at low voltages [26], as well as by implementing the critical control and datapath components, such as Program Counter, using high-reliability logic with low fan-in gates and, possibly, with dual-rail control logic for added resilience against single event upsets.

Acknowledgements

This work was supported by EPSRC grants EP/I038357/1 (eFuturesXD, project PowerProp) and EP/K034448/1 (PRiME).

Conflicts of Interest

The authors declare no conflict of interest.

References

1. The Zettabyte Era. In *Cisco Visual Networking Index: Forecast and Methodology, 2010–2015*; Cisco Systems Inc: San Jose, CA, USA, 2011.
2. Meisner, D.; Sadler, C.M.; Barroso, L.A.; Weber, W.D.; Wenisch, T.F. Power Management of Online Data-Intensive Services. In Proceedings of the 38th Annual International Symposium on Computer Architecture, San Jose, CA, USA, 4–8 June 2011; pp. 319–330.
3. Saponara, S.; Fanucci, L.; Terreni, P. Architectural-level power optimization of microcontroller cores in embedded systems. *IEEE Trans. Ind. Electron.* **2007**, *54*, 680–683.
4. Trending Back To ASICs. Available online: http://semiengineering.com/trending-asics/ (accessed on 30 October 2013).

5. Nowick, S. Automatic Synthesis of Burst-Mode Asynchronous Controllers. Ph.D. Thesis, Stanford University, Stanford, CA, USA, March 1993.
6. Baranov, S.I. *Logic Synthesis for Control Automata*; Kluwer Academic Publishers: Dordrecht, The Netherlands, 1994.
7. Cortadella, J.; Kishinevsky, M.; Kondratyev, A.; Lavagno, L.; Yakovlev, A. *Logic Synthesis of Asynchronous Controllers and Interfaces*; Advanced Microelectronics; Springer-Verlag: New York, NY, USA, 2002.
8. Mokhov, A.; Yakovlev, A. Conditional partial order graphs: Model, synthesis and application. *IEEE Trans. Comput.* **2010**, *59*, 1480–1493.
9. Mokhov, A.; Iliasov, A.; Sokolov, D.; Rykunov, M.; Yakovlev, A.; Romanovsky, A. Synthesis of processor instruction sets from high-level ISA specifications. *IEEE Trans. Comput.* **2013**, *PP*, 1.
10. Chang, K.L.; Chang, J.S.; Gwee, B.H.; Chong, K.S. Synchronous-logic and asynchronous-logic 8051 microcontroller cores for realizing the internet of things: A comparative study on dynamic voltage scaling and variation effects. *IEEE J. Emerg. Sel. Top. Circuits Syst.* **2013**, *3*, 23–34.
11. De Micheli, G. *Synthesis and Optimization of Digital Circuits*; McGraw-Hill Higher: Stanford, CA, USA, 1994.
12. Sparsoe, J.; Furber, S. *Principles of Asynchronous Circuit Design: A Systems Perspective.*; Kluwer Academic Publishers: Dordrecht, The Netherlands, 2001.
13. Kinniment, D.J. *Synchronization and Arbitration in Digital Systems*; John Wiley and Sons: Cornwall, UK, 2008. ISBN: 978-0-470-51082-7.
14. Hoffmann, A.; Schliebusch, O.; Nohl, A.; Braun, G.; Wahlen, O.; Meyr, H. A Methodology for the Design of Application Specific Instruction Set Processors (ASIP) Using the Machine Description Language LISA. In Proceedings of the ACM/IEEE International Conference on Computer-Aided Design, San Jose, CA, USA, 4–8 November 2001, pp. 625–630.
15. Fauth, A.; Freericks, M. Describing Instruction Set Processors Using nML. In Proceedings of the 1995 European Conference on Design and Test, Paris, France, 6–9 March 1995, p. 503.
16. Intel 8051 Instruction Set. Availabel online: http://www.keil.com/support/man/docs/is51 (accessed on 30 October 2013).
17. Huffman, D.A. A method for the construction of minimum-redundancy codes. *Proc. IRE* **1952**, *40*, 1098–1101.
18. Mokhov, A.; Alekseyev, A.; Yakovlev, A. Encoding of processor instruction sets with explicit concurrency control. *IET Comput. Digit. Tech.* **2011**, *5*, 427–439.
19. The Workcraft Framework Homepage. Available online: http://www.workcraft.org (accessed on 30 October 2013).
20. Price, W.J. A benchmark tutorial. *IEEE Micro* **1989**, *9*, 28–43.
21. Van Gageldonk, H.; van Berkel, K.; Peeters, A.; Baumann, D.; Gloor, D.; Stegmann, G. An Asynchronous Low-Power 80C51 Microcontroller. In Proceedings of the Fourth International Symposium on Advanced Research in Asynchronous Circuits and Systems, San Deigo, CA, USA, 30 March–2 April 1998; pp. 96–107.
22. *HMS99C52 Datasheet*; Hynix Co.: Gyeonggi, South Korea, 2003. Available online: http://www.gaw.ru/pdf/Hynix/mcs51/HMS99C52.pdf (accessed on 30 October 2013).

23. Martin, A.J.; Nystrom, M.; Papadantonakis, K.; Penzes, P.I.; Prakash, J.T.; Tura, A. The Lutonium: A Sub-Nanojoule Asynchronous 8051 Microcontroller. In Proceedings of the Ninth International Symposium on Asynchronous Circuits and Systems, Vancouver, BC, Canada, 12–16 May 2003, pp. 14–23.
24. *DS89C420 Microcontroller*; Maxim Integrated Products: San Jose, CA, USA, 2000.
25. Chang, K.L.; Gwee, B.H. A Low-Energy Low-Voltage Asynchronous 8051 Microcontroller Cores. In Proceedings of the IEEE International Symposium on Circuits and Systems (ISCAS), Island of Kos, Greece, 21–24 May 2006.
26. Baz, A.; Shang, D.; Xia, F.; Yakovlev, A. Self-timed SRAM for Energy Harvesting Systems. In Proceedings of the 20th International Workshop on Power and Timing Modeling, Grenoble, France, 7–10 September 2010, pp. 105–115.

8

A Practical Framework to Study Low-Power Scheduling Algorithms on Real-Time and Embedded Systems

Jian (Denny) Lin [1],*, Albert M. K. Cheng [2] and Wei Song [2]

[1] Department of Management Information Systems, University of Houston—Clear Lake, Houston, TX 77058, USA

[2] Department of Computer Science, University of Houston, Houston, TX 77204, USA; E-Mails: cheng@cs.uh.edu (A.M.K.C.); song.songwei@gmail.com (W.S.)

* Author to whom correspondence should be addressed; E-Mail: linjian@uhcl.edu

Abstract: With the advanced technology used to design VLSI (Very Large Scale Integration) circuits, low-power and energy-efficiency have played important roles for hardware and software implementation. Real-time scheduling is one of the fields that has attracted extensive attention to design low-power, embedded/real-time systems. The dynamic voltage scaling (DVS) and CPU shut-down are the two most popular techniques used to design the algorithms. In this paper, we firstly review the fundamental advances in the research of energy-efficient, real-time scheduling. Then, a unified framework with a real Intel PXA255 Xscale processor, namely real-energy, is designed, which can be used to measure the real performance of the algorithms. We conduct a case study to evaluate several classical algorithms by using the framework. The energy efficiency and the quantitative difference in their performance, as well as the practical issues found in the implementation of these algorithms are discussed. Our experiments show a gap between the theoretical and real results. Our framework not only gives researchers a tool to evaluate their system designs, but also helps them to bridge this gap in their future works.

Keywords: real-time/embedded systems; low-power; scheduling

1. Introduction

Recently, there has been an explosive growth in the popularity of portable and embedded devices. While power demands are increasing rapidly, battery capacity cannot keep up [1]. Power has been considered as the most important constraint in embedded systems, and how to manage it plays important roles. Managing energy consumption introduces a variety of benefits, such as the prolonged life of the battery in mobile devices, the reduced cost of energy and a low operating temperature. Today, low-power design is a widely discussed topic from the low circuit level to the high system synthesis level. This work is a study of low-power and energy-efficient design at the system level, more specifically embedded and real-time systems.

With the advanced technology of VLSI circuit designs, the dynamic voltage scaling (DVS), which adjusts the supply voltage and frequency dynamically, becomes an important power management technique in computer architecture. Modern well-known CPUs in embedded systems (Intel StrongARM [2], Xscale [3]) have adopted the DVS technique. The main sources of the power consumption of the CMOS circuits in a DVS processor are dynamic power, short circuit power and leakage power consumption. The dynamic power results from the charging and discharging of gate switching, and is dependent upon the processor's speed, s, which can be approximated by:

$$P_{dyn}(s) = C_{ef} V_{dd}^2 s \tag{1}$$

where $s = K\frac{(V_{dd}-V_t)^2}{V_{dd}}$. The C_{ef}, V_t, V_{dd} and K denote the effective switch capacitance, the threshold voltage, the supply voltage and a hardware-design-specific constant, respectively. The short-circuit power is proportional to the supply voltage. As a result, the dynamic power consumption of the DVS processor has a quadratic dependency on the supply voltage, which can be reduced by lowering the supply voltage and, thus, the speed. The leakage power is independent of the processor's speed, which is the power consumed by unintended leakage that does not contribute to the processor's function. When a processor is idle, it may be temporarily shut down to reduce the overall power consumption.

Many applications running on embedded systems are real-time tasks in which the task response time is an important requirement. A real-time task is expected to complete its execution before its deadline to maintain the system stability (e.g., the control tasks). If the timing resource is not 100% utilized on a system, there are some idle intervals in CPU cycles. These CPU idle cycles, termed as slack, can then be used to slow down the execution of a real-time task (or shut down the CPU), and its deadline can still be met. It is not hard to see that minimizing the energy consumption, as well as satisfying the multiple tasks' real-time requirements need a careful calculation to schedule the tasks. Figure 1 demonstrates a general idea in power-aware scheduling in real-time systems. In the figure, Schedule 1 is a schedule of the three tasks in Table 1 running at full speed (s = 1.0), and Schedule 2 is a schedule where the speed of the processor is slowed down by 50%. The idle intervals in Schedule 1 may be used to shut down the processor if they are long enough. Generally, a processor is temporarily shut down only if the idle interval is sufficiently long. This is because a processor's shut-down and wake-up will incur overheads.

Considering the potential ability in power saving, power-aware scheduling using DVS or/and CPU shut-down has received a lot of attention in the past few decades. A large number of algorithms have been proposed to complete the goal of power-saving using different models and considering different scenarios. Although most algorithms are shown to be quite effective under their own experimental settings, almost all of their performance results are obtained solely via simulations. The major restriction of using simulations is the difficulty in determining the real performance, because many unpredictable and non-modeled factors may affect the results. The minor restriction could be the hardness in comparing different designs, making it difficult for developers to select an appropriate energy-efficient algorithm for a given application/system.

Figure 1. An example of power-aware scheduling using a slowed-down speed.

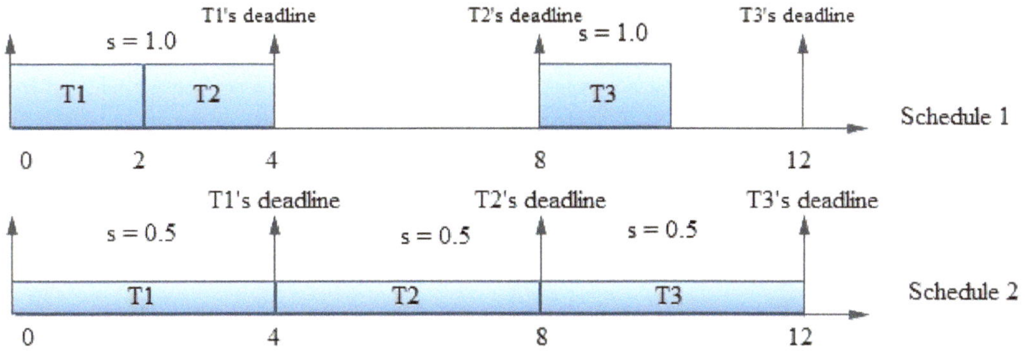

Table 1. Three real-time tasks.

Task no.	Ready time and deadline	Execution time
T1	(0, 4)	2
T2	(2, 8)	2
T3	(7, 12)	2

This paper extends and enhances our prior work in [4], which designs a complete framework, including hardware settings and software implementation, to evaluate power-aware scheduling algorithms on a real platform. Our work shows a gap between the theoretical results and the practical results. We not only evaluate low-power system designs, but also provide a way to bridge the gap in one's future work. The rest of the paper is organized as follows. In Section 2, we review the important advances in real-time, power-aware scheduling research works. Then, a novel, unified framework, called Real-Energy, is proposed in Section 3 to implement the power-aware algorithms and evaluate their performance. Section 4 presents a case study of the real performance about several classical algorithms using DVS or/and CPU shut-down. The energy efficiency and the quantitative difference in their performance, as well as the practical issues found in the implementation of these algorithms are discussed. The last section gives a summary.

2. Advances in Power-Aware Scheduling

2.1. Fundamental Techniques for Power-Aware Scheduling

The first important work in power-aware scheduling can be traced back to 1995 in a paper done by Yao *et al.* [5]. In the work, the authors study a scheduling problem of minimizing the CPU's energy consumption for a set of real-time, single-instance or non-repeating tasks, when the earliest deadline first (EDF) algorithm is used to schedule tasks. Yao's algorithm is a greedy, off-line algorithm, which recursively calculates the highest intensity in a set of time intervals where an interval $[t_1, t_2]$ defines the executions of tasks released not earlier than t_1 and have a deadline not later than t_2. After the heaviest interval is identified, the intensity ratio, workload/interval's length, is used as the speed of the processor in this interval to execute tasks, and this interval is removed from the set of intervals. This process repeats, until all tasks are assigned a processor speed to run. For example, in Figure 1, there are six intervals, (0, 4), (2, 8), (0, 8), (2, 12), (7, 12) and (0, 12). The highest intensity ratio is the one of (0, 12), which is 0.5. Since the interval includes all of the three tasks, a processor speed of 0.5 is used to run all of them.

A periodic task is another important task model for real-time systems in which a series of job instances of a task arrive continuously at a certain rate. An important type of real-time task running on embedded systems, control tasks, can be modeled as periodic tasks. Three fundamental parameters generally define a periodic task, the maximum (or worst case) execution time, c; the period or rate, p; and utilization, u, where $u = c/p$. Aydin *et al.* prove that for all tasks scheduled by EDF, using a common speed, $s = U$, where $U = \sum u$, minimizes the CPU energy consumption of a set of periodic tasks if the power consumption function is convex [6].

EDF is a dynamic priority scheduling algorithm, such that the priorities of each task to be scheduled change during execution. There is another type of scheduling algorithm, called fixed-priority, such that each task's priority for scheduling does not change during the execution of the system. Rate-monotonic (RM) [7] is the optimal fixed-priority algorithm in which a task with a shorter period is assigned with a higher priority in scheduling. The common processor speed used for a set of periodic tasks scheduled by a fixed-priority, for example RM, is calculated in [8,9]. The calculation utilizes the schedulability analysis for fixed-priority scheduling based on the critical instant theorem, which says that if a task meets its deadline whenever the task is requested simultaneously with requests for all higher priority tasks, then the deadline will always be met for all task phasings [10,11]. A lowest speed ratio is calculated for every task using RM, such that the task and all of its higher priority tasks are schedulable to meet their deadlines using that speed. The highest ratio among the calculated speed ratio of each task is used as the common speed for all tasks.

The above techniques calculate a static speed off-line by assuming the worst case execution time (WCET) of real-time tasks. In real-time systems, the use of the WCET is to guarantee the deadline requirement, but it is conservative, because during the run-time, most job instances usually spend less time than their WCET in completing the execution. An idle interval that is generated by the early completion of a task, called on-line slack, can be used to further reduce the speed of the processor. A good review is done to survey the techniques to exploit on-line slack in [12]. The stretching-to-NTA (Next Task Arrival time) technique, used in the algorithms of *lppsEDF/RM* in [9], is based on a slack up to the arrival time of the next task. While in this technique, the slack can be exploited only if one task is in the ready queue, another one, The *ccRM* algorithm [8], theoretically improves the result by distributing the slack to multiple tasks. Utilization updating is used in *ccEDF* and *laEDF* [8], based on an observation that the actual processor utilization during runtime is usually lower than the worst case. The algorithm, *laEDF*, takes a more aggressive approach by estimating the amount of work required to be completed before NTA. The *DRA* algorithm [13] exploits the priority-based slack stealing, such that when a higher-priority task completes its execution earlier than its WCET, the following lower-priority tasks can use the slack time from the completed higher-priority task. Figure 2 shows an example of how on-line slack can be used in a power-aware schedule. Assume that after the common speed, 0.5, is used, task T1's actual execution time is one unit under the full speed. T1 completes at Time 2, and the dynamic slack of two time units can be used to reduce the speed of T2 to 0.33, as in Schedule 1. If the on-line slack is distributed to be used by both T2 and T3, the two tasks both execute at a speed of 0.4. Considering that the power function is convex, Schedule 2 is theoretically better than Schedule 1 in energy minimization.

Figure 2. An example of power-aware scheduling using on-line slack.

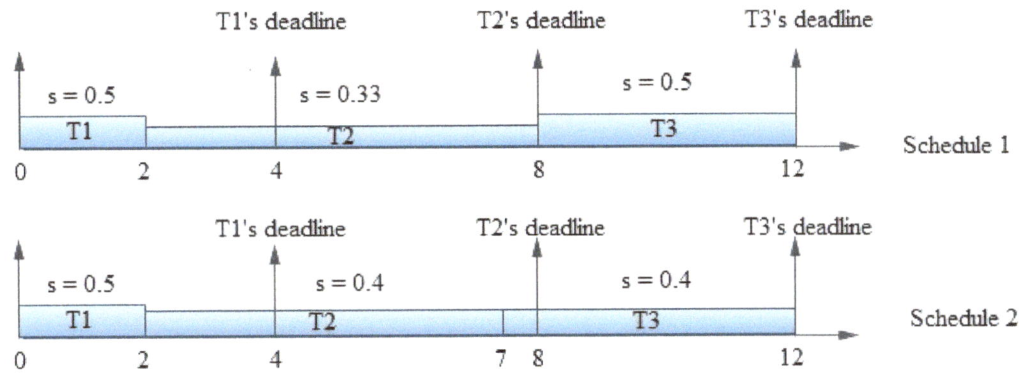

There are some other, additional considerations when scheduling a set of real-time tasks. For examples, there exist jobs that are non-preemptive. In [14], the authors describe a scheduling technique that provides an optimal schedule in terms of power consumption on a uniprocessor platform running single-instance, non-preemptive jobs. The problem of scheduling periodic tasks is studied in [15] in which an algorithm that can produce lower speeds for a variety of non-preemptive task sets is proposed. Furthermore, computer jobs can be executed on multiprocessors in order to enhance the throughput. The low-power design problem has been extensively studied for homogeneous multiprocessor systems [16–18] and heterogeneous multiprocessor systems [19–21].

While the DVS algorithms focus on the reduction of dynamic power consumption, the static one, leakage power, is no longer negligible today. When a processor is idle for execution, it may be temporarily shut down to reduce the static power consumption and wake up later to execute other jobs. In real-time, power-aware scheduling, another fundamental technique is to delay future jobs' execution as much as possible in order to create a long idle interval to shut down the processor. This technique, called procrastination, can reduce the frequency of shut-down and wake-up and, hence, reduce the overhead, including energy consumption. Leakage-aware algorithms proposed in [22–25] address procrastination scheduling to safely delay the execution of tasks in order to increase the idle interval for CPU shut-down. The general idea in these works is first to calculate an allowed idle interval for each task, and then update the wake-up time according to the dynamic arrivals of tasks. Some other works [26,27] combine the DVS with the shut-down technique. They assume that there exists a critical speed, in which executing any task higher or lower than this speed, both consume more energy. If a CPU speed is calculated as lower than the critical speed, the critical speed is used as the CPU's speed. While there is no job ready, the longest duration is calculated and used to shut down the CPU, so that the incoming jobs will not miss their deadlines.

Recently, energy harvesting, a technology that allows one to capture otherwise unused ambient energy and convert it into electrical energy that can be used immediately or later, is considered in real-time power scheduling [28,29]. When using ambient energy to support a system's execution, a challenge is that the energy source may be unstable and change from time to time. The research presented in [30] develops an energy-aware DVS algorithm, the task execution of which is slowed down if the system does not have sufficient available energy; otherwise, the tasks are executed at full speed. This work is improved later in [31] in which task executions are slowed down whenever possible for energy saving and the on-line slack are exploited by all tasks in the ready task queue. So far, the power-aware scheduling with energy harvesting makes some assumptions, such as a predictable and well-defined power-gaining function, which are difficult to apply on a real platform.

2.2. Evaluations of the Performance of Power-Aware Algorithms

In order to apply the algorithms to real systems, we are curious about their true performance. There have been several existing works to evaluate the performance of power-aware algorithms. Kim *et al.* in [32] compare several DVS algorithms for hard real-time systems, but their work uses simulations to obtain the experimental results. For measuring real energy consumption, a project called PLEB 2 developed by David *et al.* is shown in [33]. The system is specialized for observing energy distribution on the board when running applications. Although PLEB 2 was used to study the DVS technique, the following reasons limit its usability in other settings. First, designing and producing such a system requires a significant modification to the integrated circuits on the board. This may be beyond the scope of many researchers' work in developing algorithms. Second, PLEB 2 does not have a software framework for implementing real-time scheduling algorithms, and thus, it cannot be used to run multiple real-time tasks. There are several other works on estimating the power consumption for the DVS technique. Miyoshi *et al.* [34] examine a set of micro-benchmarks on two different systems. They find that the lowest speed is not always the best for saving energy. In [35], on an Xscale-based computer, Weissel monitors the number of memory references and instructions executed at runtime in order to estimate the reduction in frequency. Due to a lack of a framework for programming, these works have a similar drawback of PLEB 2, that they are not suitable for evaluating real-time scheduling algorithms.

3. Real-Energy

In this section, we describe the Real-Energy framework. Although the framework is based on an Intel PXA255 XScale processor, the methodology and programming design can be easily applied to other similar platforms.

3.1. Environment Setup

Figure 3 shows the environment's outline. Our hardware platform subsystem includes a PYTHECT phyCOREr PXA255 rapid development kit, a desktop and a data acquisition (DAQ) USB-6800 from National Instruments. The rapid development kit is the target platform in our embedded development environment. The desktop is our host machine. The computer program is developed, compiled and linked on the host machine before it is downloaded to the target platform. DAQ USB-6800 is a data collection device used to measure the power consumption of the target platform. The data is stored on a computer and analyzed by the LabView software tool.

Figure 3. Real-Energy's environment setup. DAQ data acquisition; DVS, dynamic voltage scaling.

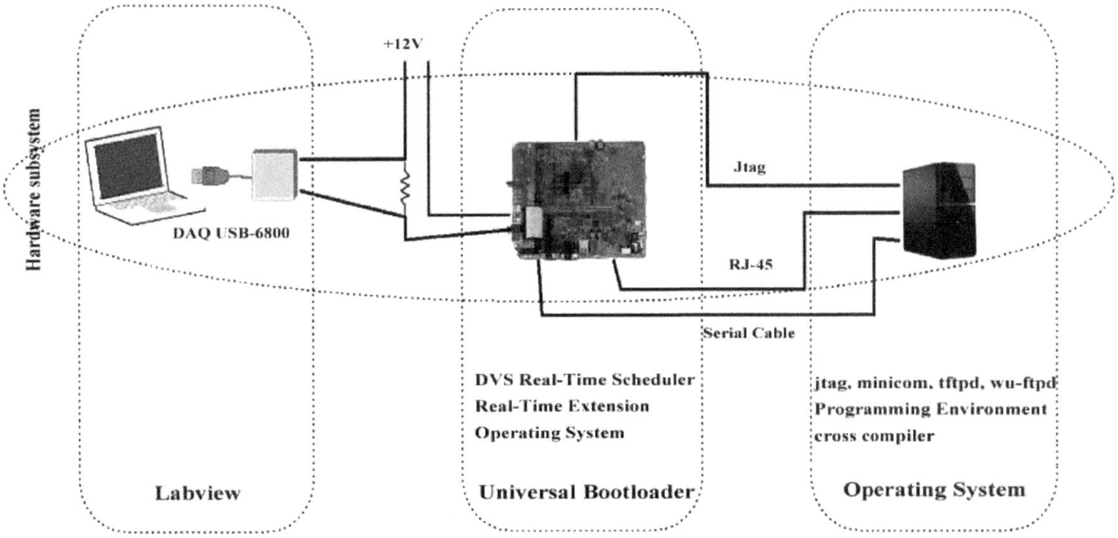

For the programming environment, we use Linux 2.14.19 as the operating system (OS). The patch RTAI-24.1.3 is used, where RTAI stands for Real-Time Application Interface, which is a real-time extension for the Linux kernel. We extend the RTAI scheduler to a power-aware enabled scheduler for accommodating various low-power scheduling algorithms. There are three connections from the host to the embedded board: A JTAG-to-parallel connection is used to download the boot loader (Universal Bootloader U-Boot-1.0.2); A serial cable connection is used to download the OS and file system image; An Ethernet connection is used to download the real-time applications.

The PYTHECT phyCORErr PXA255 rapid development kit is available on the market and designed for use in industrial applications that require high processing speed with low power. It includes a carrier board and a PXA255 computing module. All that is required for operation is a single 3.3 V supply voltage. A 128 MB SDRAM and a 64 MB flash memory are integrated onto the sub-miniature (70 mm $\times$ 57 mm) system on module. An SPI (Serial Peripheral Interface) port expander with 28 GPIOs (General-purpose input/output) for matrix keypads and a touch screen controller are also available on the PXA255 module. A 10/100 Mbps Ethernet and a CAN controller also populate the module, enabling the connection of the module to networks. To support the programming and analysis for applications running on the module, the computing module is designed to plug into a carrier board that provides I/O connectors, such as DB-9, RJ-45, USB and power jack, as well as any other interface circuitry not provided on the module itself. In order to focus on the interested portions, we have disabled the ports and controllers that we do not use in this work, such as the 10/100 Mbps Ethernet, CAN controller, USB, *etc.* The LCD touch screen is also shut off. Intel PXA255 runs at frequencies from 99.1 to 398.1 MHz. The number of supported frequency settings is fixed. Table 2 shows four supported CPU frequency settings under the same flash memory and RAM frequency, which are used in our work.

Table 2. Four supported frequency settings of Intel PXA255.

CPU frequency (MHz) at voltage (V)	Flash memory frequency (MHz)	SDRAM frequency (MHz)
99.1 MHz at 1.0 V	99.5 MHz	99.5 MHz
199.1 MHz at 1.0 V	99.5 MHz	99.5 MHz
298.6 MHz at 1.1 V	99.5 MHz	99.5 MHz
398.1 MHz at 1.3 V	99.5 MHz	99.5 MHz

3.2. Measuring Energy Consumption

The carrier board with the PXA255 computing module plugged in is connected to an external stabilized voltage supply. In order to measure the energy consumed by the computing module, we measure the energy used by the development kit (the carrier board and the computing module). To determine the energy consumed by the kit, we attach a small resistor (0.02) to the wire connecting to the external voltage supply. The voltage drop across this resistor is measured by data acquisition (DAQ). The DAQ reads and records the voltage drop 50 times per second. The instantaneous current through the system can be calculated by dividing the voltage by the resistance, and the instantaneous power consumption of the development kit can be calculated as:

$$P_{kit} \approx I \times V_{external} \tag{2}$$

where I is the recorded current used by the kit and $V_{external}$ is the external supply's voltage to the kit. Figure 4 shows an example of the current measured for running a single instance task at different CPU frequencies. We observe that the instantaneous current, therefore the instantaneous power, at a low 331 (265) MHz is higher than that at a high 398 (298) MHz. This is because the memory frequency in the setting of 331 (265) MHz is higher than the one used in the setting of 398 (298) MHz [3]. This observation reminds us of a practical issue, that a CPU frequency setting may be related to a specific memory (flash memory) frequency on a real platform, and these combination settings must be considered in the design of low-power algorithms.

3.3. CPU Energy Consumption and System-Wide Energy Consumption

The original goal of DVS is to reduce the energy consumption of the processor. This section gives the methods for evaluating the DVS algorithms for both the CPU energy and system-wide energy (the energy used by the module) consumptions. Generally, the system-wide energy consumption gives a higher hierarchical understanding in view of saving energy, so that the reason for measuring it is obvious. However, we also expect to investigate the impact of a proposed scheduling algorithm and the application workload on different components of the hardware (e.g., the CPU). Then, we can fine-tune the algorithm and application to accomplish energy savings. Since the CPU is a major energy-consuming component, evaluating the performance of scheduling algorithms and applications in reducing the CPU's energy consumption is a major step toward this goal.

Figure 4. An example of the current measured.

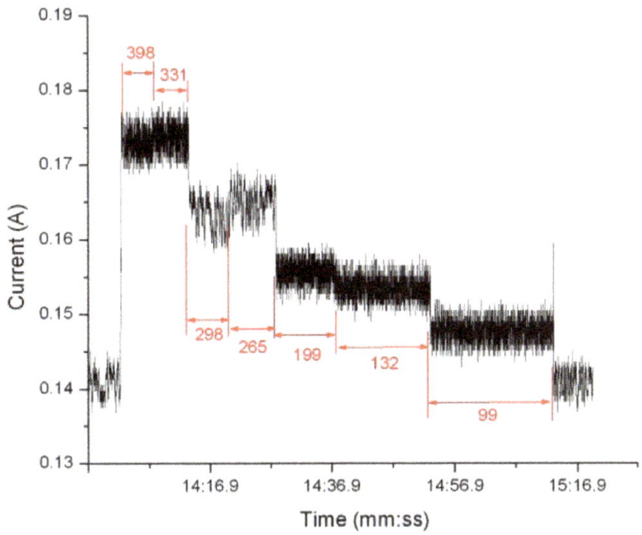

As stated earlier, the actual power we measure is the total power consumed by the development kit, which includes the computing module and the carrier board. Hence,

$$P_{kit} \approx P_{mod} + P_{board} \tag{3}$$

In Equation (3), P_{mod} is the power consumption of the computing module (considered as the target system) and P_{board} is the power consumption of the carrier board. Since most ports and controllers on the carrier board have been disabled, we can assume that the carrier board's instantaneous power consumption is stable and is a constant. P_{board} can be approximated to be the kit's energy consumption when the computing module is in the sleep mode. The sleep mode is the lowest power state with a very small energy usage, where only a real-time clock (RTC) and a power manager are active. Thus:

$$P_{board} \approx P_{kit}^{sleep} \tag{4}$$

and P_{kit}^{sleep} can be measured and known easily. For P_{mod}, the major sources of power consumption are: CPU, flash memory, SDRAM and leakage currents. Thus, P_{mod} can be defined as:

$$\begin{aligned} P_{mod} &\approx P_{CPU} + P_{flash} + P_{SD} + P_{leakage} \\ &\approx P_{kit} - P_{kit}^{sleep} \end{aligned} \tag{5}$$

In the system-wide evaluations, P_{mod} is what we want to know. However, for the algorithms targeting the CPU only, we need to find out P_{CPU}. Our solution is to run CPU-bound tasks on the system for measuring the kit's instantaneous power consumption. In a CPU-bound task, an empty loop is executed. When the task is running, there are very few memory accesses. The energy consumed by the SDRAM and flash memory is primarily from the clocks and the leakage currents. The following equations define the instantaneous power usage by the module and kit when running CPU-bound tasks.

$$P_{mod}^{CPU-Bound} \approx P_{CPU} + P_{flash}^{idle} + P_{SD}^{idle} + P_{leakage} \tag{6}$$

$$P_{kit}^{CPU-Bound} \approx P_{board} + P_{mod}^{CPU-Bound} \tag{7}$$

On the other hand, when the CPU is idle, the development kit and the module actually are in the idle mode, too. Thus,

$$P_{mod}^{idle} \approx P_{CPU}^{idle} + P_{flash}^{idle} + P_{SD}^{idle} + P_{leakage} \tag{8}$$

$$P_{kit}^{idle} \approx P_{board} + P_{mod}^{idle} \tag{9}$$

Now, subtracting Equation (9) from Equation (7), we have:

$$\begin{aligned} P_{kit}^{CPU-Bound} - P_{kit}^{idle} &\approx P_{mod}^{CPU-Bound} - P_{mod}^{idle} \\ &\approx P_{CPU} \end{aligned} \tag{10}$$

That is, the power consumption of a CPU can be obtained by measuring the power consumption of the development kit running a CPU-bound task and having the module in sleep mode. The small amount of the energy consumption of the CPU while it is idle is ignored. If the static power consumption needs to be considered, a system-wide energy measurement can be used.

In order to verify the effectiveness of our methods, a CPU-bound task and the QSort benchmark [33] are tested on our system for using DVS. The CPU's and the system-wide energy consumption are calculated according to the approach mentioned above. Figure 5 shows the normalized results for running these two tasks at the four supported frequency settings. It confirms that for Intel PXA255 using the lowest speed is not CPU energy efficient, and using a lower speed exhibits less efficiency in saving the system-wide energy.

Figure 5. Energy consumption using dynamic voltage scaling (DVS).

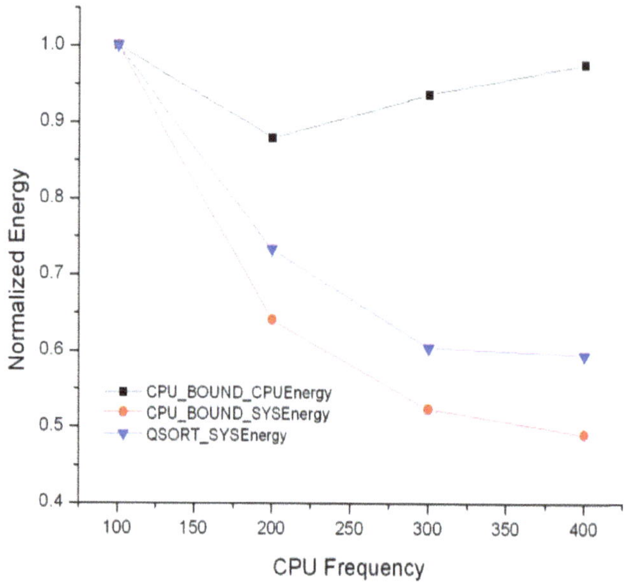

3.4. Programming Design

So far, the existing works to evaluate the real power consumption of a system do not have a programming design to evaluate real-time, power-aware scheduling algorithms. In this work, we extend

RTAI to work with DVS and CPU shut-down/wake-up. In what follows, we present several important aspects and techniques used in our framework.

Implementing a priority queue correctly is a critical part of designing a real-time scheduler. The work is easy for single-instance tasks, but for a periodic task, it requires additional efforts. In order to make a set of tasks periodically run, in our programming design, there are two task queues: time queue (TQ) and ready queue (RQ). After an initialization, a periodic task is pushed into TQ in which the tasks follow the nondecreasing order of the next job instance's arrival time. A time interrupt handler checks TQ at small, regular time intervals. If a task is ready to run, it is inserted into RQ and removed from TQ. Its position in RQ is decided by the scheduling algorithm the system uses (e.g., RM or EDF). The scheduler schedules tasks for execution based on the order in RQ. After a task is terminated, it is removed from RQ and pushed into TQ again. Figure 6 shows how these two queues work, as well as the task structure definition. The variable state in the task structure indicates the task status, such as READY, DELAYED and SUSPENDED. The policy is used to represent the energy-efficient scheduling policy. The period and resume time are release intervals of a periodic task and the release time of its next instance. Pointers *tnext, *tprev, *rnext and *rprev are pointed to the next or previous tasks in TQ and RQ, which form a double-linked list.

Figure 6. Time queue, ready queue and task structure.

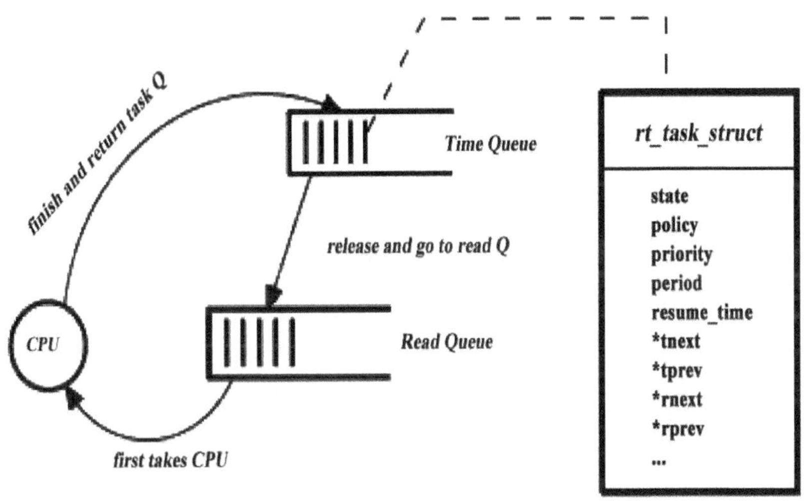

When a real-time task arrives or completes, a real-time scheduler needs to make a scheduling decision. We call this point a scheduling point. In our work, the detection of a scheduling point is done by using a timer interrupt. An interrupt handler is called when a scheduling point is found. For an inter-task, power-aware algorithm, changing the CPU's frequency or putting it into sleep mode also happen at a scheduling point. An intuitive way is to implement the code as part of the interrupt handler. However, in Linux, a change of the CPU's frequency (including putting the CPU into sleep mode) is not allowed in the interrupt context (an interrupt handler runs in the interrupt context). To overcome this obstacle, we create a special task handling the job and then make this task run in the process context. The implementation of our technique is described as in Figure 7. A special task to change the CPU's frequency is set with the highest priority, and its default status is suspended. At a scheduling point, if a change of frequency is called, the interrupt handler activates the task and inserts it into the head of the RQ. In this way, the CPU frequency can be immediately changed before the next real task is dispatched. The special task's status is set to be suspended after the change of frequency is done. This implementation is also applicable to intra-task DVS algorithms, where the change of the CPU's frequency may take place at any predefined point.

Figure 7. The process of changing the CPU's frequency.

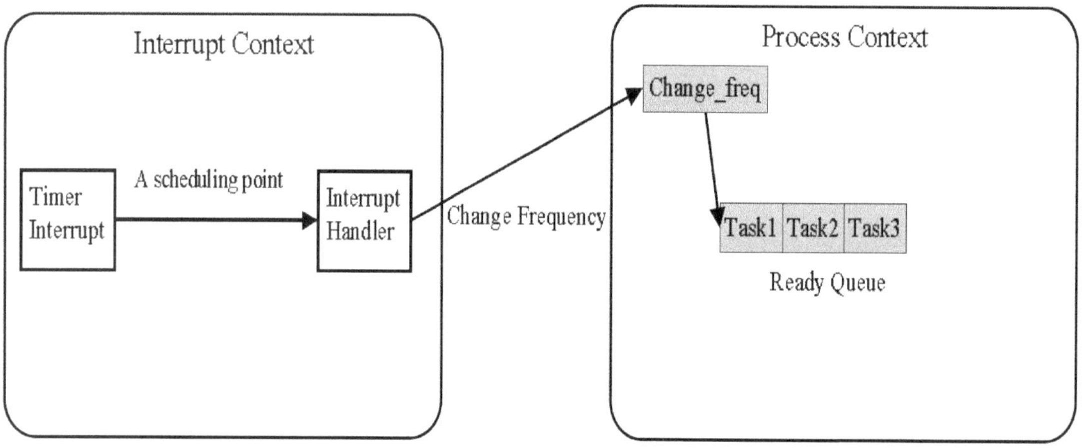

In a CPU's sleep mode, all processor and peripheral clocks are disabled, except the RTC (real-time clock). The processor does not recognize interrupts or external pin transitions, except valid wake-up signals and reset signals, *etc.* In practice, a wake-up of a system is normally done by a hardware signal, e.g., touching the screen, moving the mouse, *etc.* Apparently, this hardware mechanism cannot be used with real-time tasks. Since only an RTC is active when the system sleeps, we utilize the RTC alarm to set up the wake-up time. While RTAI blocks the RTC alarm from running, we fortunately found that the blocking mechanism actually can be dynamically mounted, as well as unmounted.

We have integrated several key DVS and shut-down algorithms into the Real-Energy framework being used in our case study. Our programming design is module oriented and fully extensible to other scheduling algorithms. When one wants to evaluate a new algorithm, the new one can be written in a new module and added to the selections for the policy.

4. A Case Study

In this section, we perform a case study using our proposed framework, Real-Energy. In the study, we restrict our attention to study preemptive, periodic tasks scheduled by EDF and RM, because these two scheduling algorithms are the most used ones in the literature. The algorithms use either DVS or CPU shut-down to save energy. In the DVS algorithms, the main difference is in how they exploit the on-line slack times. The CPU's energy consumption is used to compare their effectiveness. For the algorithms using CPU shut-down, we mainly study the so-called procrastination algorithms, which delay task execution for putting the CPU into sleep mode. A practical issue is discussed in implementing the algorithms, and we bring a solution to it.

4.1. Experimental Setting

Considering the lowest frequency setting in Intel PXA255 is not energy-efficient, as shown in Figure 5, we use the other three frequencies in Table 2 to evaluate the DVS scheduling algorithms. The calculated speeds by the algorithms are used to determine the CPU's frequency. For example, if the calculated speed is 0.38, the frequency of 199 MHz is used to execute the tasks. In our experiments, ten tasks are created to run on the system. The periods of the tasks range from 50 to 120 units, where one unit is equal to 25 ticks period (1 tick period = 2 ms). The worst case utilization of these ten tasks are randomly generated, but the total utilization is assured to be less than 100%. For evaluating the algorithms under different workloads, we use loops in a task, and the number of loops is used to control the generation of slack time. The WCET is measured off-line, and we increase it by 5% to accommodate the run-time overhead.

4.2. Experimental Results for DVS Algorithms

Six key DVS algorithms (as introduced in the previous section) listed in Table 3 are implemented and compared. These algorithms are classical in that a lot of the latter works are extensions of the former ones. The ideas of exploiting the on-line slack are still used in recent research works.

Table 3. DVS algorithms. EDF, earliest deadline first; RM, rate-monotonic.

EDF	*lppsEDF* [9]; *ccEDF* [8]; *laEDF* [8]; *DRA* [13].
RM	*lppsRM*[9]; *ccRM*[8]

The experimental results for DVS algorithms are presented in two categories, RM or EDF, as shown in Figure 8. In RM, the results of using *lppsRM* and *ccRm* are normalized and compared to the result of using a fixed static speed without utilizing an on-line slack. These results are compared at four workload conditions, where the percentage of slack means the average ratio between the actual execution time (by adjusting the number of loops) and the WCET. It is easy to see that *lppsRM* and *ccRM* both effectively reduce the CPU's energy usage. According to what is reported in [32], *ccRM* is better than *lppsRM* with the belief that distributing the on-line slack to multiple tasks enhances the power-saving performance. However, in our experiment, *lppsRM* obviously outperforms *ccRM*, which challenges the theoretical result. We believe that the disagreement is caused by the higher overhead in *ccRM*, which is not factored into the simulations in [32]. In these two algorithms, *ccRM* uses a more complicated mechanism to calculate the slack times at each scheduling point. These calculations consume energy, as well. When the system is heavily loaded, although the complicated method may be better for exploiting slack times, the slack is scarce, which could weaken this advantage. When the system is lightly loaded, since the slack times are easy to obtain, paying a high overhead may not be worthwhile. Anyway, the results of the experiment tell us that the overhead from the algorithm itself is not negligible, and a simple design sometimes beats an advanced design.

In the EDF's results, the four EDF algorithms are more aggressive than the RM algorithms in reducing the CPU's energy consumption. This is because the current technique used to guarantee tasks' deadline in RM is more conservative than EDF, so that less slack can be used to lower the speed in RM. Similar to the results of RM, the simple algorithm, *lppsEDF*, has the best performance.

Figure 8. CPU energy consumption for the RM and EDF algorithms.

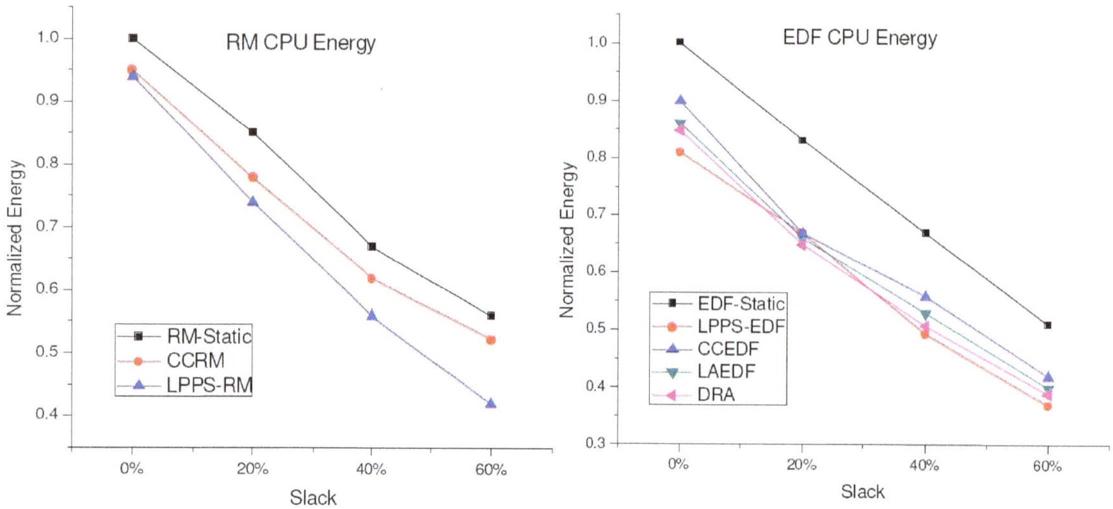

We are also interested in how much of the total energy is consumed by the system when only the CPU is working. This information is useful to analyze the energy usage contributed by the idle memory, leakage currents, *etc.* Figure 9 shows the measured system-wide energy consumption for using the RM and EDF algorithms when running CPU-bound tasks. It can be seen that the performance gaps between the algorithms and their norm (the curve of using static speed) are smaller than those in Figure 8. This is because using the DVS technique cannot reduce the energy consumption from other units. Figure 10 shows the average energy usage distribution (at 40% slack) for using the EDF and RM algorithms, where the usage for other parts is calculated by subtracting the CPU's energy usage from the total system's usage. It is apparent that the energy used by other parts on the system on chip can no longer be ignored.

Figure 9. System-wide energy consumption for the RM and EDF algorithms.

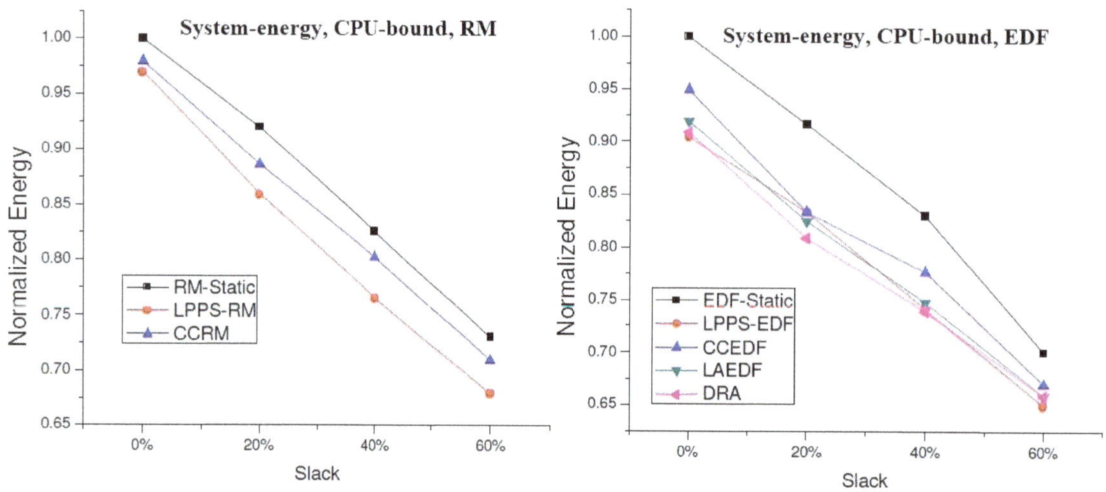

Figure 10. Energy distribution for the EDF and RM algorithms.

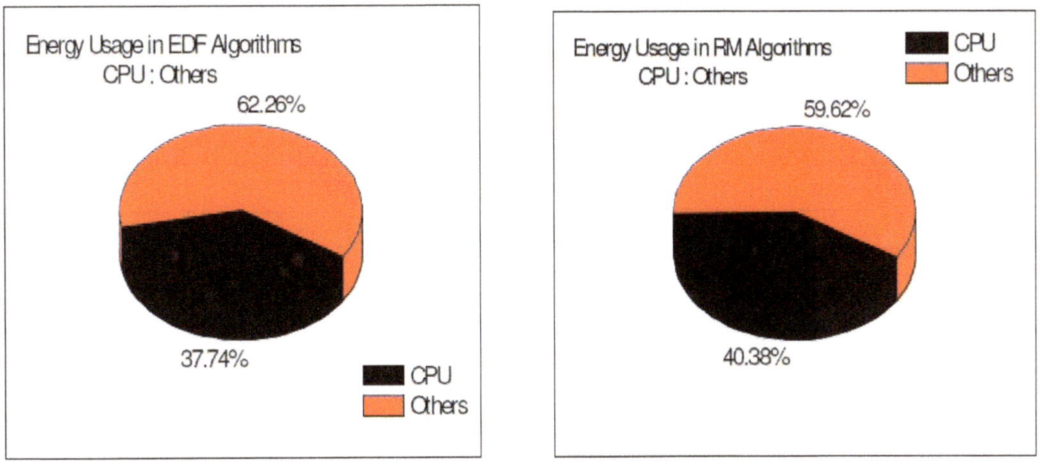

4.3. Experimental Results of Procrastination Algorithms

Procrastination algorithms [22–25] are used to temporarily shut down the processor in order to reduce the overall energy consumption, including the static one. During our study of applying the algorithms to a real hardware platform, we found that a practical issue existed in these procrastination algorithms: they all somehow rely on a calculation by a sleeping CPU. For example, in [22], the accumulated utilization needs to update during the sleeping interval (algorithm *LCEDF*) to determine a delay, Y_i, for a task, T_i; in [23–25], a function is performed to update the wake-up timer when the CPU is in sleep mode. Obviously, these computations require a CPU to be involved unless an additional computing unit undertakes the jobs, and this unit, if available, consumes energy, as well. Hence, either the additional energy usage should be studied or the algorithms should be revised to apply them on current commercial processors, where a CPU is not clocked during the sleep mode.

The revised algorithm, practical-leakage-current (PLC), of the algorithm, LC, in [22] is shown in Algorithm 1. In PLC, no computation is performed during the CPU's sleep state. Upon the completion of a task, the system keeps the CPU busy if any remaining task is in the ready queue. Otherwise, the CPU enters the sleep state, and the wake-up time is set to $NTA + Y$, where Y is the minimum value among the idle times calculated off-line for each task. Without updates during sleep mode, using the minimum value, Y, guarantees that it does not cause any task to miss its deadline.

Even with the revised algorithm, PLC, the procrastination technique can only be used for those real-time tasks for which the parameters are known in advance. For real-time tasks having dynamic arrivals, since the system cannot accept any task during the sleep mode, they may be lost due to the CPU's shut-down. Currently, in most existing works, this practical issue is not paid enough attention.

Considering the re-booting overhead, a CPU enters into the sleep state only if it is profitable to do so. In our experiment, we set this overhead to 2 ms. We compare our PLC-RM and PLC-EDF algorithms with their corresponding DVS algorithms, as shown in Figure 11. In these experiments, we use non-CPU-bound tasks in which the accessing of memory is frequent. As seen from the results, all of the DVS algorithms do not succeed in reducing the system-wide energy consumption. In some cases, the slow-down of the processor speed even increases the energy consumption of the system. This is because when the CPU is active with a longer period, the other components have to be actively working, as well. While today, the CPU's energy consumption no longer dominates in a system, using the DVS alone to slow down the execution of computer tasks may not be a good solution for saving energy. Nevertheless, the procrastination algorithms using the CPU shut-down technique are very effective at achieving this goal.

Algorithm 1: Revised procrastination algorithms.

On a task's arrival:
  If (CPU is active)
    Insert the task into the ready queue
  Else
    Do nothing
  End
On a completion of a task:
  If(the ready queue is not empty)
    Execute the highest priority task in the queue
  Else
    The CPU enters into the sleep state
    The wake-up time is set at $NTA + Y$
  End
On a CPU's wake-up:
  Begin executing the tasks in the ready

Figure 11. Comparisons of the RM and EDF procrastination algorithms.

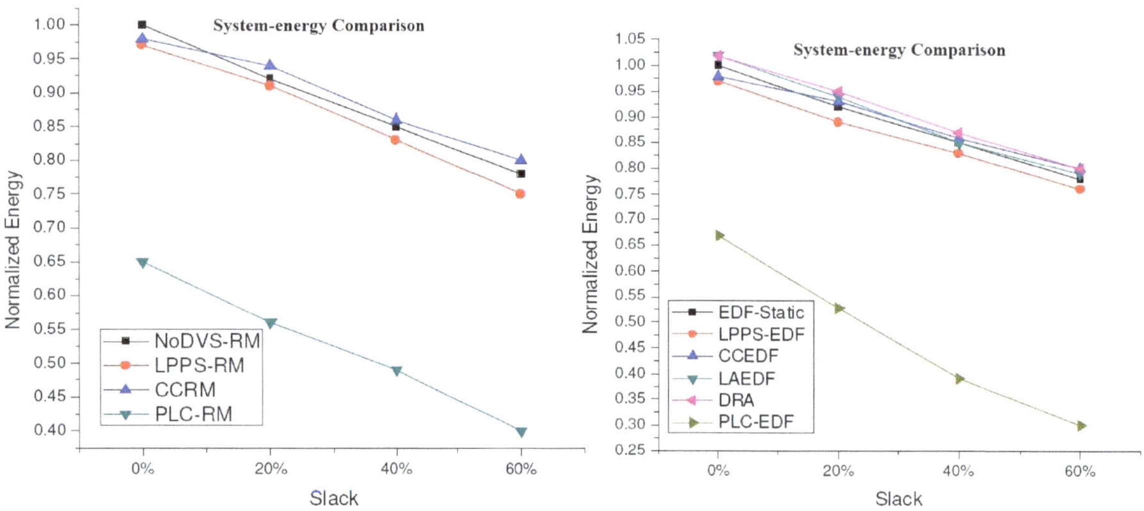

5. Conclusion

In designing power-aware algorithms, it is always important to know the real performance, because so many factors can affect the results. This motivates us to design a unified framework for the evaluations in spite of the complication. Our work does a comprehensive job of completing such a tool in hardware and software. We not only use the tool for evaluation, we also can use it to test the implementation of algorithms in order to find out practical issues. In our case study, while the effectiveness of the DVS technique for reducing the CPU's energy consumption is shown, it also confirms that it does not effectively reduce the system's energy consumption. The CPU shut-down is a simple and effective one to save the system's energy consumption, but so far, it is only applicable for running tasks with a predefined release time. Otherwise, the CPU does not know when to wake up to continue the execution. We do not implement or evaluate the algorithms using a critical speed. The use of the critical speed requires a continuous range of the CPU speed, which is unpractical today on any platform. Therefore, there has been a focus on thermal-aware scheduling recently. Our framework can be used for the topic by finding a way to measure a CPU's temperature.

The Real-Energy framework is not specifically hardware and software dependent. The OS, Linux, with programming in C is widely used in embedded systems. The development kit and data collection device and their replacements are easily found on the market. It does not need any hardware modification. When a newer generation of hardware components appears, they can be easily used in the framework, as long as they have a similar architecture and functionalities.

At the point of writing this paper, Real-Energy is being used for a single-CPU system. Because of the lack of hardware support on the market, we have not found a suitable multiprocessor system to be used in our framework. If the support in the hardware is available, we will be interested in investigating the real performance of the algorithms working with a multiprocessor/multi-core system, such as those in [36,37].

Acknowledgments

This work was supported in part by the National Science Foundation under Award No. 0720856.

Conflicts of Interest

The authors declare no conflict of interest.

References

1. Ditzel, M.; Otten, R.; Serdijn, W. *Power-Aware Architecting for Data-Dominated Dpplications.* Springer: Berlin, Germany, 2007.
2. *Intel StrongARM SA-1110 Microprocessor Developers Manual.* Intel Corporation: Santa Clara, CA, USA, 2000.
3. *Intel PXA255 Processor Developers Manual.* Intel Corporation: Santa Clara, CA, USA, 2004.
4. Lin, J.; Song, W.; Cheng. A. Real energy: A new framework and a case study to evaluate power-aware real-time scheduling algorithms. In Proceedings of the ACM/IEEE International Symposium on Low Power Electronics and Design (ISLPED), Austin, TX, USA, 18–20 August 2010.
5. Yao, F.; Demers, A.; Shenker. S. A scheduling model for reduced CPU energy. In Proceedings of the 36th Annual Symposium on Foundations of Computer Science, Milwaukee, WI, USA, 23–25 October 1995.

6. Aydin, H.; Melhem, R.; Mosse, D.; Meja-Alvarez, P. Dynamic and aggressive scheduling techniques for power-aware real-time systems. In Proceedings of the 22nd IEEE Real-Time Systems Symposium, London, UK, 2–6 December 2001.
7. Liu, J.W. *Real-Time Systems*; Prentice Hall: Englewood, Cliffs, NJ, USA, 2000.
8. Pillai, P.; Shin, K.G. Real-time dynamic voltage scaling for low power embedded operating systems. In Proceedings of the Eighteenth ACM Symposium on Operating Systems Principles, Banff, AB, Canada, 21–24 October 2001.
9. Shin, Y.; Choi, K.; Sakurai, K. Power optimization of real-time embedded systems on variable speed processors. In Proceedings of the 2000 IEEE/ACM International Conference on Computer-Aided Design, San Jose, CA, USA, 5–9 November 2000.
10. Liu, C.L.; Layland, J.W. Scheduling algorithms for multi programming in a hard real time environment. *J. ACM* **1973**, *20*, 46–61.
11. Lehoczky, J.; Sha, L.; Ding, Y. The rate monotonic scheduling algorithm: Exact characterization and average case behavior. In Proceedings of the IEEE Real-Time Systems Symposium, Santa Monica, CA, USA, 5–7 December 1989.
12. Lee, I.; Leung, J.; Son, S. *Handbook of Real-Time and Embedded Systems*; Chapman and Hall/CRC Press: Boca Raton, FL, USA, 2007.
13. Aydin, H.; Melhem, R.; Mosse, D.; Mejia-Alvarez, P. Dynamic and aggressive scheduling techniques for power-aware real-time systems. In Proceedings of the 22nd IEEE Real-Time Systems Symposium, London, UK, 2–6 December 2001.
14. Andrei, S.; Cheng, A.; Radulescu, V.; McNicholl, T. Toward an optimal power-aware scheduling technique. In Proceedings of the 14th IEEE International Symposium on Symbolic and Numeric Algorithms for Scientific Computing, Timisoara, Romania, 26–29 September 2012.
15. Li, J.; Shu, L.; Chen, J.; Li, G. Energy-efficient scheduling in non-preemptive systems with real-time constraints. *IEEE Trans. Syst. Man Cybern. Syst.* **2013**, *43*, 332–344.
16. Chen, J.; Hsu, H.; Chuang, K.; Yang, C.; Pang, A.; Kuo, T. Multiprocessor energy-efficient scheduling with task migration considerations. In Proceedings of the IEEE EuroMicro Conference on Real-Time Systems, Catania, Italy, 30 June–2 July 2004.
17. Aydin, H.; Yang, Q. Energy-aware partitioning for multiprocessor real-time systems. In Proceedings of the IEEE 17th International Parallel and Distributed Processing Symposium, Nice, France, 22–26 April 2003.
18. Lin, J.; Cheng, A.M.K. Real-time task assignment in rechargeable multiprocessor systems. In Proceedings of the IEEE 14th IEEE International Conference on Embedded and Real-Time Computing Systems and Applications, Kaohisung, Taiwan, 25–27 August 2008.
19. Schmitz, M.; Al-Hashimi, B.; Eles, P. Energy-efficient mapping and scheduling for DVS enabled distributed embedded systems. In Proceedings of the IEEE Conference on Design, Automation and Test in Europe, Paris, France, 4–8 March 2002.
20. Lin, J.; Cheng, A.M.; Kumar, R. Real-time task assignment in heterogeneous distributed systems with rechargeable batteries. In Proceedings of the IEEE International Conference on Advanced Information Networking, Bradford, UK, 26–29 May 2009.

21. Luo, J.; Jha, N. Static and dynamic variable voltage dcheduling algorithms for realtime heterogeneous distributed embedded systems. In Proceedings of the 15th International Conference on VLSI Design, Bangalore, India, 7–11 January 2002.
22. Lee, Y.; Reddy, K.P.; Krishna, C.M. Scheduling techniques for reducing leakage power in hard real-time systems. In Proceedings of the EcuroMicro Conference on Real Time Systems (ECRTS), Porto, Portugal, 2–4 July 2003.
23. Jejruikar, R.; Gupta, P.K. Procrastination scheduling in fixed priority real-time systems. In Proceedings of the 2004 ACM SIGPLAN/SIGBED Conference on Languages, Compilers and Tools for Embedded Systems, Washington, DC, USA, 11–13 June 2004.
24. Jekruikar, R.; Gupta, P.K. Dynamic slack reclamation with procrastination scheduling in real-time embedded systems. In Proceedings of the ACM IEEE Design Automation Conference, San Diego, CA, USA, 13–17 June 2005.
25. Jekruikar, R.; Pereira, C.; Gupta, P.K. Leakage aware dynamic voltage scaling for real-time embedded systems. In Proceedings of the ACM IEEE Design Automation Conference, San Diego, CA, USA, 7–11 June 2004.
26. Irani, S.; Shukla, S.; Gupta, R. Algorithms for power savings. In Proceedings of the 14th Annual ACM-SIAM Symposium on Discrete Algorithms, Baltimore, MD, USA, 12–14 January 2003.
27. Lin, J.; Cheng, A. Energy reduction for scheduling a set of multiple feasible interval jobs. *J. Syst. Archit. Embed. Softw. Des.* **2011**, *57*, 663–673.
28. Chetto, M. A note on EDF scheduling for real-time energy harvesting systems. *IEEE Trans.Comput.* **2014**, *63*, 1037–1040
29. Chetto, M. Optimal scheduling for real-time jobs in energy harvesting computing systems. *IEEE Trans. Emerg. Top. Comput.* **2014**, doi:10.1109/TETC.2013.2296537.
30. Liu, S.; Qiu, Q.; Wu., Q. Energy aware dynamic voltage and frequency selection for real-time systems with energy harvesting. In Proceedings of the Design, Automation, and Test in Europe, Munich, Germany, 10–14 March 2008.
31. Liu, S.; Lu, J.; Wu, Q.; Qiu, Q. Harvesting-aware power management for real-time systems with renewable energy. *IEEE Trans. Large Scale Integr. (VLSI) Syst.* **2011**, *20* ,1473–1486 .
32. Kim, W.; Shin, D.; Yun, H.-S.; Kim, J.; Min, S.L. Performance comparison of dynamic voltage scaling algorithms for hard real-time systems. In Proceedings of the Real-Time and Embedded Technology and Applications Symposium, San Jose, CA, USA, 24–27 September 2002.
33. Snowdon, D.; Ruocco, S.; Heiser, G. Power management and dynamic voltage scaling: Myths and facts. In Proceedings of the 2005 Workshop on Power Aware Real-time Computing, Jersey City, NJ, USA, 18–22 September 2005.
34. Miyoshi, A.; Lefurgy, C.; Hensbergen, E.V.; Rajamony, R.; Rajkumar, R. Critical power slope: Understanding the runtime effects of frequency scaling. In Proceedings of the 16th Annual ACM International Conference on Supercomputing, New York, NY, USA, 22–26 June 2002.
35. Weissel, A.; Bellosa, F. Process cruise control: Event-driven clock scaling for dynamic power management. In Proceedings of the International Conference on Compilers, Architecture and Synthesis for Embedded Systems (CASE 2002), Greenoble, France, 8–11 October 2002.

36. Yang, C.; Chen, J.; Kuo, T. Preemption control for energy efficient task scheduling in systems with a DVS processor and Non-DVS Devices. In Proceedings of the 13th IEEE International Conferences on Embedded and Real-Time Computing Systems and Applications, Daegu, Korea, 21–24 August 2007.
37. AlEnawy, T.A.; Aydin, H. Energy-aware task allocation for rate monotonic scheduling. In Proceedings of the 11th IEEE Real-time and Embedded Technology and Applications Symposium (RTAS'05), San Francisco, CA, USA, 7–10 March 2005.

9

MOS Current Mode Logic Near Threshold Circuits †

Alexander Shapiro * and Eby G. Friedman

Department of Electrical Engineering, University of Rochester, Rochester, NY 14627, USA;
E-Mail: eby.friedman@rochester.edu

† This is an extended version of the "Performance Characteristics of 14 nm Near Threshold MCML Circuits" paper that was presented at the IEEE S3S Conference 2013.

* Author to whom correspondence should be addressed; E-Mail: alexander.shapiro@rochester.edu

---

Abstract: Near threshold circuits (NTC) are an attractive and promising technology that provides significant power savings with some delay penalty. The combination of NTC technology with MOS current mode logic (MCML) is examined in this work. By combining MCML with NTC, the constant power consumption of MCML is reduced to leakage power levels that can be tolerated in certain modern applications. Additionally, the speed of NTC is improved due to the high speed nature of MCML technology. A 14 nm Fin field effect transistor (FinFET) technology is used to evaluate these combined circuit techniques. A 32-bit Kogge Stone adder is chosen as a demonstration vehicle for feasibility analysis. MCML with NTC is shown to yield enhanced power efficiency when operated above 1 GHz with a 100% activity factor as compared to standard CMOS. MCML with NTC is more power efficient than standard CMOS beyond 9 GHz over a wide range of activity factors. MCML with NTC also exhibits significantly lower noise levels as compared to standard CMOS. The results of the analysis demonstrate that pairing NTC and MCML is efficient when operating at high frequencies and activity factors.

Keywords: near threshold circuits (NTC); MOS current mode logic (MCML); high performance; power efficiency

---

1. Introduction

In the era of handheld mobile devices, the performance of integrated circuits is a primary concern. Power consumption and speed are two primary characteristics of high performance integrated circuits [1]. In this work, both of these characteristics are addressed by utilizing low power near threshold circuits (NTC) [2] in combination with high speed MOS current mode logic (MCML) [3]. In recent years, NTC has become an increasingly popular approach to lower power consumption at the expense of decreased speed. The quadratic reduction in power consumption is accomplished by operating near the threshold voltage. A similar reduction in supply voltage, however, results in a ten times reduced speed of near threshold circuits as compared to circuits operating at the nominal supply voltage. In contrast, MCML utilizes a differential circuit topology driven by a constant tail current and is generally characterized by high speed and high power consumption. The combination of MCML with NTC produces a balanced circuit methodology that compensates for the vulnerable aspects while benefiting from the advantages of each technology.

The paper is structured as follows. In Section 2, NTC and MCML are introduced. The benefits of combining MCML with NTC are described in Section 3. The simulation environment is reviewed in Section 4, and the results are summarized in Section 5. Some conclusions are offered in Section 6.

2. Background

Combining MOS current mode logic with near threshold circuits is proposed in this work. Each technology is individually described in this section to provide a basis for the combination presented in Section 3. Near threshold circuits are discussed in Section 2.1, and MCML is described in Section 2.2.

2.1. Near Threshold Circuits

Near threshold circuits consume an order of magnitude less power than circuits operating under nominal voltages while not suffering from the significant delay penalty found in subthreshold circuits. NTC has therefore become an attractive methodology for sub-30 nm CMOS circuits [4]. By operating near the threshold voltage (as compared to a much lower voltage deep within the subthreshold region), near threshold circuits represent a balanced approach to tackling the power issue, while maintaining circuit delays within a reasonable range. This concept is illustrated in Figure 1. In this figure, NTC is compared to two opposite extremes. At one extreme, subthreshold circuits represent minimal energy consumption coupled with slow speed operation. At the other extreme, nominal circuits consume significant energy coupled with fast speed of operation. With respect to these extrema, circuits operating in the near threshold region consume only two times more energy as compared to the subthreshold region while remaining energy efficient (ten times less than nominal voltage operation [5]). Alternatively, circuits operating in the near threshold region exhibit ten times longer delays as compared to circuits operating in the nominal voltage region. The delay of circuits operating in the subthreshold region can be a hundred to a thousand times greater than NTC [5].

Figure 1. Energy and delay in different operating regions.

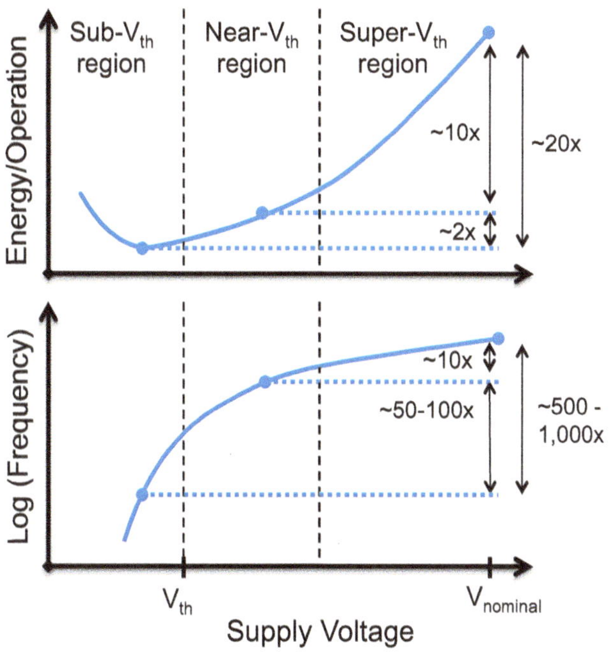

The primary difficulty of operating near the threshold voltage is the increased sensitivity to process, voltage, and temperature variations (PVT). Small variations in supply voltage can greatly affect the operating point (speed and power consumption) of NTC. Power noise in the range of 50 to 100 mV can shift the operating point from above the threshold voltage to below the threshold voltage, essentially pushing NTC either to subthreshold or above threshold operation. Alternatively, the threshold voltage of a circuit can shift due to process variations, leading to the same effect, operating a circuit either in the subthreshold region or above the threshold voltage. This behavior can lead to large shifts in gate driving capabilities of NTC transistors due to the exponential dependence of gate current on supply and threshold voltages [6]. Additionally, the low power characteristics of NTC degrade when the supply voltage is above the threshold voltage. This increased sensitivity, however, can be mitigated using the techniques discussed in Section 3.

2.2. MCML Circuits

MCML is the CMOS counterpart of bipolar emitter coupled logic (ECL), which has been in use in high speed applications since the 1970s. MCML maintains the benefits of traditional ECL, such as high speed, reduction in dI/dt noise, and common mode noise rejection, without requiring bipolar transistors [7].

An ideal MCML gate is shown in Figure 2. The gate is composed of three parts: the pull-up load resistors, the pull-down logic network, and a constant current source. The pull-up load resistors are typically PMOS transistors, as depicted in Figure 3 for an MCML universal gate. PMOS transistors are used in the pull-up network, similar to standard CMOS. During the low-to-high transition, the PMOS pull-up network charges the output to V_{DD}, unlike NMOS, which charges the output to $V_{DD} - V_{th}$.

Figure 2. Ideal MOS current mode logic (MCML) gate modeled with resistive loads and a tail current.

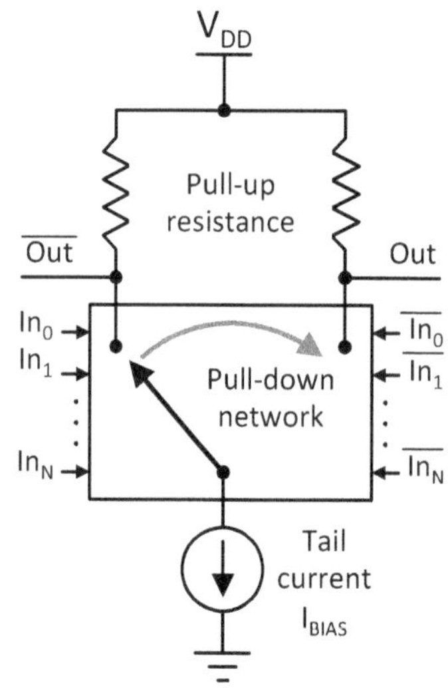

Figure 3. Basic MCML gates that share an asymmetric universal MCML gate topology: (**a**) MCML NAND gate; and (**b**) MCML NOR gate. The PMOS pull-up gate voltage, V_{pbias}, is typically connected to ground. The gate voltage, V_{nbias}, drives the NMOS transistor, providing the tail current.

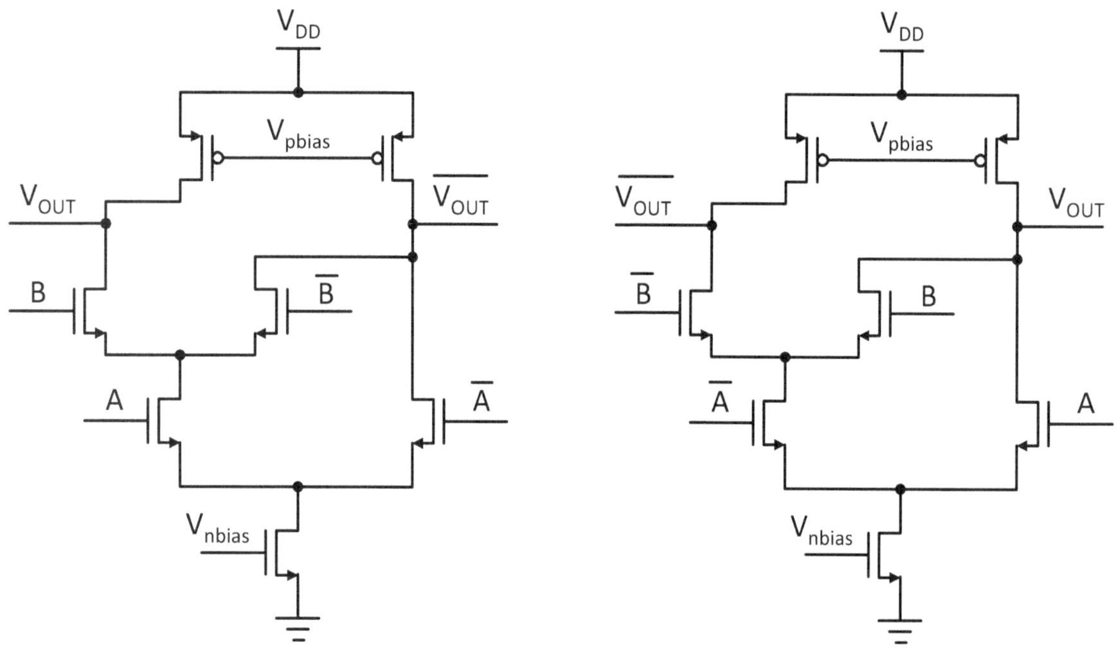

The pull-down network is fully differential and generates both the true and complementary forms of the output signal; consequently, the logic can often be simplified by eliminating inverters. The constant current source is provided by a single NMOS transistor and typically uses a separate control voltage, V_{nbias}. This constant current is steered between the differential branches (*i.e.*, the pull-up loads) to change the outputs, while the total current from V_{DD} to ground is ideally maintained as constant.

2.2.1. Power Efficiency of MCML

The power consumed by an MCML gate is

$$P_{MCML} = I_{BIAS} \times V_{DD}. \tag{1}$$

Note that the power consumed by an MCML gate does not depend on the operating frequency. In other words, an MCML gate consumes constant current (and power) from the power supply network independent of the logic activity or frequency. This behavior is in contrast to the CV^2f power dissipated by conventional CMOS, where the power consumed by a static CMOS gate exhibits a linear relationship with operating frequency. MCML is therefore more power efficient at high frequencies than static CMOS. Standard MCML circuits operating under nominal conditions exhibit enhanced power efficiency at frequencies above 5 GHz. At these frequencies, MCML, although more power efficient than standard CMOS above 5 GHz, suffers from high power densities not acceptable in modern ultra-mobile microprocessors. To reduce the frequency at which MCML dissipates less power than static CMOS, MCML circuits are operated near the threshold voltage, as suggested in this paper. By combining MCML with NTC technology, the frequency at which MCML dissipates less power than standard CMOS can be lowered to around 1 GHz, as shown in Figure 8.

2.2.2. High Speed of MCML

MOS current mode logic operates at frequencies significantly higher than standard CMOS. These frequencies are achieved due to the low delay of MCML. This property of MCML circuits is largely due to the reduced voltage swing of MCML gates. The voltage swing of an MCML gate is commonly two to ten times lower than V_{DD}, resulting in higher circuit speeds as compared to standard CMOS [8].

2.2.3. Low Noise Environment of MCML

CMOS circuits suffer from simultaneous switching noise (SSN), a significant source of on-chip noise [9]. In contrast, the near constant current of MCML (regardless of the state, *i.e.*, idle, transition, or active) produces significantly less on-chip SSN. The low noise of MCML is particularly relevant when combined with NTC due to the exponential sensitivity of NTC circuits to the power supply when operating near the threshold voltage [5]. In this paper, the simultaneous switching noise generated by MCML NTC circuits is shown to be ten times less noise than in standard CMOS with NTC circuits. A noise analysis of these circuits is described in Section 5.2.

2.2.4. Logic Gates

The design process of MCML circuits is more complex than standard CMOS. Circuit parameters, such as the supply voltage, voltage swing, pull-up equivalent resistance, tail current, and input network, need to be considered. These parameters are correlated. A change in one parameter leads to adjustments in the other parameters. For example, the voltage that determines the low output is a function of both the supply voltage and voltage swing of the gate. The voltage swing affects the pull-up resistance and tail current. If the tail current is chosen for a low power operating point, the voltage swing is affected if the pull-up resistance is not modified. This behavior is in contrast to standard CMOS gates, which have fewer design parameters (e.g., supply voltage, transistor sizes, and threshold voltage) , and each parameter independently affects the operating point. The use of high threshold voltage transistors rather than standard transistors to set a low power operating point does not change the output swing of the gate.

To overcome this limitation of MCML technology, a family of logic gates has been designed based on universal MCML gates [10]. This approach standardizes and simplifies the process of MCML logic design to a small number of universal gates. This capability is possible because basic MCML gates (*i.e.*, NAND, AND, NOR, and OR) only differ in the input and output connections. These basic MCML gates share a common circuit topology, also referred to as a universal gate structure. This universal gate structure can be either symmetric or asymmetric, depending upon the set of gates that use this universal gate structure, as well as power, speed, and area constraints. The asymmetric universal gate structure of NAND and NOR gates is illustrated in Figure 3 [1,10]. As shown in this figure, the only difference between these gates are the input and output connections. The lack of symmetry in this universal gate structure leads to asymmetric rise and fall times. An asymmetric universal gate structure, however, has two fewer input transistors and wires.

This set of basic gates can be further expanded to include an XOR gate if a symmetric topology is considered. An MCML XOR gate is shown in Figure 4b to illustrate a symmetric universal gate structure [1]. Other basic gate types (such as NAND, AND, NOR, and OR) use this symmetric universal gate topology with modified input and output connections. The drawback of a symmetric universal gate is the increased area. A symmetric universal gate structure exhibits symmetric rise and fall transitions and is simpler to design due to the symmetry of the circuit structure. The symmetry enables the use of equally sized transistors in both the left and right branches of the logic gate, eliminating the need to balance the branch currents.

3. Combination of MCML and NTC

A novel approach for combining MCML with NTC is presented in this section. The combination of MCML with NTC is proposed to exploit the mutual benefits and to offset the drawbacks of each technology. This combination is presented in Section 3.1. In Section 3.2, the sensitivity to PVT variations of MCML with NTC is discussed. The characterization of basic gates is presented in Section 3.3.

3.1. MCML with NTC

The reason for combining MCML with NTC is as follows. Standard CMOS with NTC consumes less power when operated near the threshold voltage, as discussed in Section 2.1. This low voltage operation, however, is responsible for the slower speed, as compared to the same circuit operating at a nominal supply voltage. Alternatively, MCML circuits consume greater power as compared to standard CMOS, due to the static current, as described in Section 2.2. The differential nature of MCML gates, however, requires a smaller voltage swing at the output, which significantly reduces the gate delay. CMOS with NTC therefore dissipates less power, but operates at a lower speed, while standard MCML technology provides enhanced speed, but consumes a constant high power during both active and idle periods.

Figure 4. Symmetric universal MCML gate structure used as a topology for basic MCML gates: (**a**) Symmetric universal MCML gate; and (**b**) MCML XOR gate. The PMOS pull-up gate voltage, V_{pbias}, is typically connected to ground. The gate voltage, V_{nbias}, drives the NMOS transistor, providing the tail current.

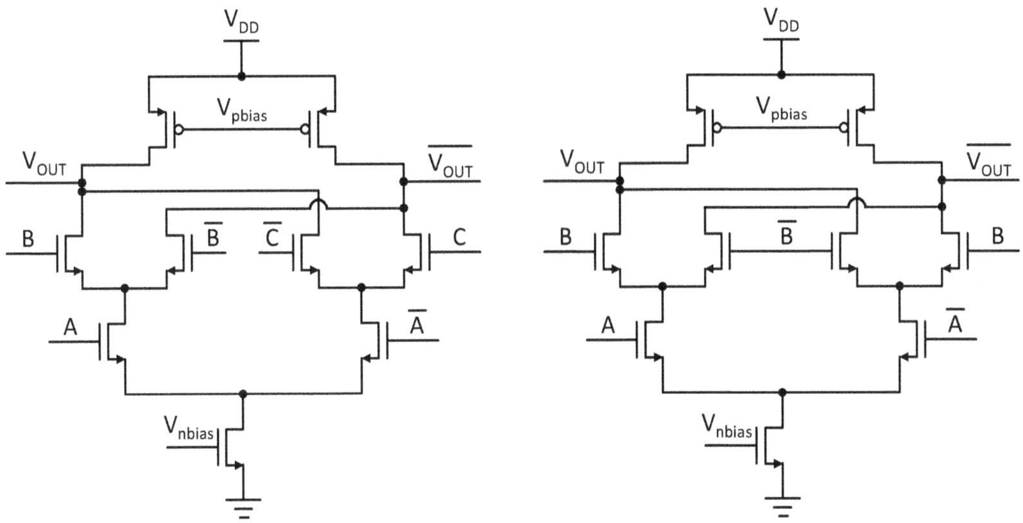

(**a**) (**b**)

MCML and NTC uncombined therefore either dissipate excessive power or are too slow. When combined, the constant power consumption of MCML is reduced to much lower levels, producing an effective circuit topology. Additionally, the low noise advantages of MCML, as described in Section 2.2, are maintained in the combined circuit topology.

One issue, however, remains partly unresolved by this combination: the high sensitivity of MCML NTC to PVT variations. This difficulty is discussed with greater detail in Section 3.2.

The advantages and disadvantages of combining both circuit approaches are summarized in Table 1. In this table, as discussed in Sections 2.1, 2.2, and 3, the speed of MCML with NTC is comparable to the speed of standard CMOS. The energy is approximately one order of magnitude less than standard CMOS; however, the same energy is consumed during idle periods. The simultaneous switching noise induced on the power network is up to two orders of magnitude lower, than standard CMOS, and is one order of magnitude lower than in CMOS with NTC. Finally, MCML with NTC is primarily sensitive to V_{th} mismatch among the PMOS pull-up transistors; however, the sensitivity can be reduced using the techniques described in this section.

Table 1. Combination of NTC and MCML.

	Standard CMOS	**CMOS with NTC**	**Standard MCML**	**MCML with NTC**
Speed	S	$S/10$	Up to $10 \times S$	Up to S
Energy consumption	E	$E/10$	$\approx E$ (also when idle)	$\approx E/10$ (also when idle)
Power network inductive noise	$\approx 10 \times N$	$\approx N$	$\approx N$	$\approx N/10$
Variations	Standard sensitivity	V_{th} variations can cause timing failures	Sensitivity to mismatch	V_{th} mismatch can cause logical failures

3.2. Sensitivity to Process Variation of MCML with NTC

There are three aspects to this issue, voltage variations, process mismatch, and temperature variations. MCML circuits provide lower SSN noise, which reduces voltage variations by about ten times, as compared to standard CMOS. This low noise environment significantly limits fluctuations near the threshold voltage, reducing variations caused by sensitivity to noise. MCML can suffer from process mismatch between the two differential branches. The PMOS pull-ups, however, are located in close proximity, allowing the transistors to be aligned to alleviate this effect [11,12]. Additionally, process variations can adversely affect the threshold voltage of the PMOS transistors within the pull-up network. This effect, however, is significantly reduced in sub-20 nm FinFET technologies due to the light doping of the transistor channel and improved gate control [13]. Finally, local temperature variations have minimal effect on the differential branches due to the physical proximity and the aforementioned layout techniques.

A Monte Carlo analysis is used to demonstrate the effect of process mismatch on MCML with NTC. In this analysis, a universal MCML with NTC gate is characterized with 1000 iterations. For each iteration, the threshold voltage of the pull-up PMOS transistors in the gate is assigned with an independently generated and normally distributed value; resulting in process mismatch between threshold voltages of up to 41 mV, which is equal to approximately 11% of nominal threshold voltage. The analyzed MCML with NTC gate exhibited correct logic values and did not fail under mismatch of up to 11%. Histograms of delay and power are presented, respectively, in Figure 5a and 5b.

Figure 5. Monte Carlo simulation of MCML with NTC gate: (**a**) Delay variation; and (**b**) Variation of power consumption. The mean delay is $\mu = 110$ ps with $\sigma = 24$ ps, while the mean power is $\mu = 827$ nW with $\sigma = 2.8$ nW.

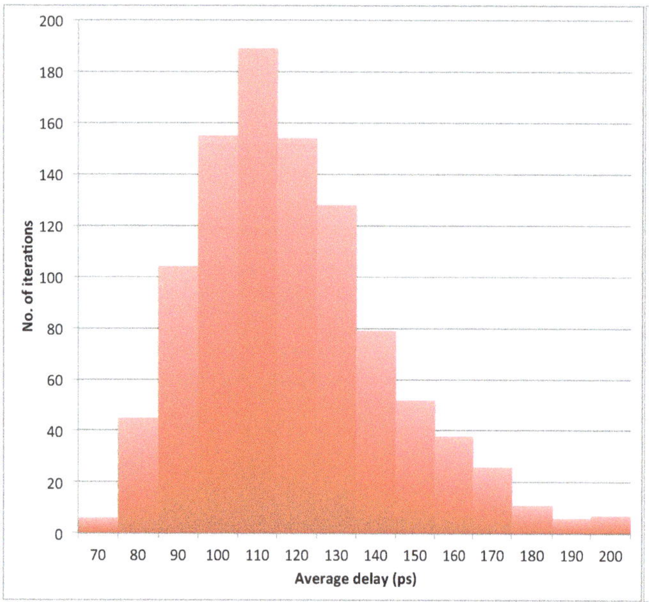

(a)

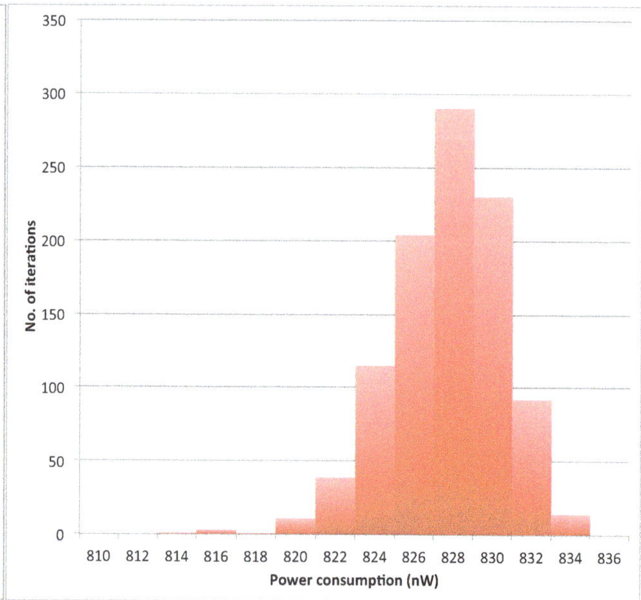

(b)

3.3. Characterization of Basic MCML with NTC Gates

The basic gates described in Section 2.2 are characterized in terms of power consumption and delay, as summarized in Table 2. The combination of MCML with NTC exhibits promising characteristics at the gate level. For basic gates, the combination of MCML with NTC exhibits lower delay, achieving higher operating frequencies as compared to standard CMOS with NTC. Additionally, the dynamic power dissipated by MCML gates operating near the threshold voltage is significantly lower than the dynamic power dissipated by standard CMOS gates operating near the threshold voltage. As described in Section 2.2, however, MCML technology exhibits the same power dissipation during idle periods, which is significantly higher than the static power of standard CMOS. This behavior is less significant for the combination of MCML and NTC when operating at high frequencies where the idle time is less, as described in Section 5.

Table 2. Performance comparison of basic logic gates using standard CMOS with NTC, and MCML with NTC.

Gate type	Technology	Delay (ps)	Dynamic power (nW)	Static power (nW)	Supply voltage (mV)
NAND gate	CMOS with NTC	120	2,270	0.150	400
	MCML with NTC	99	800	800	400
NOR gate	CMOS with NTC	112	1,600	0.090	400
	MCML with NTC	89	1,200	1,200	400
XOR gate	CMOS with NTC	267	2,600	1.225	400
	MCML with NTC	147	800	800	400

4. Simulation Setup

In this section, standard CMOS circuits and MCML-based NTC circuits are compared. The analysis is based on 14 nm low power (LP) FinFET predictive technology models [14]. A standard threshold voltage of $V_{th} = 350$ mV is assumed. The supply voltage is set to 400 mV to operate near the threshold voltage with an MCML voltage swing of 100 mV. A Kogge Stone adder is used to evaluate this proposed circuit topology and is described in Subsection 4.1. Power and noise simulation setups are described, respectively, in Subsections 4.2 and 4.3.

4.1. Description of Test Circuit

A 32-bit Kogge Stone adder [15] is evaluated in both standard CMOS, and MCML with NTC. The Kogge Stone adder is a parallel carry look ahead adder [15]. The choice of a 32-bit Kogge Stone adder as a test circuit is due to the high speed nature of this circuit topology. The structure of the 32-bit test circuit is demonstrated with an 8-bit Kogge Stone adder, as presented in Figure 6. The 8-bit Kogge Stone adder has the same periodic structure as the 32-bit adder. The adder is composed of three building blocks: bit propagate, group generate, and group propagate cells, with a 32-bit input and 32-bit output. The critical delay path is highlighted in red, as shown in Figure 6. This critical path is used for evaluating the worst case delay to determine the maximum operating frequency of the circuit. Two versions of a 32-bit Kogge Stone adder are compared. One version is based on MCML with NTC logic, while the other version is based on CMOS with NTC logic.

4.2. Power Simulation Setup

The power *versus* frequency characteristics of the Kogge Stone adder are illustrated in Figure 8. Both MCML and standard CMOS circuits are stimulated with the same 32-bit input with a duty cycle equivalent to 1 GHz. The input with the longest propagation delay, T_D, sets the maximum possible operating frequency $F_{max} = 1/T_D$ of the circuit. The different power consumption levels are dependent on the supply voltage for standard CMOS, varying from subthreshold operation (200 mV) to nominal voltage (800 mV). For MCML NTC circuits, however, the power consumption is primarily dependent on the tail current. The power consumption is the average with different inputs, and consists of both dynamic and static power consumption. The dynamic power consumption is the average power consumed by the test circuits during a signal transition. The static power consumption is the average power consumed during the remaining portion of the input cycle when the logic is idle. Note that the static power consumption is a key difference between standard CMOS and MCML. Therefore, to produce a fair comparison, standard CMOS is compared to MCML with activity factors of 10%, 20%, and 100%. These activity factors represent a wide range of circuits, from active each cycle (*e.g.*, clock distribution signals) to active only every tenth cycle (common data paths).

Figure 6. Eight-bit Kogge Stone adder within a 32-bit Kogge Stone adder. The white blocks represent the bit propagate (BP) cells, solid gray blocks represent the group propagate (GP) cells, and doted gray blocks represent the group generate (GG) cells. The critical delay path is highlighted by a bold red line.

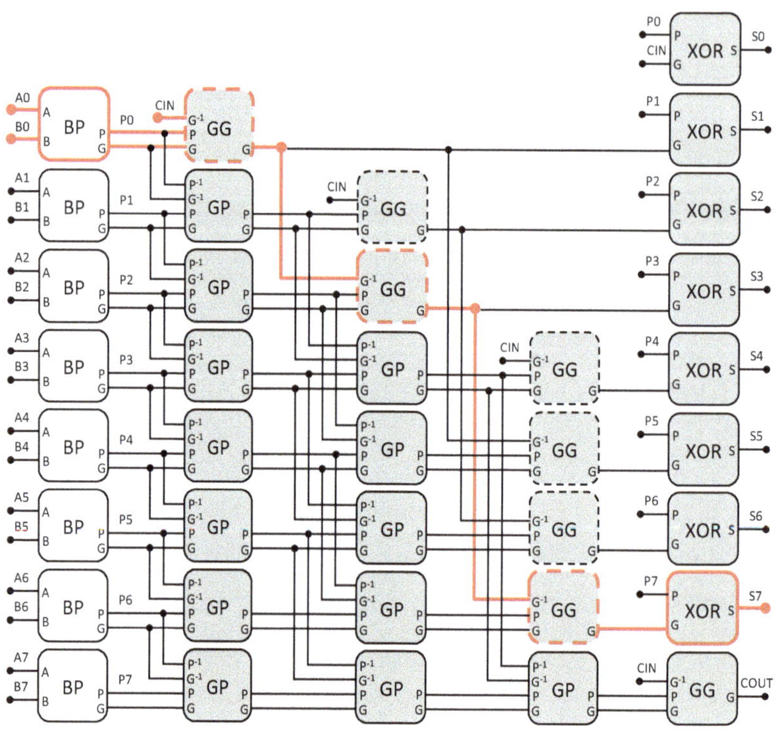

For standard CMOS, frequencies above 9 GHz cannot be achieved with minimum sized gates in 14 nm FinFET CMOS. The exponential increase in the power consumed by standard CMOS beyond 9 GHz is caused by the increased size of the gates with a nominal voltage supply of 1 volt. For MCML with NTC, the supply voltage is set constant near a threshold voltage of 400 mV.

4.3. Noise Simulation Setup

The same circuit structure is used to analyze the noise induced within the power network. An equivalent lump model of the power network is illustrated in Figure 7. The resistive, capacitive, and inductive impedances are based on the international technology roadmap for semiconductors (ITRS) guidelines [16]. The results are discussed in Section 5.

Figure 7. Test circuit with a lumped impedance model for evaluating noise in power and ground networks.

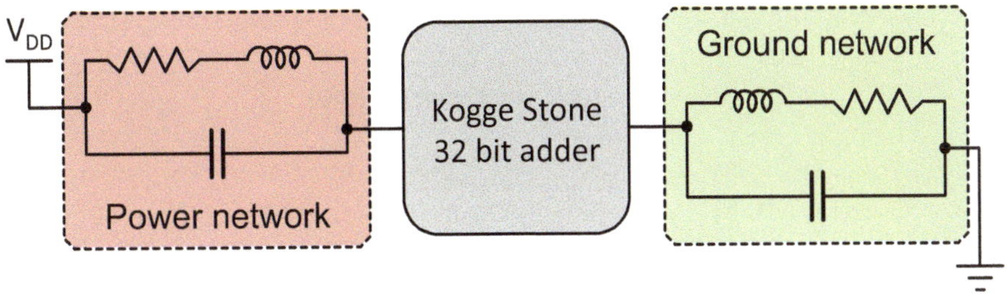

5. Simulation Results

The power, speed, and noise characteristics of standard CMOS and MCML with NTC are presented in the following subsections.

5.1. Power/Speed

The power consumed by a 32-bit Kogge Stone adder is shown in Figure 8, as described in Section 4. As expected from the gate performance characteristics listed in Table 2, MCML with NTC is more power efficient than standard CMOS with NTC when operating at high frequencies.

Three speed/power behaviors of interest are exhibited, as illustrated in Figure 8. At 1 GHz, the power consumed by MCML with NTC is less than standard CMOS at a 100% activity factor. In other words, the combination of MCML with NTC is more power efficient for circuits operating at frequencies above 1 GHz and switching every cycle. These types of circuits are not limited to clock distribution networks.

At around 9 GHz, the CMOS circuit reaches the maximum operating frequency at a nominal supply voltage with minimum sized gates. To further lower the delay (to increase the frequency), the CMOS gates need to be significantly larger. A sharp increase in power consumption is noted for standard CMOS operating at multi-gigahertz frequencies switching at 10% and 20% activity factors. Due to this sharp rise, at around 9 GHz, MCML with NTC is more power efficient than standard CMOS switching at a 20% activity factor.

Figure 8. Power *vs.* maximum frequency of MCML with near threshold circuits (NTC) and standard CMOS for activity factors of 10%, 20%, and 100%.

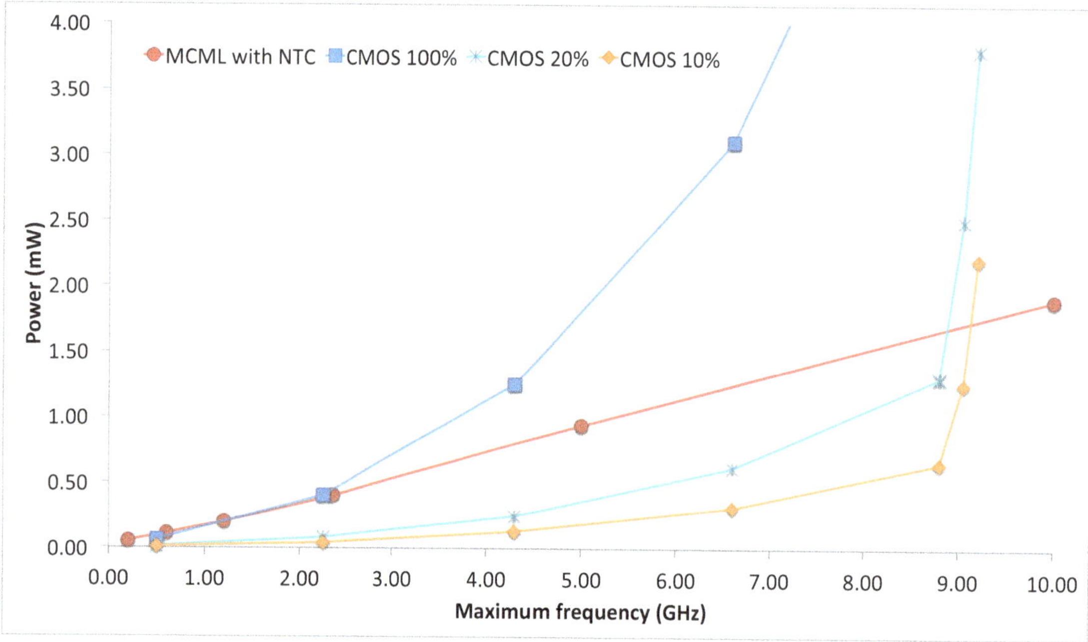

Additionally, above 9 GHz, MCML with NTC is more power efficient than standard CMOS switching at a 10% activity factor. At these frequencies, MCML with NTC is the methodology of choice for general high performance circuits. Activity factors below 25% represent general switching characteristics of typical data paths.

These characteristics of MCML with NTC position this technology as a significant competitor to standard CMOS when operating at high activity factors, or multi-gigahertz frequencies regardless of the activity factor. Unlike standard CMOS, MCML is capable of power efficient operation at frequencies beyond 9 GHz (in 14 nm technology), enabling power efficient operation at multi-gigahertz frequencies.

5.2. Noise

The maximum induced noise in a power network due to switching activity is listed in Table 3. Noise values are presented for standard CMOS and MCML as a function of power network resistance, capacitance, and inductance. As noted in this table, the SSN in MCML circuits is, on average, an order of magnitude lower than in static CMOS. The low noise environment is particularly beneficial for deeply scaled circuits operating near the threshold voltage that suffer from high sensitivity to process and environment variations.

This capability supports heterogeneous systems that integrate noise sensitive analog circuits with digital logic and memory. In contrast to standard CMOS circuits with significant simultaneous switching noise, extensive effort to isolate sensitive circuits from switching noise is not required in NTC MCML. Additionally, the low noise environment enables lower noise margins, resulting in more power efficient and/or higher speed circuits. This low noise characteristic represents a significant advantage of MCML combined with NTC, particularly in heterogeneous systems [17].

Table 3. Comparison of noise in CMOS and MCML circuits.

Power network parasitic impedances			Noise induced on power network (mV)		
Resistance (ohm)	Capaitance (fF)	Inductance (nH)	MCML absolute value	CMOS absolute value	Ratio
2	50	1	0.56	6.27	11
2	50	2	0.94	9.92	11
2	50	4	0.70	14.64	21
2	100	1	1.28	6.19	5
2	100	2	0.95	9.14	10
2	100	4	1.81	13.51	7
2	200	1	0.55	6.21	11
2	200	2	0.93	9.96	11
2	200	4	0.66	12.56	19
5	50	1	1.32	6.50	5
5	50	2	0.84	9.75	12
5	50	4	1.71	14.63	9
5	100	1	0.51	6.52	13
5	100	2	0.93	9.29	10
5	100	4	0.72	13.24	18
5	200	1	1.25	6.60	5
5	200	2	0.83	10.02	12
5	200	4	1.61	12.16	8

6. Conclusions

The combination of NTC and MCML exhibits high performance by exploiting the advantages of each technology. The proposed combination of MCML with NTC is shown to be best suitable for two types of applications. The first type are high activity circuits operating at frequencies above 1 GHz (assuming 14 nm FinFET CMOS). This behavior is in contrast to standard CMOS, which dissipates excessive power at high activity factors. The second type are low activity circuits operating at frequencies above 9 GHz. At these high frequencies, CMOS is inefficient due to the linear dependence of dynamic power with frequency. The combination of MCML with NTC, therefore, provides an effective high performance, power efficient circuit technology for high speed and/or high activity applications.

Acknowledgements

This research is supported in part by a grant from Qualcomm. The authors would like to thank Jeff Fischer, Francois Atallah, Kyungseok Kim, and Jihoon Jeong for their technical support and feedback.

Conflicts of Interest

The authors declare no conflict of interest.

References

1. Salman, E.; Friedman, E.G. *High Performance Integrated Circuit Design*; McGraw-Hill Publisher: New York, NY, USA, 2012.
2. Kaul, H.; Anders, M.; Hsu, S.; Agarwal, A.; Krishnamurthy, R.; Borkar, S. Near-Threshold Voltage (NTV) Design: Opportunities and Challenges. In Proceedings of the ACM/IEEE Design Automation Conference, San Francisco, CA, USA, 3–7 June 2012; pp. 1149–1154.
3. Alioto, M.; Palumbo, G. Feature-Power-Aware Design Techniques for Nanometer MOS Current-Mode Logic Gates: A Design Framework. *IEEE Circuits Syst. Mag.* **2006**, *6*, 42–61.
4. Jain, S.; Khare, S.; Yada, S.; Ambili, V.; Salihundam, P.; Ramani, S.; Muthukumar, S.; Srinivasan, M.; Kumar, A.; Gb, S.K.; *et al.* A 280 mV-to-1.2 V Wide-Operating-Range IA-32 Processor in 32 nm CMOS. In Proceedings of the IEEE Solid-State Circuits Conference, San Francisco, CA, USA, 19–23 February 2012; pp. 66–68.
5. Dreslinski, R.; Wieckowski, M.; Blaauw, D.; Sylvester, D.; Mudge, T. Near-Threshold Computing: Reclaiming Moore's Law through Energy Efficient Integrated Circuits. *IEEE Proc.* **2010**, *98*, 253–266.
6. Islam, A.; Akram, M.; Imran, A.; Hasan, M. Energy Efficient and Process Tolerant Full Adder Design in Near Threshold Region Using FinFET. In Proceedings of the International Symposium on Electronic System Design, Bhubaneswar, India, 20–22 December 2010; pp. 56–60.
7. Hassan, H.; Anis, M.; Elmasry, M. MOS Current Mode Logic: Design, Optimization, and Variability. In Proceedings of the IEEE International SOC Conference, Santa Clara, CA, USA, 12–15 September 2004; pp. 247–250.
8. Hassan, H.; Anis, M.; Elmasry, M. MOS Current Mode Circuits: Analysis, Design, and Variability. *IEEE Trans. Very Large Scale Integr. (VLSI) Syst.* **2005**, *13*, 885–898.
9. Tang, T.; Friedman, E. Simultaneous Switching Noise in on-Chip CMOS Power Distribution Networks. *IEEE Trans. Very Large Scale Integr. (VLSI) Syst.* **2002**, *10*, 487–493.
10. Abdulkarim, O.M.; Shams, M. A Symmetric MOS Current-Mode Logic Universal Gate for High Speed Applications. In Proceedings of the ACM Great Lakes Symposium on VLSI, Lago Maggiore, Italy, 11–13 March 2007; pp. 212–215.
11. Mandapati, V.S.R.; Nishanth, P.V.; Paily, R. Study of Transistor Mismatch in Differential Amplifier at 32 nm CMOS Technology. *Int. J. Comput. Sci. Issues* **2011**, *1*, 109–115.
12. Kuhn, K. Reducing Variation in Advanced Logic Technologies: Approaches to Process and Design for Manufacturability of Nanoscale CMOS. In Proceedings of the IEEE International Electron Devices Meeting, Washington, DC, USA, 10–12 December 2007; pp. 471–474.
13. Tawfik, S.; Kursun, V. FinFET Technology Development Guidelines for Higher Performance, Lower Power, and Stronger Resilience to Parameter Variations. In Proceedings of the IEEE International Midwest Symposium on Circuits and Systems, Cancun, Mexico, 2–5 August 2009; pp. 431–434.
14. Cao, Y.K. Predictive Technology Models. Available online: http://ptm.asu.edu/ (accessed on 29 April 2014).

15. Kogge, P.M.; Stone, H.S. A Parallel Algorithm for the Efficient Solution of a General Class of Recurrence Equations. *IEEE Trans. Comput.* **1973**, *C-22*, 786–793.
16. The International Technology Roadmap for Semiconductors. Available online: http://www.itrs.net/ (accessed on 29 April 2014).
17. Pavlidis, V.F.; Friedman, E.G. *Three-Dimensional Integrated Circuit Design*; Morgan Kaufmann: Burlington, MA, USA, 2009.

10

Comparative Study of Charge Trapping Type SOI-FinFET Flash Memories with Different Blocking Layer Materials †

Yongxun Liu [1,*], Toshihide Nabatame [2], Takashi Matsukawa [1], Kazuhiko Endo [1], Shinichi O'uchi [1], Junichi Tsukada [1], Hiromi Yamauchi [1], Yuki Ishikawa [1], Wataru Mizubayashi [1], Yukinori Morita [1], Shinji Migita [1], Hiroyuki Ota [1], Toyohiro Chikyow [2] and Meishoku Masahara [1]

[1] National Institute of Advanced Industrial Science and Technology (AIST), 1-1-1 Umezono, Tsukuba, Ibaraki 305-8568, Japan; E-Mails: t-matsu@aist.go.jp (T.M.); endo.k@aist.go.jp (K.E.); shinichi.ouchi@aist.go.jp (S.O.); junichi-tsukada@aist.go.jp (J.T.); hiro-yamauchi@aist.go.jp (H.Y.); yuki.ishikawa@aist.go.jp (Y.I.); w.mizubayashi@aist.go.jp (W.M.); y.morita@aist.go.jp (Y.M.); s-migita@aist.go.jp (S.M.); hi-ota@aist.go.jp (H.O.); m.masahara@aist.go.jp (M.M.)

[2] National Institute for Materials Science (NIMS), 1-1 Namiki, Tsukuba, Ibaraki 305-0044, Japan; E-Mails: NABATAME.Toshihide@nims.go.jp (T.N.); CHIKYO.toyohiro@nims.go.jp (T.C.)

† This is an extended version of paper "Charge Trapping Type FinFET Flash Memory with Al_2O_3 Blocking Layer" that was presented at the IEEE S3S Conference 2013.

* Author to whom correspondence should be addressed; E-Mail: yx-liu@aist.go.jp

---

Abstract: The scaled charge trapping (CT) type silicon on insulator (SOI) FinFET flash memories with different blocking layer materials of Al_2O_3 and SiO_2 have successfully been fabricated, and their electrical characteristics including short-channel effect (SCE) immunity, threshold voltage (V_t) variability, and the memory characteristics have been comparatively investigated. It was experimentally found that the better SCE immunity and a larger memory window are obtained by introducing a high-k Al_2O_3 blocking layer instead of a SiO_2 blocking layer. It was also confirmed that the variability of V_t before and after one program/erase (P/E) cycle is almost independent of the blocking layer materials.

Keywords: charge trapping (CT); flash memory; silicon on insulator (SOI); FinFET; blocking layer; high-k metal gate; variability

1. Introduction

Device scaling is very effective for the fabrication of high-density and low-cost flash memories. However, further scaling of conventional bulk planar MOSFET type flash memories becomes very difficult because of the increased short-channel effect (SCE) and the lowered source-drain (SD) breakdown voltage (BV_{DS}) with scaling down device size [1–3]. Especially, in the NOR-type flash memory, further scaling of device size faces the theoretical limit of BV_{DS} which corresponds to the silicon (Si) and silicon dioxide (SiO_2) conduction band difference (3.2 eV). This indicates that channel hot electron (CHE) programming cannot be guaranteed in the scaled NOR-type flash memories with gate length (L_g) smaller than 100 nm [2,3]. On the other hand, three-dimensional (3D) channel devices, such as fin-type double-gate (DG) MOSFET (FinFET) and fin-channel tri-gate (TG) device, provide excellent SCE immunity owing to the strong controllability of channel potential by the multiple gates [4–13]. Moreover, threshold voltage (V_t) variability in the FinFETs and TG devices is much smaller than that in the conventional bulk planar MOSFETs because the random dopant fluctuation (RDF) induced V_t variation is negligible in the FinFETs and TG devices due to the undoped fin-channels [14–22]. Therefore, the scaled charge trapping (CT) type fin-channel flash memories using silicon on insulator (SOI)-based fin-channels and body-tied bulk Si fin-channels have actively been developed [23–32]. However, a high-k blocking layer is strongly required in the ultimately scaled CT type flash memory fabrication to overcome the gate coupling area decrease with scaling down the device size [33]. As a high-k blocking layer, an Al_2O_3 layer has been used in the planar MOSFET type and body-tied bulk FinFET type flash memories [28,34]. By introducing such a high-k blocking layer, the gate injection current is effectively suppressed during program/erase (P/E) operations because the electric field across the blocking layer is proportionally reduced owing to its high dielectric constant, which is useful for the enlarging of memory window. However, the blocking layer material effect on the electrical characteristics of SOI-FinFET flash memories has not been studied sufficiently. Very recently, we have demonstrated floating gate (FG) type split-gate fin-channel flash memories with a highly suppressed over erase, and experimentally confirmed that nanosize triangular cross-section tunnel areas are useful for the fabrication of the low operating voltage flash memories owing to the enhanced local electric field at the tips of triangular tunnel areas [35–38]. We have also fabricated and investigated FG type crystalline and polycrystalline Si fin-channel flash memories with DG and TG structures, and confirmed that TG structured flash memory shows the better SCE immunity and a larger memory window than the DG structured one owing to the additional top gate and recessed buried oxide (BOX) region [39–41]. Moreover, we have also investigated the gate material effect on the electrical characteristics of the CT type SOI-FinFET flash memories by introducing different gate materials of physical vapor deposited (PVD) titanium nitride (TiN) and n^+-poly-Si [42]. It was experimentally found that a larger memory window is obtained in the PVD-TiN metal gate flash memories than the n^+-poly-Si gate ones owing to the higher work function of PVD-TiN metal gate, which is efficient to suppress electron back tunneling during erase operation [28,34].

As a further study, in this work, we fabricate CT type SOI-FinFET flash memories with different blocking layer materials of atomic layer deposited (ALD) Al_2O_3 and chemical vapor deposited (CVD) SiO_2, and comparatively investigate their electrical characteristics including SCE immunity, V_t variability and memory characteristics [43].

2. Device Fabrication

In order to investigate the blocking layer material effect on the electrical characteristics and memory properties of the CT type SOI-FinFET flash memories, we fabricated MANOS (PVD-TiN Metal-Al_2O_3-Nitride-Oxide-Silicon) type and MONOS (PVD-TiN Metal-Oxide-Nitride-Oxide-Silicon) type TG structured SOI-FinFET flash memories. Figure 1 shows the schematic three-dimensional (3D) device structure for the CT type SOI-FinFET flash memory. Except for the deposition of blocking layer materials, the same process was used in the fabrication for both types of devices. To fabricate an ideal rectangular cross-section Si-fin-channel by using the orientation-dependent wet etching, we used lightly doped p-type (110)-oriented SOI wafers as the starting material [7,8]. The initial thicknesses of the top silicon and BOX layers were 70 and 145 nm, respectively.

Figure 1. Schematic three-dimensional (3D) device structure for the CT type SOI-FinFET flash memory with different blocking layer materials of Al_2O_3 and SiO_2.

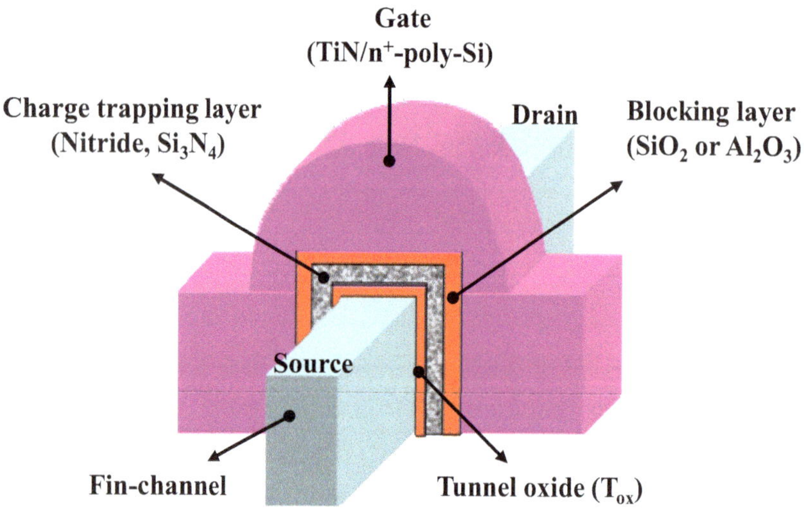

The schematic device fabrication process flow for the CT type SOI-FinFET flash memory is shown in Figure 2. First, the lightly doped p-type (110)-SOI wafers were thermally oxidized to form a 30-nm-thick SiO_2 layer as shown in Figure 2a, and fin-patterns were delineated in parallel to the <112> direction by using electron-beam (EB) lithography as shown in Figure 2b. The trimming of EB-resist was performed using oxygen plasma to reduce the fin width. The fin-patterns were then transferred to the SiO_2 layer on the (110)-SOI wafers by reactive ion etching (RIE). After removing the EB-resist, further narrowing of the width of SiO_2 hard-mask was carried out with dilute hydrofluoric acid (DHF) solution. By using these techniques, the width of SiO_2 hard-mask was controlled to around 20 nm.

Figure 2. Schematic device fabrication process flow for the CT type SOI-FinFET flash memory. (**a**) Thermal oxidation; (**b**) fin-pattern formation by electron-beam (EB) lithography; (**c**) fin-channel formation by the orientation-dependent wet etching; (**d**) SiO_2 hard-mask removing by RIE; (**e**) tunnel oxide (T_{ox}) formation by thermal oxidation; (**f**) nitride (Si_3N_4) layer deposition by LPCVD; (**g**) blocking layer (Al_2O_3 for MANOS and SiO_2 for MONOS) deposition; (**h**) PVD-TiN deposition; (**i**) n^+-poly-Si deposition.

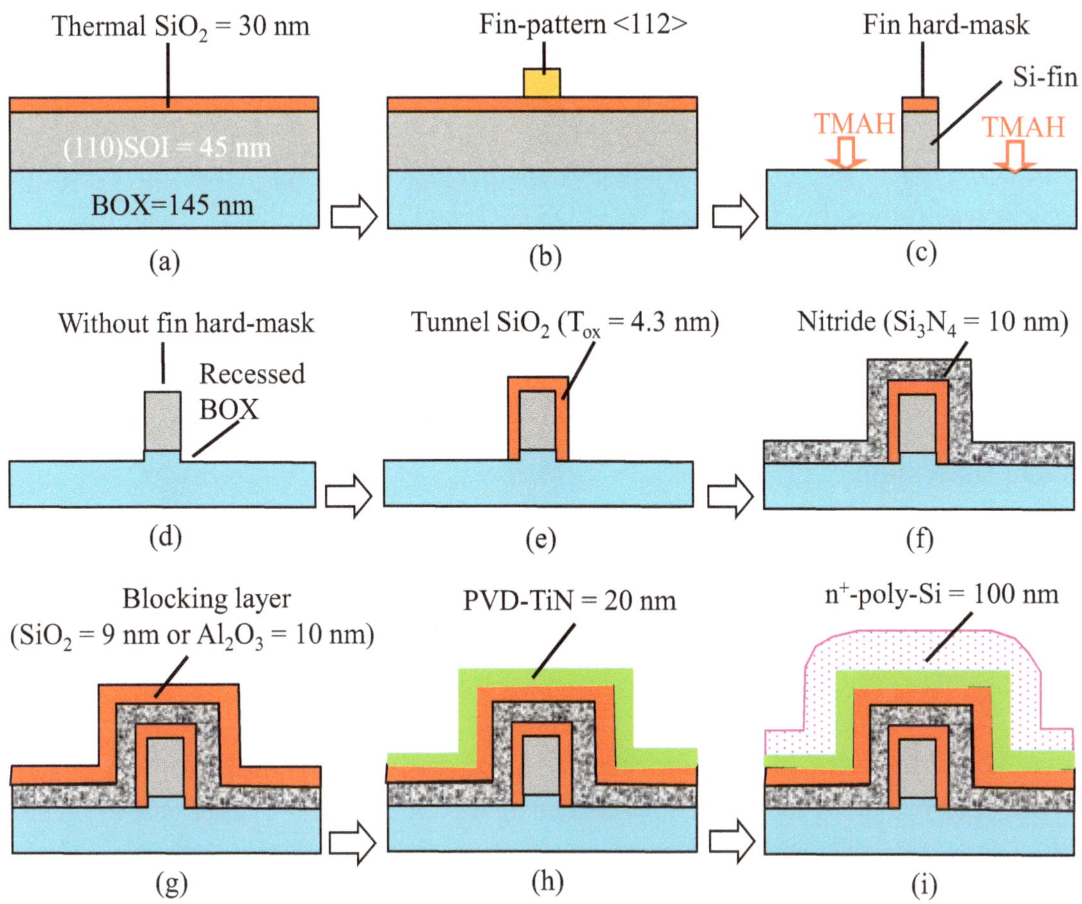

Through the SiO_2 hard-masks, fin-channels were fabricated by the orientation-dependent wet etching with a 2.38% tetramethylammonium hydroxide (TMAH) solution at 50 °C for 30 s, as shown in Figure 2c. Since the sidewalls of the fin-channels have a (111)-oriented plane with an extremely low etching rate in TMAH compared with other planes, very narrow and straight fin-channels can easily be fabricated. It was experimentally confirmed that the etching rates for (100)-, (110)-, and (111)-oriented Si wafers were 214, 359, and 9 nm/min, respectively [44]. This indicates that the (111) plane is successfully retained in the etching process. To fabricate TG structure, the SiO_2 hard-mask layer was removed by RIE as show in Figure 2d, which results in a slight reduction of fin-height. A recessed BOX region was also formed in this RIE process, which is useful for the suppression of SCE [39–41].

After the fin-channel formation, a 4.3-nm-thick tunnel oxide (T_{ox}) layer was formed by thermal oxidation at 850 °C as shown in Figure 2e, followed by the deposition of a 10-nm-thick Si_3N_4 layer as the charge trapping layer by low-pressure chemical vapor deposition (LPCVD) at 790 °C as shown in Figure 2f. As the blocking layer, a 9-nm-thick tetraethylorthosilicate (TEOS)-SiO_2 layer was deposited for MONOS type devices by plasma CVD at 350 °C, and a 10-nm-thick Al_2O_3 layer was deposited for

MANOS type device by atomic layer deposition (ALD) at 300 °C as shown in Figure 2g. In the ALD deposition, we used trimetylaluminum (TMA) and water gas at pressure of 14 hPa. As the same gate material, a combination of a 20-nm-thick PVD-TiN layer and a 100-nm-thick n^+-poly-Si layer, was deposited on all wafers as shown in Figure 2h,i. The gate electrodes were also patterned by EB lithography and fabricated by using combination of inductively coupled plasma (ICP) RIE for n^+-poly-Si and wet etching for PVD-TiN. In the PVD-TiN wet etching, we used an ammonium hydroxide (NH_4OH):hydrogen peroxide (H_2O_2):deionized water (H_2O) = 1:2:5 (APM) solution at 60 °C, which provides a high etching selectivity of PVD-TiN to SiO_2 [45,46].

After etching the top ONA and ONO layers on the fin extension and SD electrode regions by RIE, arsenic (As) ion implantation (I/I) was performed for SD-extension with a dose (D) of 4×10^{14} cm^{-2} and a tilting angle (θ) of 60° at a fixed implant energy of 5 keV [47]. Then, an 80-nm-thick gate sidewall spacer was formed by deposition of TEOS-SiO_2 and RIE. For the SD-region I/I, phosphorus ion (P^+) was used with $D = 1.5 \times 10^{15}$ cm^{-2} and $\theta = 7°$, which was followed by the deposition of a 100-nm-thick TEOS-SiO_2 layer on all wafers. To activate the implanted impurities, rapid thermal annealing (RTA) was performed at 830 °C for 2 s. Finally, contact holes and aluminum electrodes were formed, and all wafers were sintered in forming gas ambient at 450 °C for 30 min.

3. Results and Discussion

The scanning electron microscopy (SEM) images of the fabricated MANOS type SOI-FinFET flash memory after fin-channel formation and gate formation are shown in Figure 3a,b, respectively. Note that a straight 22-nm-thick Si-fin channel is successfully fabricated thanks to the orientation-dependent wet etching. Moreover, a scaled 26-nm gate is also clearly confirmed.

Figure 3. SEM images of the fabricated CT type SOI-FinFET flash memory (**a**) after fin-channel formation by the orientation-dependent wet etching and (**b**) after gate formation.

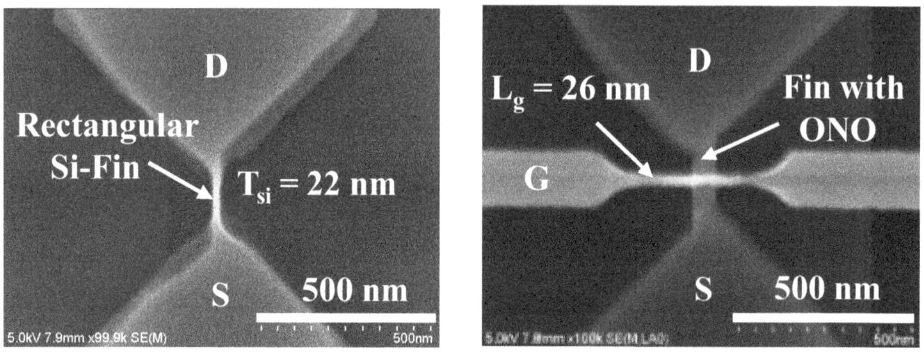

(**a**) (**b**)

Figure 4a,b show the cross-sectional scanning transmission electron microscopy (STEM) images of the fabricated MANOS and MONOS type SOI-FinFET flash memories, respectively. Note that ultrathin and ideal rectangular cross-sectional Si fin-channels are fabricated uniformly. It should be mentioned that the fin-width and fin-height were slightly narrowed and lowered during wafer cleaning and tunnel oxide formation. Moreover, it is clearly confirmed that a uniform tunnel oxide layer is formed on top and sidewalls of the fin-channels without a thick SiO_2 hard-mask on top of Si fin-channel. This indicates that TG structure is fabricated successfully. Furthermore, it can be seen from STEM images that the charge trapping ONA and ONO layers are not only formed around the top and sidewalls of the fin-channels but also extended to the recessed BOX region. Such a gate all around (GAA) like structure is useful to the suppression of SCE as mentioned before. The observed a thin SiO_2 blocking layer (<9 nm) in Figure 4b probably depends on the pick-up position of STEM sample owing to the poor uniformity of TEOS-SiO_2 layer thickness on a wafer although a 9-nm-thick TEOS-SiO_2 layer was confirmed on a dummy wafer as mentioned before.

Figure 4. Cross-sectional STEM images of the fabricated (**a**) MANOS type SOI-FinFET flash memory with an Al_2O_3 blocking layer; and (**b**) MONOS type SOI-FinFET flash memory with a SiO_2 blocking layer.

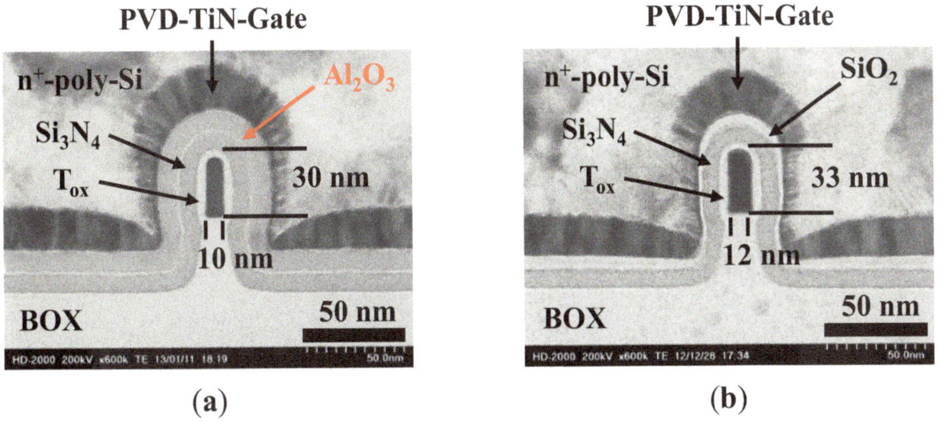

At first, we evaluated the V_t variability and SCE immunity for the fabricated MANOS and MONOS type SOI-FinFET flash memories with different gate length (L_g) values from 26 to 103 nm. Figure 5a,b show the measured initial I_d-V_g characteristics of the fabricated MANOS type flash memories with different L_g values of 26 and 38 nm, respectively. In this measurement, we used 40 cell transistors for each type to investigate statistical V_t variations. Note that an excellent S-slope of 105-mV/decade is obtained even L_g is scaled down to 26 nm due to the GAA like structure. However, the uniformity of I_d-V_g curve is slightly deteriorated with scaling down L_g from 38 to 26 nm.

To evaluate V_t variations quantitatively, we measured the initial V_t values at a constant drain current of $I_d = 1$ μA for all devices with different L_g values from 26 to 103 nm. Figure 6 shows the measured standard deviations of V_t (σV_t) and S-slope (σS) as a function of L_g for the fabricated MANOS and MONOS type SOI-FinFET flash memories. It is clear that almost the same σV_t and σS are obtained for both types of devices although they increase slightly with scaling down L_g. This result indicates that the variations of V_t and S-slope are almost independent of the blocking layer materials. The measured average V_t ($\langle V_t \rangle$) and S-slope ($\langle S \rangle$) values are summarized as a function of L_g as shown in Figure 7. Note that the smaller V_t roll-off and the better S-slope are obtained in the MANOS devices than the MONOS ones at whole range of L_g due to the high-k effect of Al_2O_3 in MANOS type devices.

Figure 5. Initial I_d-V_g characteristics of the fabricated MANOS type SOI-FinFET flash memories with different L_g values of (**a**) L_g = 26 nm and (**b**) L_g = 38 nm.

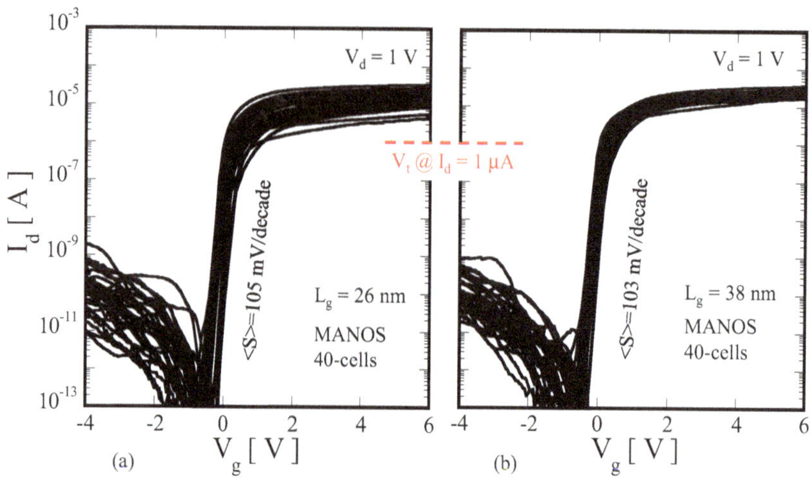

Figure 6. σV_t and σS as a function of L_g for the fabricated MANOS and MONOS type SOI-FinFET flash memories.

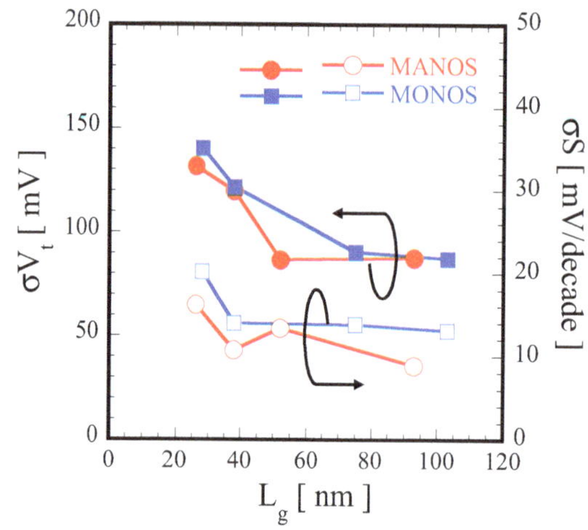

Figure 7. Average V_t (<V_t>) and S-slope (<S>) values as a function of L_g for the fabricated MANOS and MONOS type SOI-FinFET flash memories.

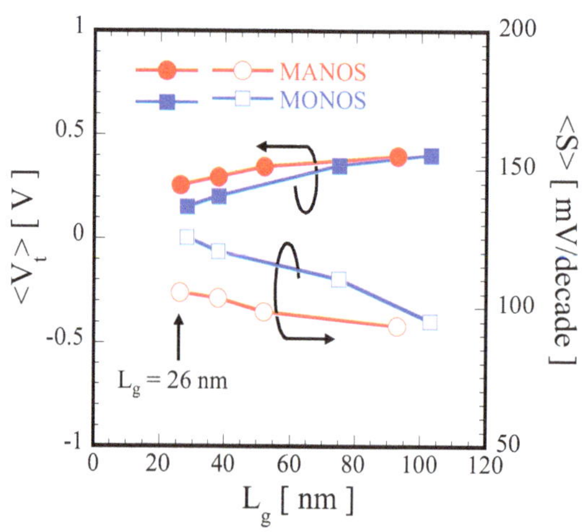

Before evaluation of memory characteristics, gate breakdown voltage was measured for the fabricated MANOS and MONOS type SOI-FinFET flash memory cell transistors. It was found that the gate breakdown voltage is around 17 V. According to this experimental result, P/E bias conditions were determined as $V_g = 16$ V, pulse-time (t) = 50 μs for programming, $V_g = -16$ V, t = 20 ms for erasing. In order to evaluate the dependence of the memory characteristics on the blocking layer materials, including V_t variability and memory window, the I_d-V_g characteristics of the fabricated MANOS and MONOS type devices with different L_g values were systematically investigated after one P/E cycle. Figure 8a,b show the measured I_d-V_g characteristics of the fabricated MANOS type flash memory cell transistors with different L_g values of 26 and 38 nm after one P/E cycle, respectively. In this measurement, we used 40 cell transistors for each type. It is clear that a larger memory window is reasonably obtained in the long channel devices as compared to the short channel devices. A similar behavior was also observed in the MONOS type devices.

To compare the V_t variations between MANOS and MONOS type devices, V_t values for all fabricated devices were also evaluated at a constant drain current of $I_d = 1$ μA. As an example, Figure 9 shows the measured cumulative probability of V_t and σV_t values for the MANOS and MONOS type devices with the same L_g of 38 nm after one P/E cycle. Note that a larger memory window is obtained in the MANOS type devices than the MONOS type ones due to the high-k effect of Al_2O_3 blocking layer in MANOS type devices. However, it is clear that σV_t values for both devices are almost the same although slightly smaller σV_t values are observed in the MONOS type device than MANOS type one. Such comparison was also carried out for all fabricated MANOS and MONOS type devices with different L_g values, and it was confirmed that actually a larger memory window is obtained in the MANOS type devices than the MONOS type ones at whole range of L_g as shown in Figure 10. The deep erase in MANOS type devices should be resulted from the high-k effect of an Al_2O_3 (k ~ 9) blocking layer, which is efficient to enhance the electric field across the tunnel oxide and to reduce the electric field across the layer its self. Therefore, the electron back tunneling is effectively suppressed during erase operation which contributes to the deep erase in MANOS type devices [28,34].

Figure 8. I_d-V_g characteristics of the fabricated MANOS type SOI-FinFET flash memory cell transistors (40 cells) with different L_g values of (**a**) $L_g = 26$ nm and (**b**) $L_g = 38$ nm. P/E conditions: $V_g = 16$ V, t = 50 μs for program, and $V_g = -16$ V, t = 20 ms for erase.

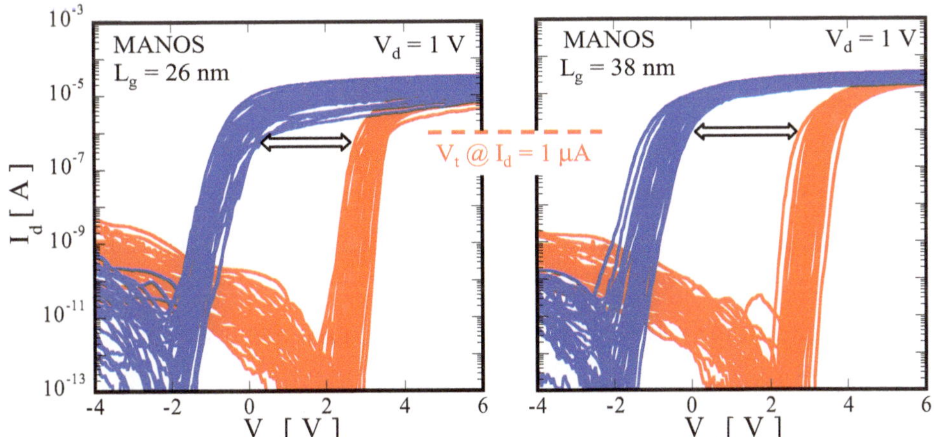

(a) (b)

Figure 9. (**a**) Cumulative probability of V_t and (**b**) σV_t of the fabricated MANOS and MONOS type SOI-FinFET flash memories with the same L_g of 38 nm after one P/E cycle. P/E conditions: $V_g = 16$ V, t = 50 μs for program, and $V_g = -16$ V, t = 20 ms for erase.

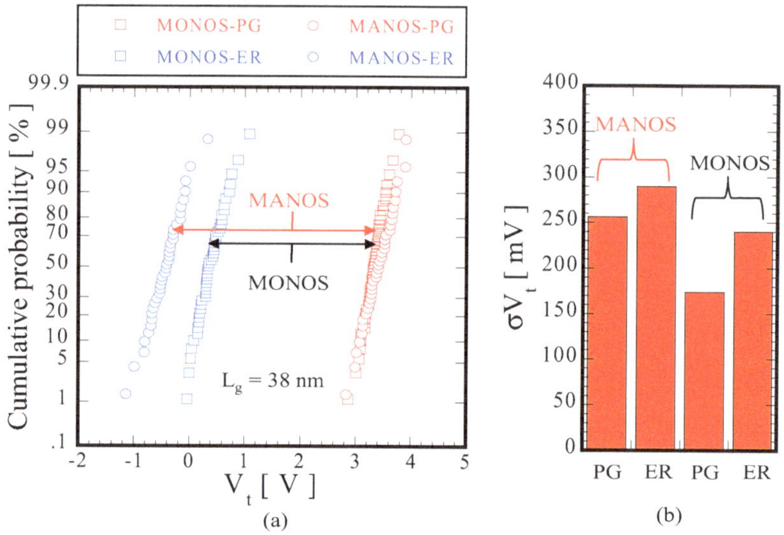

Figure 10. Average V_t ($<V_t>$) values as a function of L_g for the fabricated MANOS and MONOS type SOI-FinFET flash memory cell transistors after one P/E cycle. P/E conditions: $V_g = 16$ V, t = 50 μs for program, and $V_g = -16$ V, t = 20 ms for erase.

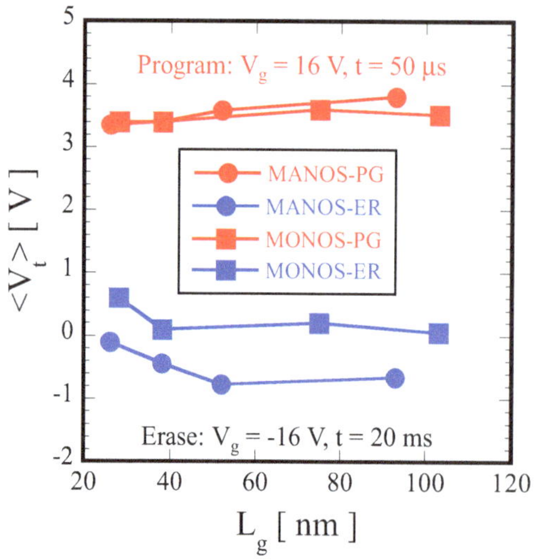

Figure 11a shows the measured P/E characteristics of the fabricated MANOS type device with L_g of 93 nm at different $|V_g|$ values, and the program-time dependence of the I_d-V_g characteristics at V_g of 14 V is shown in Figure 11b. It is clear that a large memory window of 5.6 V is obtained keeping an excellent S-slope of 100-mV/decade due to the GAA like structure. Figure 12a,b show the measured endurance characteristics of the fabricated MANOS and MONOS type devices, respectively. Note that these two kinds of device can operate over 100 k cycles although somewhat remarkable degradation is observed in the MONOS type device after 10 k cycles probably due to the insufficient uniformity of CVD-SiO_2 blocking layer. Moreover, the better data retention is obtained in the MANOS type device than MONOS type one due to the uniform ALD-Al_2O_3 blocking layer as shown in Figure 13.

Figure 11. (**a**) P/E characteristics and (**b**) program-time dependence of the I_d-V_g characteristics of the fabricated MANOS type SOI-FinFET flash memory with L_g = 93 nm.

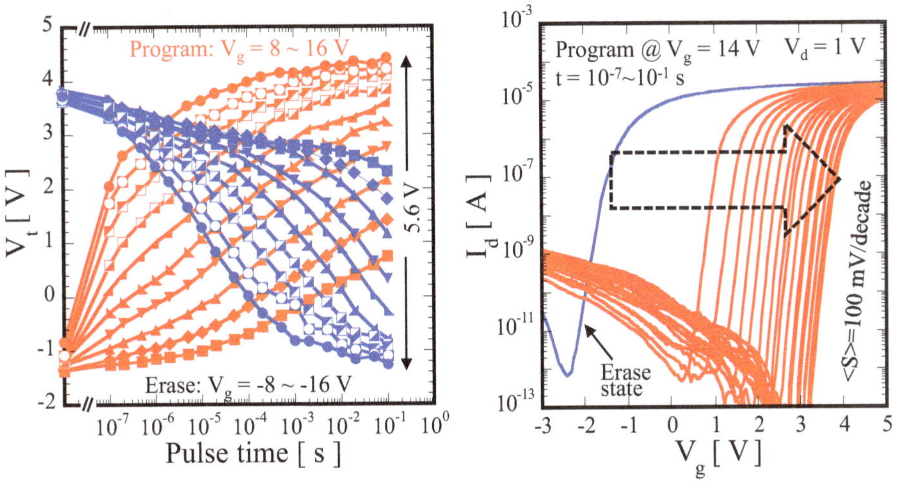

(a) (b)

Figure 12. Endurance characteristics of the fabricated (**a**) MANOS type SOI-FinFET flash memory with L_g = 93 nm; and (**b**) MONOS type SOI-FinFET flash memory with L_g = 103 nm.

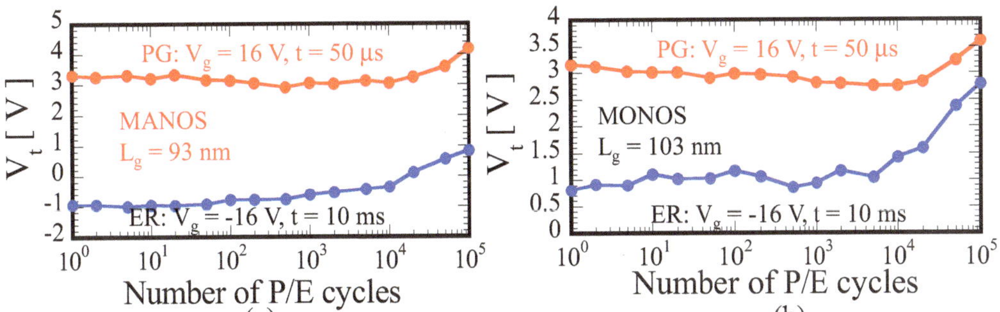

(a) (b)

Figure 13. Retention characteristics of the fabricated MANOS type SOI-FinFET flash memory with L_g = 93 nm and MONOS type SOI-FinFET flash memory with L_g = 103 nm at room temperature.

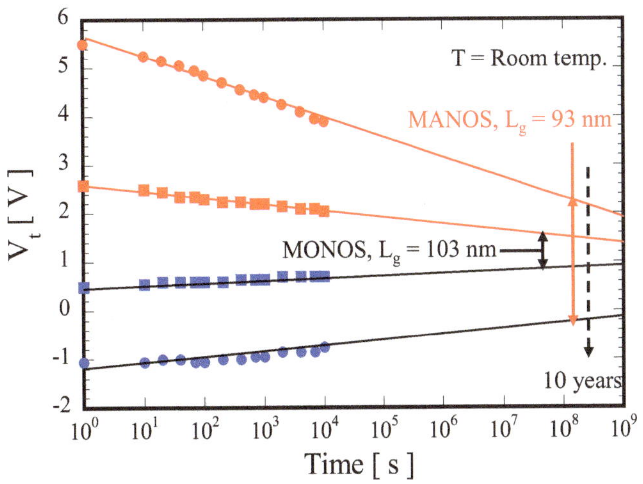

4. Conclusions

We have comparatively investigated the electrical characteristics and memory properties of the fabricated charge trapping type SOI-FinFET flash memories with different blocking layer materials of Al_2O_3 and SiO_2. It was experimentally found that introducing a high-k Al_2O_3 blocking layer instead of a SiO_2 one is very efficient for the enlarging of memory window owing to the high-k effect of Al_2O_3. It was also confirmed that V_t variability before and after one P/E cycle is almost independent of blocking layer materials.

Acknowledgments

This work was supported in part by the Nanotechnology Project of NEDO, Japan. The authors would like to thank Num Nguyen and Takeshi Inoshita of National Institute for Materials Science (NIMS) for fruitful discussions.

Conflicts of Interest

The authors declare no conflict of interest.

References

1. Xuan, P.; She, M.; Harteneck, B.; Liddle, A.; Bokor, J.; King, T.-J. FinFET SONOS Flash Memory for Embedded Applications. In Proceedings of the IEEE International Electron Devices Meeting, Washington, DC, USA, 8–10 December 2003; pp. 609–612.
2. Kim, T.-Y.; Cho, E.S.; Cho, H.J.; Sung, S.-K.; Lee, C.-H.; Choi, B.Y.; Cho, B.K.; Park, D.G. 256 Mb FinFET NOR Flash Memory for sub-70 nm Technology Breakthrough. In Proceedings of the International Silicon Nanoelectronics Workshop, Kyoto, Japan, 12–13 June 2005; pp. 98–99.
3. Sim, S.-P.; Kim, K.S.; Lee, H.K.; Han, J.I.; Kwon, W.H.; Han, J.H.; Lee, B.Y.; Jung, C.; Park, J.H.; Kim, D.J.; *et al.* Fully 3-Dimensional NOR Flash Cell with Recessed Channel and Cylindrical Floating Gate-A Scaling Direction for 65nm and Beyond. In Proceedings of the Symposium on VLSI Technology, Honolulu, HI, USA, 13–15 June 2006; pp. 22–23.
4. Hisamoto, D.; Lee, W.-C.; Kedzierski, J.; Takeuchi, H.; Asano, K.; Kuo, C.; Anderson, E.; King, T.-J.; Boker, J.; Hu, C. FinFET—A self-aligned double-gate MOSFET scalable to 20 nm. *IEEE Trans. Electron Devices* **2000**, *47*, 2320–2325.
5. Choi, Y.-K.; Lindert, N.; Xuan, P.; Tang, S.; Ha, D.; Anderson, E.; King, T.-J.; Bokor, J.; Hu, C. Sub-20 nm CMOS FinFET Technologies. In Proceedings of the IEEE International Electron Devices Meeting, Washington, DC, USA, 2–5 December 2001; pp. 421–424.
6. Yu, B.; Chang, L.; Ahmed, S.; Wang, H.; Bell, S.; Yang, C.-Y.; Tabery, C.; Ho, C.; Xiang, Q.; King, T.-J.; *et al.* FinFET Scaling to 10 nm Gate Length. In Proceedings of the IEEE International Electron Devices Meeting, San Fransico, CA, USA, 8–11 December 2002; pp. 251–254.
7. Liu, Y.X.; Ishii, K.; Tsutsumi, T.; Masahara, M.; Takashima, H.; Suzuki, E. Ideal Rectangular Cross-Section Si-Fin Channel Double-Gate MOSFETs Fabricated Using Orientation-Dependent Wet Etching. *IEEE Electron Device Lett.* **2003**, *24*, 484–486.

8. Liu, Y.X.; Masahara, M.; Ishii, K.; Tsutsumi, T.; Sekigawa, T.; Takashima, H.; Yamauchi, H.; Suzuki, E. Flexible Threshold Voltage FinFETs with Independent Double Gates and an Ideal Rectangular Cross-Section Si-Fin Channel. In Proceedings of the IEEE International Electron Devices Meeting, Washington, DC, USA, 8–10 December 2003; pp. 986–988.
9. Masahara, M.; Liu, Y.X.; Sakamoto, K.; Endo, K.; Matsukawa, T.; Ishii, K.; Sekigawa, T.; Yamauchi, H.; Tanoue, H.; Kanemaru, S.; *et al.* Demonstration, Analysis, and Dvice Design Cosiderations for Independent DG MOSFETs. *IEEE Trans. Electron Devices* **2005**, *52*, 2046–2053.
10. Liu, Y.X.; Matsukawa, T.; Endo, K.; Masahara, M.; Ishhi, K.; O'uchi, S.; Yamauchi, H.; Tsukada, J.; Ishikawa, Y.; Suzuki, E. Advanced FinFET CMOS Technology: TiN-Gate, Fin-Height Control and Asymmetric Gate Insulator Thickness 4T-FinFETs. In Proceedings of the IEEE International Electron Devices Meeting, San Fransico, CA, USA, 11–13 December 2006; pp. 989–992.
11. Doyle, B.; Boyanov, B.; Datta, S.; Doczy, M.; Hareland, S.; Jin, B.; Kavalieros, J.; Linton, T.; Rios, R.; Chau, R. Tri-Gate Fully-Depleted CMOS Transistors: Fabrication, Design and Layout. In Proceedings of the Symposium on VLSI Technology, Kyoto, Japan, 12–14 June 2003; pp. 133–134.
12. Yang, F.-L.; Chen, H.-Y.; Chen, F.-C.; Chan, Y.-L.; Yang, K.-N.; Chen, C.-J.; Tao, H.-J.; Choi, Y.-K.; Liang, M.-S.; Hu, C. 35 nm CMOS FinFET. In Proceedings of the Symposium on VLSI Technology, Honolulu, HI, USA, 13–15 June 2002; pp. 104–105.
13. Endo, K.; O'uchi, S.; Ishikawa, Y.; Liu, Y.X.; Matsukawa, T.; Sakamoto, K.; Tsukada, J.; Ishii, K.; Yamauchi, H.; Suzuki, E.; *et al.* Enhancing SRAM Cell Performance by Using Independent Double-Gate FinFET. In Proceedings of the IEEE International Electron Devices Meeting, San Fransico, CA, USA, 15–17 December 2008; pp. 857–860.
14. O'uchi, S.; Matsukawa, T.; Endo, K.; Liu, Y.X.; Sekigawa, T.; Tsukada, J.; Ishikawa, Y.; Yamauchi, H.; Ishii, K.; Suzuki, E.; *et al.* Characterization of Metal-Gate FinFET Variability Based on Measurements and Compact Model Analyses. In Proceedings of the IEEE International Electron Devices Meeting, San Fransico, CA, USA, 15–17 December 2008; pp. 709–712.
15. Matsukawa, T.; O'uchi, S.; Endo, K.; Ishikawa, Y.; Yamauchi, H.; Liu, Y.X.; Tsukada, J.; Sakamoto, K.; Masahara, M. Comprehensive Analysis of Variability Sources of FinFET Characteristics. In Proceedings of the Symposium on VLSI Technology, Kyoto, Japan, 15–17 June 2009; pp. 118–119.
16. Endo, K.; O'uchi, S.; Ishikawa, Y.; Liu, Y.X.; Matsukawa, T.; Sakamoto, K.; Tsukada, J.; Yamauchi, H.; Masahara, M. Variability analysis of TiN metal-gate FinFETs. *IEEE Electron Device Lett.* **2010**, *6*, 546–548.
17. Liu, Y.X.; Endo, K.; O'uchi, S.; Kamei, T.; Tsukada, J.; Yamauchi, H.; Ishikawa, Y.; Hayashida, T.; Sakamoto, K.; Matsukawa, T.; *et al.* On the Gate-Stack Origin Threshold Voltage Variability in Scaled FinFETs and Multi-FinFETs. In Proceedings of the Symposium on VLSI Technology, Honolulu, HI, USA, 13–15 June 2010; pp. 101–102.
18. Matsukawa, T.; Liu, Y.X.; O'uchi, S.; Endo, K.; Tsukada, J.; Yamauchi, H.; Ishikawa, Y.; Ota, H.; Migita, S.; Morita, Y.; *et al.* Comprehensive Analysis of I_{on} Variation in Metal Gate FinFETs for 20 nm and Beyond. In Proceedings of the IEEE International Electron Devices Meeting, Washington, DC, USA, 5–7 December 2011; pp. 517–520.

19. Endo, K.; O'uchi, S.; Ishikawa, Y.; Liu, Y.X.; Matsukawa, T.; Sakamoto, K.; Tsukada, J.; Yamauchi, H.; Masahara, M. Correlative Analysis between Characteristics of 30-nm L_G FinFETs and SRAM Performance. In Proceedings of the IEEE International Electron Devices Meeting, Washington, DC, USA, 8–10 December 2011; pp. 761–764.
20. Liu, Y.X.; Kamei, T.; Matsukawa, T.; Endo, K.; O'uchi, S.; Tsukada, J.; Yamauchi, H.; Ishikawa, Y.; Hayashida, T.; Sakamoto, K.; *et al.* Varaiability Analysis of Scaled Crystal Channel and Poly-Si Channel FinFETs. *IEEE Trans. Electron Devices* **2012**, *59*, 573–581.
21. Matsukawa, T.; Liu, Y.X.; Mizubayashi, W.; Tsukada, J.; Yamauchi, H.; Endo, K.; Ishikawa, Y.; O'uchi, S.; Ota, H.; Migita, S.; *et al.* Suppressing V_t and G_m Variability of FinFETs Using Amorphous Metal Gates for 14 nm and Beyond. In Proceedings of the IEEE International Electron Devices Meeting, San Fransico, CA, USA, 10–12 December 2012; pp. 175–178.
22. Endo, K.; O'uchi, S.; Matsukawa, T.; Liu, Y.X.; Sakamoto, K.; Mizubayashi, W.; Migita, S.; Morita, Y.; Ota, H.; Suzuki, E.; *et al.* Enhancing SRAM Performance by Advanced FinFET Device and Circuit Technology Collaboration for 14 nm Node and Beyond. In Proceedings of the Symposium on VLSI Technology, Kyoto, Japan, 11–13 June 2013; pp. 214–215.
23. Specht, M.; Dorda, U.; Dreeskornfeld, L.; Kretz, J.; Hofmann, F.; Stadele, M.; Luyken, R.J.; Rosner, W.; Reisinger, H.; Landgraf, E.; *et al.* 20 nm Tri-Gate SONOS Memory Cells with Multi-Level Operation. In Proceedings of the IEEE International Electron Devices Meeting, San Fransico, CA, USA, 13–15 December 2004; pp. 1083–1085.
24. Cho, E.-S.; Lee, C.-H.; Kim, T.-Y.; Sung, S.-K.; Cho, B.K.; Lee, C.; Cho, H.J.; Roh, Y.; Park, D.; Kim, K.; *et al.* Hf-Silicate Inter-Poly Dielectric Technology for Sub 70 nm Body Tied FinFET Flash Memory. In Proceedings of the Symposium on VLSI Technology, Kyoto, Japan, 14–16 June 2005; pp. 208–209.
25. Kim, S.; Kim, W.; Hyun, J.; Byun, S.; Koo, J.; Lee, J.; Cho, K.; Lim, S.; Park, J.; Yoo, I.K.; *et al.* Paired FinFET Charge Trap Flash Memory for Vertical High Density Storage. In Proceedings of the Symposium on VLSI Technology, Honolulu, HI, USA, 13–15 June 2006; pp. 104–105.
26. Cho, E.-S.; Kim, T.Y.; Cho, B.-K.; Lee, C.-H.; Lee, J.-J.; Fayrushin, A.; Lee, C.; Park, D.; Ryu, B.-I. Technology Breakthrough of Body-Tied FinFET for Sub 50 nm NOR Flash Memory. In Proceedings of Symposium on VLSI Technology, Honolulu, HI, USA, 13–15 June 2006; pp. 110–111.
27. Lee, S.H.; Lee, J.J.; Choe, J.D.; Cho, E.S.; Ahn, Y.J.; Hwang, W.; Kim, T.; Kim, W.J.; Yoon, Y.B.; Jang, D.; *et al.* Improved Post-Cycling Characteristics of FinFET NAND Flash. In Proceedings of the IEEE International Electron Devices Meeting, San Fransico, CA, USA, 11–13 December 2006; pp. 33–36.
28. Sung, S.-K.; Lee, S.-H.; Choi, B.Y.; Lee, J.J.; Choe, J.-D.; Cho, E.S.; Ahn, Y.J.; Choi, D.; Lee, C.-H.; Kim, D.H.; *et al.* SONOS-Type FinFET Device Using P^+ Poly-Si Gate and High-k Blocking Dielectric Integrated on Cell Array and GSL/SSL for Multi-Gigabit NAND Flash Memory. In Proceedings of the Symposium on VLSI Technology, Honolulu, HI, USA, 13–15 June 2006; pp. 106–107.

29. Ahn, Y.-J.; Choe, J.-D.; Lee, J.J.; Choi, D.; Cho, E.S.; Choi, B.Y.; Lee, S.-H.; Sung, S.-K.; Lee, C.-H.; Cheong, S.H.; *et al.* Trap Layer Engineered FinFET NAND Flash with Enhanced Memory Window. In Proceedings of the Symposium on VLSI Technology, Honolulu, HI, USA, 13–15 June 2006; pp. 108–109.
30. Friederich, C.; Specht, M.; Lutz, T.; Hofmann, F.; Dreeskornfeld, L.; Weber, W.; Kretz, J.; Melde, T.; Rosner, W.; Landgraf, E.; *et al.* Multi-Level p^+ Tri-Gate SONOS NAND String Arrays. In Proceedings of the IEEE International Electron Devices Meeting, San Fransico, CA, USA, 11–13 December 2006; pp. 963–966.
31. Razafindramora, J.; Perniola, L.; Jahan, C.; Scheiblin, P.; Gely, M.; Vizioz, C.; Carabasse, C.; Boulanger, F.; Salvo, B.D.; Lombardo, S.; *et al.* Low Voltage Hot-Carrier Programming of Ultra-Scaled SOI Fin Flash Memories. In Proceedings of the European Solid-State Device Research Conference, Munich, Germany, 11–13 September 2007; pp. 414–417.
32. Endo, T.; Kinoshita, K.; Tanigami, T.; Wada, Y.; Sato, K.; Kazuya, K.; Yamada, K.; Yokoyama, T.; Takeuchi, N.; Tanaka, K.; *et al.* Novel Ultra High Density Flash Memory with A Stacked-Surrounding Gate Transistor (S-SGT) Structured Cell. In Proceedings of the IEEE International Electron Devices Meeting, Washington, DC, USA, 2–5 December 2001; pp. 33–35.
33. Goveoreanu, B.; Wellekens, D.; Haspeslagh, L.; Vos, J.D.; Houdt, J.V. Investigation of the Low-Field Leakage Through High-k Interpoly Dielectric Stacks and Its Impact on Nonvolatile Memory Data Retention. In Proceedings of the IEEE International Electron Devices Meeting, San Fransico, CA, USA, 11–13 December 2006; pp. 479–482.
34. Lee, C.H.; Choi, K.I.; Cho, M.K.; Song, Y.H.; Park, K.C.; Kim, K. A Novel SONOS Structure of SiO_2/SiN/Al_2O_3 with TaN Metal Gate for Multi-Giga Bit Flash Memories. In Proceedings of the IEEE International Electron Devices Meeting, Washington, DC, USA, 8–10 December 2003; pp. 613–616.
35. Kamei, T.; Liu, Y.X.; Matsukawa, T.; Endo, K.; O'uchi, S.; Tsukada, J.; Yamauchi, H.; Ishikawa, Y.; Hayashida, T.; Sakamoto, K.; *et al.* Comparative Study of Tri-Gate Flash Memories with Split and Stack Gates. In Proceedings of the IEEE International SOI Conference, Tempe, AZ, USA, 3–6 October 2011; pp. 294–295.
36. Kamei, T.; Liu, Y.X.; Matsukawa, T.; Endo, K.; O'uchi, S.; Tsukada, J.; Yamauchi, H.; Ishikawa, Y.; Hayashida, T.; Sakamoto, K.; *et al.* Demonstration of Split-Gate Type Trigate Flash Memory with Highly Suppressed Over-Erase. *IEEE Electron Device Lett.* **2012**, *33*, 345–347.
37. Kamei, T.; Liu, Y.X.; Matsukawa, T.; Endo, K.; O'uchi, S.; Tsukada, J.; Yamauchi, H.; Ishikawa, Y.; Hayashida, T.; Sakamoto, K.; *et al.* Fabrication and Characterization of NOR-Type Tri-Gate Flash Memory with Improved Inter-Poly Dielectric Layer by Rapid Thermal Oxidation. *Jpn. J. Appl. Phys.* **2012**, *51*, doi:10.1143/JJAP.51.06FE19.
38. Liu, Y.X.; Guo, R.F.; Kamei, T.; Matsukawa, T.; Endo, K.; O'uchi, S.; Tsukada, J.; Yamauchi, H.; Ishikawa, Y.; Hayashida, T.; *et al.* Experimental Study of Floating-Gate-Type Metal-Oxide-Semiconductor Capacitors with Nanosize Triangular Cross-Sectional Tunnel Areas for Low Operating Voltage Flash Memory Application. *Jpn. J. Appl. Phys.* **2012**, *51*, doi:10.1143/JJAP.51.06FF01.

39. Liu, Y.X.; Kamei, T.; Matsukawa, T.; Endo, K.; O'uchi, S.; Tsukada, J.; Yamauchi, H.; Ishikawa, Y.; Hayashida, T.; Sakamoto, K.; *et al*. Direct Comparison of Electrical Characteristics for Double-Gate and Tri-Gate Flash Memories. In Proceedings of the Extended Abstracts of International Conference on Solid State Devices and Materials, Nagoya, Japan, 28–30 September 2011; pp. 985–986.
40. Liu, Y.X.; Kamei, T.; Matsukawa, T.; Endo, K.; O'uchi, S.; Tsukada, J.; Yamauchi, H.; Ishikawa, Y.; Hayashida, T.; Sakamoto, K.; *et al*. Fabrication of Floating-Gate-Type Fin-Channel Double- and Tri-Gate Flash Memories and Comparative Study of Their Electrical Characteristics. *Jpn. J. Appl. Phys.* **2012**, *51*, doi:10.1143/JJAP.51.04DD03.
41. Liu, Y.X.; Kamei, T.; Matsukawa, T.; Endo, K.; O'uchi, S.; Tsukada, J.; Yamauchi, H.; Ishikawa, Y.; Hayashida, T.; Sakamoto, K.; *et al*. Gate Structure Dependence of Variability in Polycrystalline Silicon Fin-Channel Flash Memories. *Jpn. J. Appl. Phys.* **2013**, *52*, doi:10.7567/JJAP.52.06GE01.
42. Liu, Y.X.; Matsukawa, T.; Endo, K.; O'uchi, S.; Tsukada, J.; Yamauchi, H.; Ishikawa, Y.; Mizubayashi, W.; Morita, Y.; Migita, S.; *et al*. Comparative Study of TiN Metal Gate and Poly-Si Gate Charge-Trapping Type FinFET Flash Memories. In Proceedings of the International Silicon Nanoelectronics Workshop, Kyoto, Japan, 11–14 June 2013; pp. 9–10.
43. Liu, Y.X.; Matsukawa, T.; Endo, K.; O'uchi, S.; Tsukada, J.; Yamauchi, H.; Ishikawa, Y.; Mizubayashi, W.; Morita, Y.; Migita, S.; *et al*. Charge Trapping Type FinFET Flash Memory with Al_2O_3 Blocking Layer. In Proceedings of the IEEE SOI-3D-Subthreshold Microelectronics Technology Unified Conference, Monterey, CA, USA, 7–10 October 2013.
44. Liu, Y.X.; Ishii, K.; Masahara, M.; Tsustumi, T.; Takashima, H.; Yamauchi, H.; Suzuki, E. Cross-Sectional Channel Shape Dependence of Short-Channel Effects in Fin-Type Double-Gate Metal Oxide Semiconductor Field-Effect Transistors. *Jpn. J. Appl. Phys.* **2004**, *43*, 2151–2155.
45. Liu, Y.X.; Kijima, S.; Sugimata, E.; Masahara, M.; Endo, K.; Ishii, K.; Matsukawa, T.; Takashima, H.; Yamauchi, H.; Takanashi, Y.; *et al*. Investigation of the TiN Gate Electrode with Tunable Work Function and Its Application for FinFET Fabrication. *IEEE Trans. Nanotechnol.* **2006**, *5*, 723–730.
46. Liu, Y.X.; Kamei, T.; Endo, K.; O'uchi, S.; Tsukada, J.; Yamauchi, H.; Hyashida, T.; Ishikawa, Y.; Mtsukawa, T.; Sakamoto, K.; *et al*. Nanoscale Wet Etching of Physical-Vapor-Deposited Titanium Nitride and Its Application to Sub-30-nm-Gate-Length Fin-Type Double-Gate Metal-Oxide-Semiconductor Field-Effect Transistor Fabrication. *Jpn. J. Appl. Phys.* **2010**, *49*, doi:10.1143/JJAP.49.06GH18.
47. Liu, Y.X.; Matsukawa, T.; Endo, K.; O'uchi, S.; Sakamoto, K.; Tsukada, J.; Ishikawa, Y.; Yamauchi, H.; Masahara, M. Investigation of Low-Energy Tilted Ion Implantation for Fin-Type Double-Gate Metal-Oxide-Semiconductor Field-Effect Transistor Extension Doping. *Jpn. J. Appl. Phys.* **2010**, *49*, doi:10.1143/JJAP.49.04DC18.

11

SOTB Implementation of a Field Programmable Gate Array with Fine-Grained Vt Programmability [†]

Masakazu Hioki [1,*], Chao Ma [1], Takashi Kawanami [2], Yasuhiro Ogasahara [1], Tadashi Nakagawa [1], Toshihiro Sekigawa [1], Toshiyuki Tsutsumi [3] and Hanpei Koike [1]

[1] Electroinformatics Group, Nanoelectronics Research Institute, National Institute of Advanced Industrial Science and Technology (AIST), 1-1-1 Umezono, Tsukuba, Ibaraki 305-8568, Japan; E-Mails: c.ma@aist.go.jp (C.M.); ys.ogasahara@aist.go.jp (Y.O.); nakagawa.tadashi@aist.go.jp (T.N.); t.sekigawa@aist.go.jp (T.S.); h.koike@aist.go.jp (H.K.)

[2] Department of Information and Computer Science, College of Engineering, Kanazawa Institute of Technology, 7-1 Ohgigaoka, Nonoichi, Ishikawa 921-8501, Japan; E-Mail: t-kawanami@neptune.kanazawa-it.ac.jp

[3] Computer Science Course, Fundamental Science and Technology, Graduate School of Science and Technology, Meiji University, 1-1-1 Higashi-Mita, Tama-ku, Kawasaki-shi, Kanagawa 214-8571, Japan; E-Mail: tsutsumi@cs.meiji.ac.jp

† This is an extended version of paper that was presented at the IEEE S3S Conference 2013.

* Author to whom correspondence should be addressed; E-Mail: m.hioki@aist.go.jp

---

Abstract: Field programmable gate arrays (FPGAs) are one of the most widespread reconfigurable devices in which various functions can be implemented by storing circuit connection information and logic values into configuration memories. One of the most important issues in the modern FPGA is the reduction of its static leakage power consumption. Flex Power FPGA, which has been proposed to overcome this problem, uses a body biasing technique to implement the fine-grained threshold voltage (Vt) programmability in the FPGA. A low-Vt state can be assigned only to the component circuits along the critical path of the application design mapped on the FPGA, so that the static leakage power consumption can be reduced drastically. Flex Power FPGA is an important application target for the SOTB (silicon on thin buried oxide) device, which features a wide-range body biasing ability and the high sensitivity of Vt variation by body biasing, resulting in a drastic subthreshold leakage current reduction caused by static leakage power. In this paper, the Flex Power FPGA test chip is fabricated in SOTB technology, and the functional test and performance evaluation of a mapped 32-bit binary counter circuit are performed successfully. As a result, a three orders of magnitude static leakage reduction with a bias range of 2.1 V demonstrates the excellent Vt controllability of the SOTB transistors, and the 1.2 V bias difference achieves a 50× leakage reduction without degrading speed.

Keywords: field programmable gate array (FPGA); static leakage power reduction; fine-grained body biasing; silicon on thin buried oxide (SOTB)

---

1. Introduction

Field programmable gate arrays (FPGAs) are one of the most widespread reconfigurable devices in which various functions can be implemented by storing circuit connection information and logic values into configuration memories. By this reconfigurability, various applications, such as prototyping, automotive, consumer products, mobile devices, industrial equipment, military equipment, aerospace applications, video/picture processing and telecommunication, can be performed on FPGAs. Another attractive point of FPGA is that the logic implementation on FPGAs needs no mask production for front-end design in ASICs. The system embedding FPGAs, promising an overall lower non-recurring engineering (NRE) cost, has cost benefits in comparison with ASIC-based systems. Moreover, FPGAs can be fabricated by advanced semiconductor process technologies, resulting in the reduction of chip costs and the improvement of device performance by rapidly shrinking the silicon die area, because FPGAs ensure a high-volume production capability, due to the implementability of many types of applications. However, the silicon area of FPGAs becomes lager than that of ASICs generally because FPGAs integrate many silicon resources, e.g., programmable interconnects, logic elements and configuration memories, to maintain the feasibility of many types of applications. While the increase of the silicon area, which lengthens interconnection wires, reduces the operating speed of mapped circuits on FPGAs, longer interconnection wires consume more dynamic power. This is one of the reasons why FPGAs are applied to advanced semiconductor process technologies aggressively.

One of the most important issues in the modern FPGA is the reduction of its static leakage power consumption [1]. Static leakage is a serious problem in FPGAs, partly because of the rich and redundant circuit resources provided for flexible programmability and partly because of the lack of a multi-threshold voltage (Vt) optimization capability for the sake of its field-programmability. Previously, techniques to reduce the static leakage current in FPGAs were aggressively examined. It is shown in [2] that the look-up-table (LUT) leakage is reduced by setting a higher threshold voltage (Vt) to static random access memories (SRAMs). In [3], various low-leakage techniques, such as redundant SRAM design, dual Vt design, body biasing and gate biasing, are evaluated. As described in [4], cutting off the power supply of an unused region by sleep transistors aims at reducing the logic slice leakage, which occupies 45% of the total leakage in their FPGA, while the assignment of a higher Vt to the configuration SRAM reduces 98% of the SRAM leakage, while increasing the configuration time by 20%. A routing switch in [5] can reduce the leakage current by 40% and 61% in low-power mode and sleep mode, respectively. The interconnection with fine-grained power-gating technique reduces the total power by 38% in [6]. It is shown in [7] that the heterogeneous routing architecture reduces standby power by 33% without any area penalty and at the cost of less than a 5% performance degradation. A standby power reduction by 99% by using low-leakage memories and the power-gating technique is presented in [8]. On the other hand, our Flex Power FPGA technique [9] uses a body biasing technique to implement the fine-grained programmability of the Vt of the FPGA building block circuits, such as LUT and multiplexer (MUX). A low-Vt state is assigned only to the building blocks along the critical path of the user application design based on the analysis of the design, while the high-Vt state is assigned to the greater part of the FPGA. As a result, a drastic reduction of the static power consumption can be expected. The key factors to attain good static leakage power reduction performance are: (1) a higher sensitivity of the Vt in the transistor to the back gate bias; (2) a fine grain size of the Vt control domain; and (3) the nature of the critical path of the application design.

The SOTB (silicon on thin buried oxide) transistor [10] has an excellent Vt controllability. Compared with the bulk MOS transistor, a wider variation range of the threshold voltage is promised in principle, because it is formed using a thinner buried oxide under a lightly-doped and fully-depleted body region. The SOTB structure is ideal for Flex Power FPGA to attain better static power reduction performance than bulk technology.

After the development of generations of the conventional bulk transistor implementation of Flex Power FPGA test chips [11], the SOTB version of the Flex Power FPGA test chip is fabricated, and its functional test and performance evaluation are performed successfully by mapping the 32-bit binary counter circuit on the test chip. This paper is organized as follows. Section 2 gives overviews of the SOTB and Flex Power FPGA architecture. In Section 3, evaluation results of the operating speed, static leakage current and area-overhead in the fabricated chip are shown. Finally, in Section 4, this paper is concluded.

2. Overviews of the SOTB Transistor and Flex Power FPGA Architecture

2.1. Silicon on Thin Buried Oxide (SOTB) Transistor

SOTB transistors, as shown in Figure 1a, have some advantages compared with the bulk MOS transistors. SOTB transistors allow a wide back-bias control range for low-power and high-performance applications, because a thinner buried oxide under the lightly-doped and fully-depleted body region is formed as an SOI structure. As a result, by applying this to the SRAM cells, the static noise margins (SNMs) are improved by adding back feedback from the back gate to the front gate [10]. Moreover, SOTB has the smallest threshold voltage variability among bulk MOS, due to its low-dose channel [12].

Figure 1b shows the SPICE simulation results of on-currents and off-currents to the back-bias voltage for the range from 0 V to −1.0 V in a bulk NMOS transistor and an SOTB NMOS transistor in 65-nm technology. Each on-current value is normalized by the on-current value in the case that a back-bias voltage of 0 V is applied. The same manner holds for off-currents. The off-current of the SOTB transistor shifts more than an order of magnitude by applying −1 V to the back gate, while that of the bulk MOS transistor reduces by about 1/5 in the same condition as the SOTB transistor. The on-current variation of the SOTB transistor is also larger than that of the bulk MOS transistor. These results show that the SOTB transistor has a higher sensitivity for the threshold voltage by back gate biasing compared with the bulk MOS transistor, and further, static power reduction efficiency can be expected by applying SOTB transistors to ICs.

Figure 1. (**a**) The structure of the silicon on thin BOX (SOTB) transistor; and (**b**) SPICE simulation results of on- and off-currents in a bulk MOS transistor and an SOTB transistor.

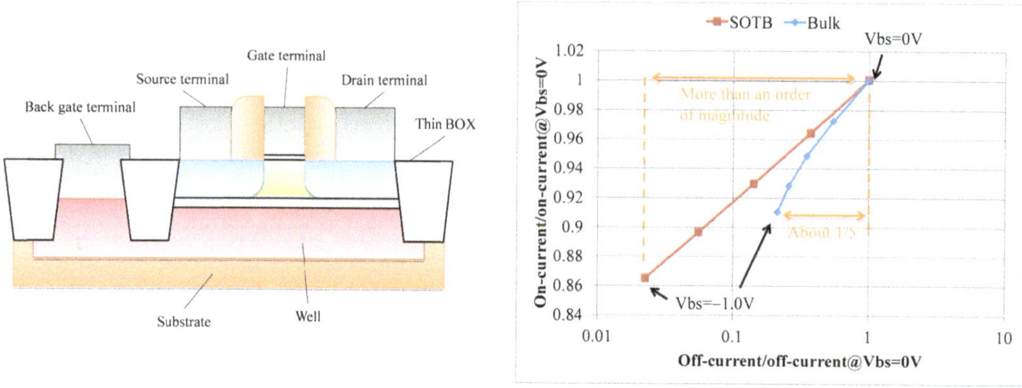

(**a**) (**b**)

2.2. Flex Power FPGA Architecture

Figure 2 shows the die photo of the SOTB version test chip and the FPGA tile. The test chip described here is a direct conversion from the latest test chip fabricated in 90-nm bulk CMOS technology [11] to the 65-nm SOTB technology. The test chip has a typical island-style FPGA structure with 11 × 11 FPGA tiles. An FPGA tile is connected to other tiles on the left, right, top and bottom of itself via wire segments, which span only four FPGA tiles.

Figure 2. Die photo of (**a**) the test chip and (**b**) the FPGA tile.

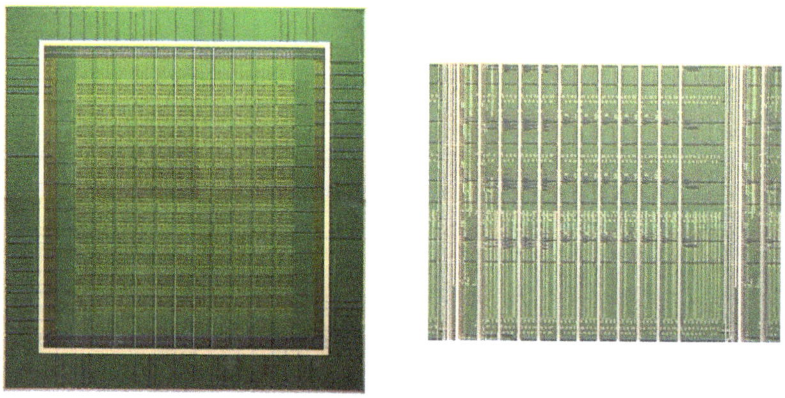

(**a**) (**b**)

Figure 3a shows the FPGA tile architecture. An FPGA tile has two circuit blocks, a switch block and a logic block. Switch matrixes, input multiplexers (IMUX) and local multiplexers (LMUX), for interconnection of a logic block and a switch block, are contained in the switch block. The switch matrix in the switch block is illustrated in Figure 3b. The switch matrix can interconnect wire segments in the four directions. A signal from one direction can be propagated selectively to the other three directions by switch multiplexers (SMUX), depicted as four trapezoid symbols in Figure 3b.

Figure 3. (**a**) FPGA tile architecture; (**b**) Switch matrix in the switch block.

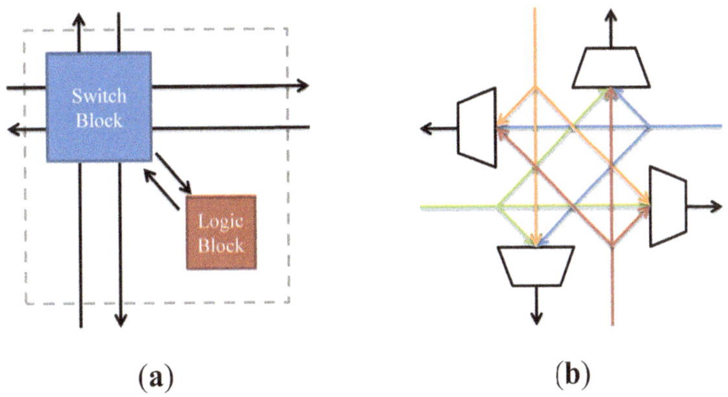

Figure 4a shows the block diagram of a logic block, which is often called a configurable logic block or CLB. A logic block includes IMUXs, LMUXs and four basic logic elements (BLE). IMUXs select 12 signals from signals on the 32 wire segments, and LMUXs input four signals to each basic logic element after selecting from 12 signals as outputs of the IMUXs. The outputs of basic logic elements connect to the switch matrix directly, while these can be fed back via local multiplexers to the inputs of the basic logic elements. Figure 4b depicts a basic logic element. The basic logic element comprises a look-up-table (LUT), a d-flip-flop (DFF) and a 2:1 multiplexer (2:1 MUX). The look-up-table has four inputs and an output. The basic logic element outputs an LUT output signal asynchronously or synchronously via d-flip-flop, which is selected by a 2:1 multiplexer.

Figure 4. (**a**) Logic block architecture; (**b**) Basic logic element. MUX, multiplexer; DFF, d-flip-flop.

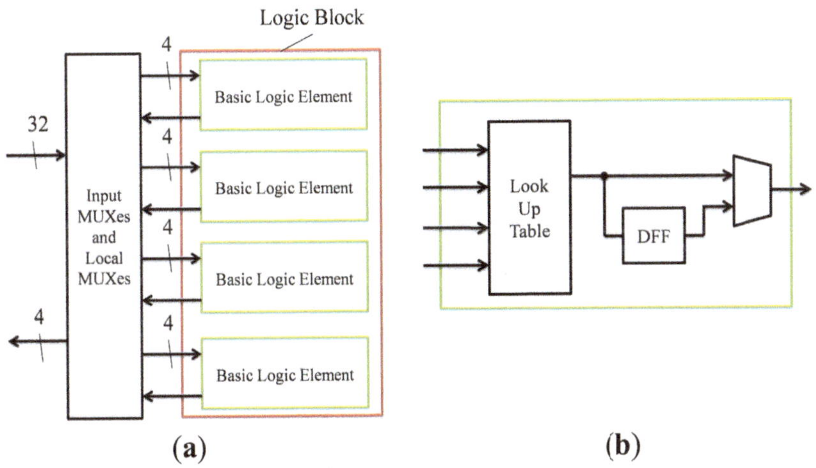

All of the routing through the MUXs and all of the logic functions of the LUTs are defined by the contents of the configuration memories in the FPGA. Figure 5 shows the configuration memory system of the Flex Power FPGA. A word line shift register for the word line strobe and a data line shift register for loading the row by row configuration data are placed on the left side and upper side of the configuration memory array, respectively. A user application design is converted to the configuration information in the bit stream format and is downloaded to the FPGA. The number of word lines is 88, and the number of data lines is 660. Each tile uses 358 configuration bits to represent logic functions, routing and Vt mapping, so that the total number of configuration bits in our test chip is 43,318.

Figure 5. Configuration memory system of the 11 × 11 Flex Power FPGA tile.

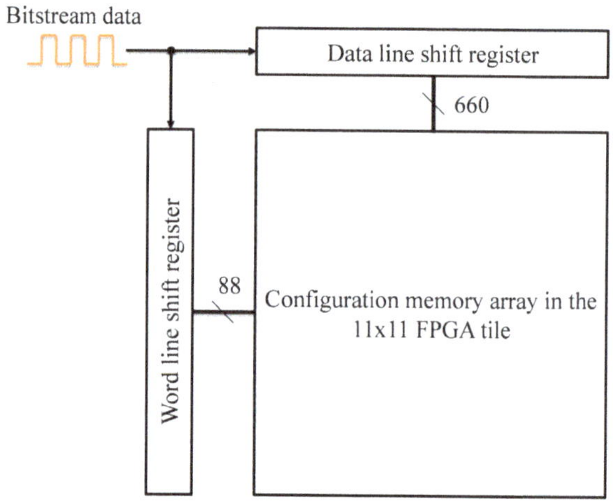

For each FPGA building block, such as the LUT and MUX, the associated Vt configuration memory bit and body bias control circuit, shown in Figure 6, are provided. Four different bias voltages (VBNL, VBPL, VBNH and VBPH for low-threshold voltage (LVT)/high-threshold voltage (HVT) to NMOS/PMOS) are supplied from the external voltage sources, and the content of the Vt configuration memory bit controls the Vt state (*i.e.*, LVT or HVT) of the corresponding building block circuit by selecting the appropriate bias voltages. As each substrate of the FPGA building block is separated by the triple-well structure, which plays the role of the back gate, individual control of the Vt in FPGA building blocks can be done by the programmable body bias control circuit.

Figure 6. Programmable body bias control circuit. VBPH, voltage bias for high-threshold voltage to PMOS; VBPL, voltage bias for low-threshold voltage to PMOS; VBNL, voltage bias for low-threshold voltage to NMOS; VBNH, voltage bias for high-threshold voltage to NMOS.

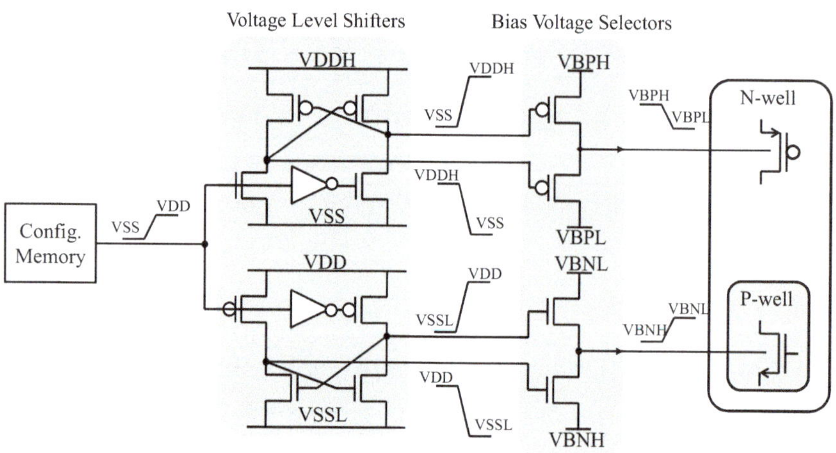

Figure 7 shows the SPICE simulation result of the programmable body bias control circuit. The blue (upper), green (middle) and red (lower) waveforms on the figure show the input signal from a configuration memory along with the resulting n-well bias voltage for PMOS and p-well bias voltage for NMOS. The green line, blue line and red line indicate the signal from a configuration memory, n-well bias voltage for PMOS and p-well bias voltage for NMOS, respectively. When the signal from a configuration memory is 1.2 V, 1.2 V and 0 V are applied to the n-well of PMOS and the p-well of NMOS, respectively, resulting in lower Vt conditions for PMOS and NMOS. In the case that the signal from a configuration memory is 0 V, the body bias control circuit outputs 2.15 V for the n-well and −1.05 V for the p-well. A higher Vt is assigned to PMOS and NMOS.

Figure 7. SPICE simulation results of the programmable body bias control circuit.

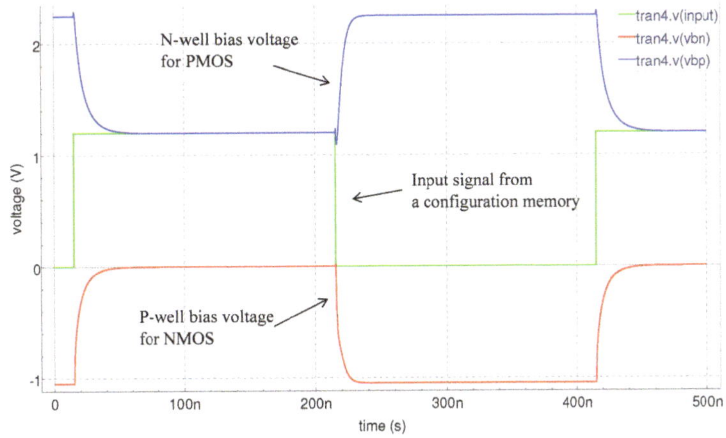

The grain size of the Vt control domains is set to extremely fine, *i.e.*, multiplexer level, in order to fully demonstrate the maximum power reduction performance of Flex Power FPGA. The number of the Vt control domains in a tile is 57, and the total number in the test chip reaches 6897 bits of the Vt configuration memory. Programmable body bias control circuits, as well as well separation areas are the overheads for the fine-grained Vt control, which will be optimized in the later versions. The FPGA architecture in the test chip is summarized in Table 1.

Table 1. Summary of the FPGA architecture. BLE, basic logic elements; Vt, threshold voltage.

Parameter	Value
# of tiles	121 (11 × 11)
Wire length	4
Switch matrix topology	Disjoint
Routing architecture	Unidirectional and single driver programmable interconnect
Routing channel width	16
# of BLEs per a CLB (configurable logic block)	4
# of CLB inputs	13
# of CLB outputs	4
# of BLE inputs	7
# of Vt control domains per a tile	57

As shown in Figure 8, the dedicated CAD toolchain for Flex Power FPGA [9,11] includes newly introduced software, called Vt mapper (VTM), in addition to the conventional FPGA tools, such as placer and router. GGEN generates a routing resource graph that reflects the FPGA architecture. Then, logic synthesis and packing need to be performed beforehand for a target circuit. CNV translates and generates information for placement and routing phases. Next, placement and routing are subsequently performed by VP and VR, respectively. After that, VTM analyzes all signal paths and obtains the timing slack information in the user application design mapped on the Flex Power FPGA. Following this, timing slack is reduced on non-critical paths by slowing down the blocks in those paths. This is achieved by assigning the HVT state to as many Vt control domains as possible, while assigning the LVT state to the Vt control domains along the critical signal paths to keep the operation speed constant.

Figure 8. The dedicated CAD toolchain.

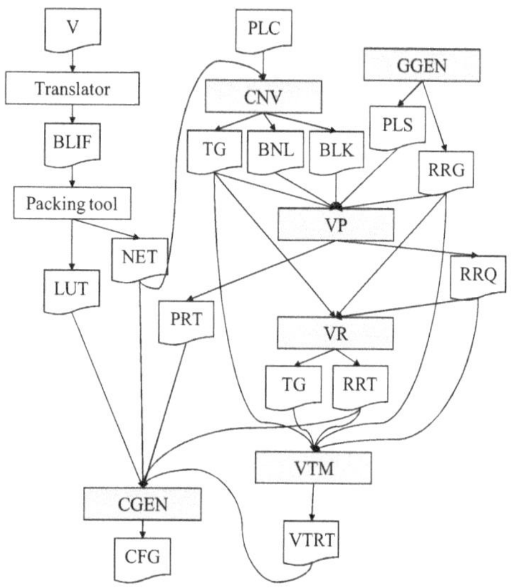

Software	
GGEN	RRG and PLS generator
CNV	NETto BNL transformer TG and BLK generator
VP	Placer
VR	Router
VTM	Vth mapper
CGEN	Config. bit stream generator

File	
V	Verilog RTL netlist
BLIF	Gate level netlist
NET	CLB netlist
LUT	LUT truth table
PLC	Pad location
TG	Timing graph
BNL	VP inputtable netlist
BLK	Fix block location
PLS	Function block location
RRG	Routing resource graph
RRQ	Routing resource request
PRT	Placement result
RRT	Routing result
VTRT	Vt map result
CFG	Config. bit stream

Figure 9 shows an example of the VTM behavior [9]. As shown in Figure 9a, the Vt of all nodes is set to LVT at the initial state. Then, the first step of the back tracing algorism in VTM is done, as shown in Figure 9b. The Vt of node on the critical path remains LVT, while unused DFF is set to HVT. Next, HVT is assigned to nodes having a large slack in Figure 9c. Finally, the VTM algorism is completed, as shown in Figure 9d.

Figure 9. Example of the Vt mapper (VTM) behavior [9] (**a**) initial state; (**b**) first step of the back trace; (**c**) HVT assignment; (**d**) VTM is completed.

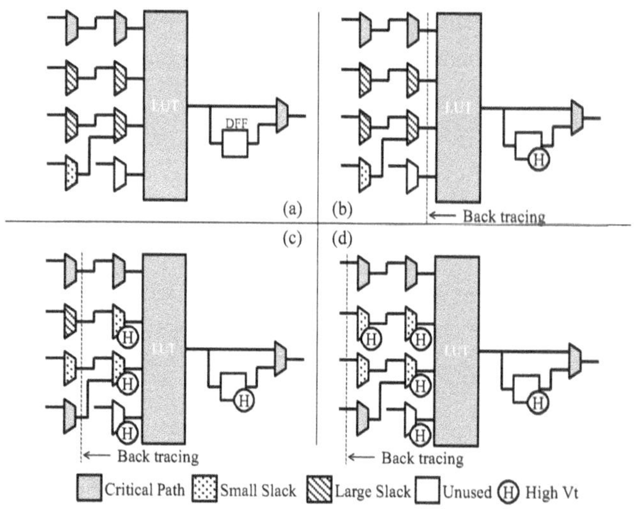

3. Evaluation Results

3.1. Operating Speed and Static Leakage Current

A 32-bit binary counter circuit is used as a sample application mapped onto the FPGA. The maximum operating frequency of the counter circuit and the static leakage current of the FPGA are measured in various bias voltage conditions. Bias voltages are chosen as follows: (1) the NMOS bias voltage for LVT (VBNL) is chosen from {−1.3, −1.2, −0.8, −0.4, 0.0, 0.4, 0.8}; (2) the NMOS bias voltage for HVT (VBNH) is chosen, so that VBNH ≤ VBNL; and (3) the PMOS bias voltages are appropriately determined from the corresponding NMOS bias voltages by subtracting from VDD. As a result, a total of 28 separate conditions are examined. The Vt mapper is re-executed condition by condition, so that the correct Vt mapping is produced based on the speed ratio of LVT and HVT.

Figure 10a shows the measured maximum operating frequencies of the 32-bit binary counter circuit on various back-gate bias conditions. The core voltage of FPGA is set to 1.2 V. The horizontal axis shows the NMOS back-gate bias voltage for HVT (VBNH), and each line corresponds to the NMOS back-gate bias voltage for LVT (VBNL). The result shows that the operating speed can be widely controlled by the LVT bias voltage, which changes in the range from 0.8 V to −1.3 V when the HVT bias voltage is set to −1.3 V. Furthermore, the operating speed keeps a constant value against the change of the HVT bias voltages in the conditions of any LVT bias voltage.

Figure 10b shows the measured static leakage current on the various back-bias voltages. The supply voltage VDD is set to 1.2 V. Diagonal conditions or a contour line (dashed line) show the leakage characteristics of a single-Vt FPGA, where this means the condition in which the LVT bias is equal to the HVT bias. The single-Vt FPGA characteristic shows a three orders of magnitude static leakage reduction with a bias range of 2.1 V. This is a demonstration of the excellent Vt controllability of the SOTB transistor.

Figure 10. (**a**) Maximum operating frequency and (**b**) the static leakage measurement results in the various bias voltage conditions (VDD = 1.2 V).

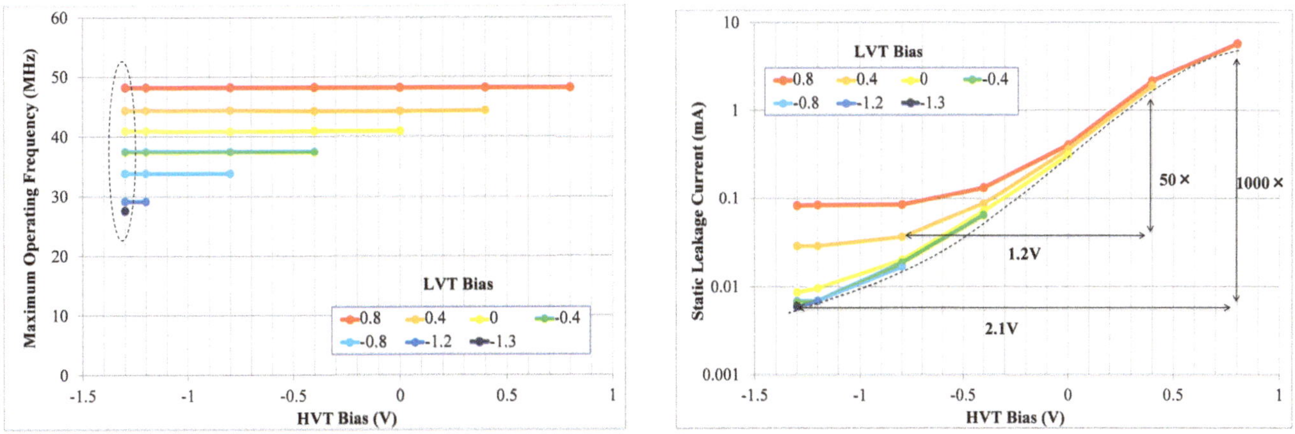

(a) (b)

With lowering of the HVT bias, each line for a specific LVT bias gradually departs from the contour line and is saturated. This is because the leakage current of HVT domains becomes negligibly small by lowering the HVT bias sufficiently, while the leakage current of LVT domains becomes dominant. When the HVT bias is relatively high, the leakage current of HVT domains, which accounts for a large percentage in the chip, is dominant. By comparing two points, for example, a 1.2-V bias difference achieves more than a 50× leakage reduction without degrading speed. A huge improvement of the static leakage reduction over 10× to 20× with bulk version [11] is obtained.

Figure 11 shows the maximum operating frequency and static leakage measurement results in the various bias voltage conditions. The core voltage of FPGA is lowered to 0.8 V to evaluate the behavior of the chip characteristics. As a result of the maximum operating frequency, the speed variation range becomes almost four times wider compared with Figure 8, in which VDD is set to 1.2 V. Another interesting point is that the static leakage current increases again after saturation, as indicated by the portion surrounded by a dashed line in the right figure. This is because the LVT is assigned to domains to be assigned originally as HVT, as discussed in [13]. First, the timing slack information in each building block is obtained after carrying out static timing analysis by the router included in the dedicated CAD toolchain, in the case that LVT is assigned to all building blocks. After that, in order, from the sink nodes, Vt mapper assigns HVT to the build blocks, satisfying that the timing slack is larger than the propagation delay difference between building blocks assigned as LVT and HVT. The Vt mapping process proceeds with decreasing the timing slack and is finally completed at the condition that the timing slack becomes smaller than the above-mentioned propagation delay difference. When the core voltage of the FPGA is set to 1.2 V, HVT is assigned to all building blocks on non-critical paths. However, when the supply voltage VDD is set to 0.8 V, LVT building blocks remain, because the timing slack is exhausted before assigning HVT to all building blocks on non-critical paths.

Figure 11. (**a**) Maximum operating frequency and (**b**) static leakage measurement results in the various bias voltage conditions (VDD = 0.8 V).

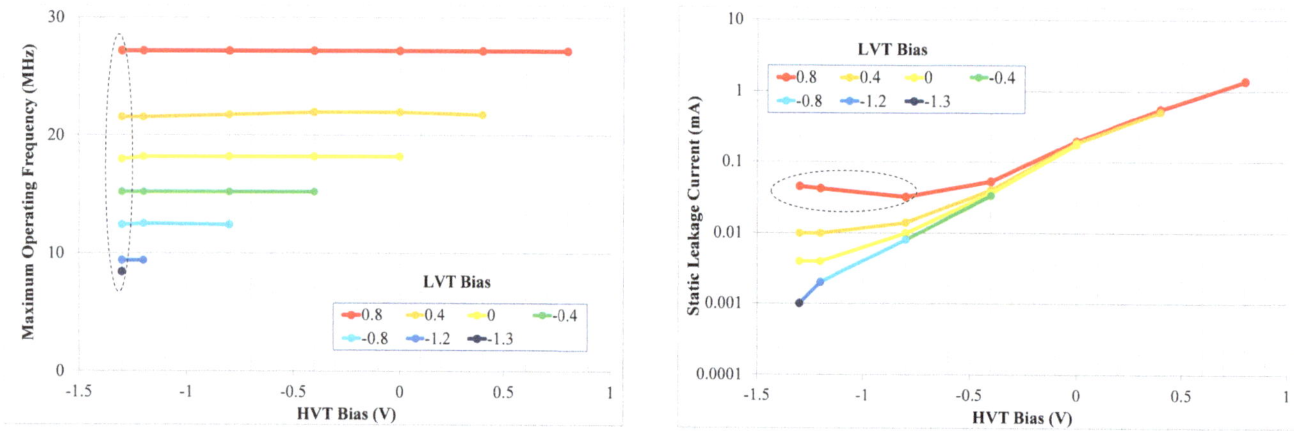

(a) (b)

3.2. Overhead

In order to program the Vt of elemental circuit components in an FPGA with a fine granularity, additional circuits, such as a body bias circuit, a bias voltage selector and configuration memory to store the Vt state information, and an extra separation margin between elemental circuits are needed. Figure 12a shows the area breakdown in the FPGA tile. The elemental circuits in FPGA, such as the MUX, buffer, inverter, LUT and DFF, occupy 54% of an FPGA tile area. The area of the configuration memory is only 3%. The Vt configuration memory, body bias circuit and separation margin occupies 1%, 26% and 16% of an FPGA tile area, respectively. An area overhead to control the Vt of an elemental circuit with a fine granularity becomes 43% of an FPGA tile area. The grain resolution for the Vt control is necessary to examine, because it directly affects the FPGA tile area. Figure 12b shows the breakdown of the static leakage current. The elemental circuits are applied to a zero body bias, resulting in the threshold voltage of the built-in state. The elemental circuits and configuration memory consume 93% of the static leakage current in an FPGA tile. On the other hand, the extra leakage current caused by the Vt configuration memory and body bias circuit is only 7%. This is because the body bias circuit is always applied to the reverse body bias, resulting in the higher Vt state, while the bias voltage selector is designed by a high-voltage transistor.

Figure 12. The breakdown of (**a**) the area and (**b**) the static leakage current in an FPGA tile.

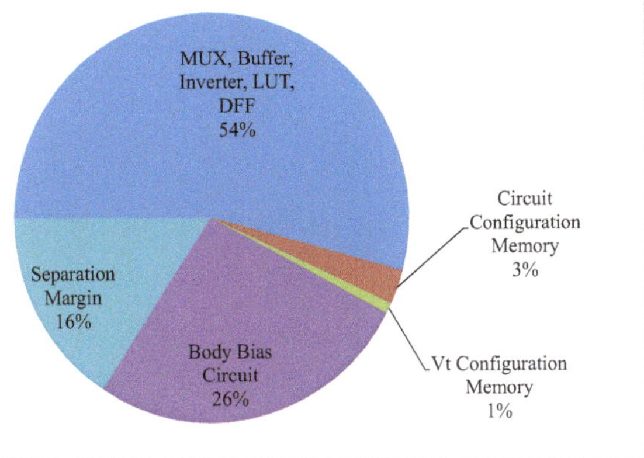

(**a**)

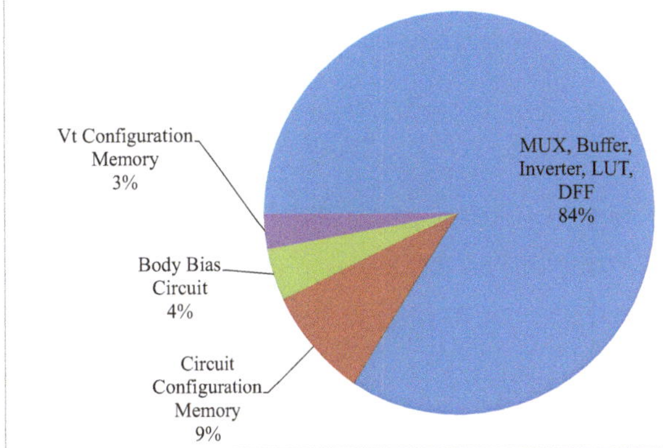

(**b**)

4. Conclusions

The field programmable gate array (FPGA) is a promising target for advanced semiconductor technologies. However, static leakage is a serious problem in FPGAs. We have proposed an FPGA with fine-grained programmable body biasing, called the Flex Power FPGA, for static leakage power reduction. For further static power reduction, an excellent Vt controllability of the SOTB transistors can help a lot. The SOTB implementation of the Flex Power FPGA test chip is reported in this paper. The maximum operating frequency measurement results show that the operating speed can be widely controlled by the LVT bias, while the operating speed remains constant against the change in the HVT bias. Moreover, in the static leakage current measurement, the three orders of magnitude static leakage reduction with a bias range of 2.1 V demonstrates the excellent Vt controllability of the SOTB transistors. A 1.2 V bias difference achieves a 50× leakage reduction without degrading speed. The huge improvement of the static leakage reduction over 10× to 20× with the bulk version is obtained. The effectiveness of the Flex Power FPGA and the excellent Vt controllability of the SOTB transistor are clearly shown. The Flex Power FPGA is an important application target for this novel device.

Acknowledgments

This work was performed in the "Ultra-Low Voltage Device Project" of the Low-power Electronics Association and Project (LEAP) funded and supported by the Ministry of Economy, Trade and Industry (METI) and the New Energy and Industrial Technology Development Organization (NEDO).

Author Contributions

Masakazu Hioki designed the Flex Power FPGA architecture. Chao Ma measured the fabricated chips. Takashi Kawanami, Yasuhiro Ogasahara, Tadashi Nakagawa, Toshihiro Sekigawa and Toshiyuki Tsutsumi discussed the results of this paper. Hanpei Koike coordinated this research project.

Conflicts of Interest

Thc authors declare no conflict of interest.

References

1. Kuon, I.; Rose, J. Measuring the Gap between FPGAs and ASICs. In Proceedings of the 2006 ACM/SIGDA 14th International Symposium on Field Programmable Gate Arrays, Monterey, CA, USA, 22–24 February 2006; pp. 21–30.
2. Li, F.; Lin, Y.; He, L.; Cong, J. Low-Power FPGA Using Predefined Dual-Vdd/Dual-Vt Fabrics. In Proceedings of the 2004 ACM/SIGDA 12th International Symposium on Field Programmable Gate Arrays, Monterey, CA, USA, 22–24 February 2004; pp. 42–50.
3. Rahman, A.; Polavarapuv, V. Evaluation of Low-Leakage Design Techniques for Field Programmable Gate Arrays. In Proceedings of the 2004 ACM/SIGDA 12th International Symposium on Field Programmable Gate Arrays, Monterey, CA, USA, 22–24 February 2004; pp. 23–30.
4. Gayasen, A.; Tasi, Y.; Vijaykrishnan, N.; Kandemir, M.; Irwin, M.J.; Tuan, T. Reduced Leakage Energy in FPGAs Using Region-Constrained Placement. In Proceedings of the 2004 ACM/SIGDA 12th International Symposium on Field Programmable Gate Arrays, Monterey, CA, USA, 22–24 February 2004; pp. 51–58.
5. Anderson, J.H.; Najim, F.N. Low-Power Programmable Routing Circuitry for FPGAs. In Proceedings of the 2004 IEEE/ACM International Conference on Computer-aided Design, Washington, DC, USA, 7–11 November 2004; pp. 602–609.
6. Lin, Y.; Li, F.; He, L. Routing Track Duplication with Fine-Grained Power-Gating for FPGA Interconnect Power Reduction. In Proceedings of the 2005 Asia and South Pacific Design Automation Conference, Shanghai, China, 18–21 January 2005; pp. 645–650.
7. Rahman, A.; Das, S.; Tuan, T.; Rahut, A. Heterogeneous Routing Architecture for Low-Power FPGA Fabric. In Proceedings of the IEEE Custom Integrated Circuits Conference 2005, San Jose, CA, USA, 21 September 2005; pp. 183–186.

8. Tuan, T.; Kao, S.; Rahman, A.; Das, S.; Trimberger, S. A 90 nm Lowpower FPGA for Battery-Powered Applications. In Proceedings of the 2006 ACM/SIGDA 14th International Symposium on Field Programmable Gate Arrays, Monterey, CA, USA, 22–24 February 2006; pp. 3–11.
9. Kawanami, T.; Hioki, M.; Nagase, H.; Tsutsumi, T.; Nakagawa, T.; Sekigawa, T.; Koike, H. Preliminary evaluation of flex power FPGA: A power reconfigurable architecture with fine granularity. *IEICE Trans. Inf. Syst.* **2004**, *8*, 2004–2010.
10. Tsuchiya, R.; Horiuchi, M.; Kimura, S.; Yamaoka, M.; Kawahara, T.; Maegawa, S.; Ipposhi, T.; Ohji, Y.; Matsuoka, H. Silicon on Thin BOX: A New Paradigm of the CMOSFET for Low-Power High-Performance Application Featuring Wide-Range Back-Bias Control. In Proceedings of the IEEE International Electron Devices Meeting 2004, San Francisco, CA, USA, 13–15 December 2004; pp. 631–634.
11. Hioki, M.; Sekigawa, T.; Nakagawa, T.; Kawanami, T.; Matsumoto, Y; Tsutsumi, T.; Koike, H. Fully-Functional FPGA prototype with Fine-Grain Programmable Body Biasing. In Proceedings of the ACM/SIGDA International Symposium on Field Programmable Gate Arrays, Monterey, CA, USA, 11–13 February 2013; pp. 73–80.
12. Sugii, N.; Tsuchiya, R.; Ishigaki, T.; Morita, Y.; Yoshimoto, H.; Kimura, S. Local V_{th} Variability and Scalability in Silicon-on-Thin-Box (SOTB) CMOS with Small Random-Dopant Fluctuation. *IEEE Trans. Electron Devices* **2010**, *57*, 835–845.
13. Kawanami, T.; Hioki, M.; Matsumoto, Y.; Tsutsumi, T.; Nakagawa, T.; Sekigawa, T.; Koike, H. Optimal Set of Body Bias Voltages for an FPGA with Field-Programmable Vth Components. In Proceedings of the 2006 IEEE International Conference on Field Programmable Technology, Bangkok, Thailand, 13–15 December 2006; pp. 329–332.

12

Embedded Memory Hierarchy Exploration Based on Magnetic Random Access Memory

Luís Vitório Cargnini [1,2,*], Lionel Torres [2,*], Raphael Martins Brum [2], Sophiane Senni [2] and Gilles Sassatelli [2]

[1] HGST Inc., 3403 Yerba Buena Road, San Jose, CA 95135, USA
[2] LIRMM - UMR CNRS 5506 - University of Montpellier 2, 161 Rue Ada, Montpellier 34095, France; E-Mails: raphael.brum@lirmm.fr (R.M.B.); sophiane.senni@lirmm.fr (S.S.); gilles.sassatelli@lirmm.fr (G.S.)

* Authors to whom correspondence should be addressed; E-Mails: lvcargnini@ieee.org (L.V.C.) lionel.torres@lirmm.fr (L.T.)

Abstract: Static random access memory (SRAM) is the most commonly employed semiconductor in the design of on-chip processor memory. However, it is unlikely that the SRAM technology will have a cell size that will continue to scale below 45 nm, due to the leakage current that is caused by the quantum tunneling effect. Magnetic random access memory (MRAM) is a candidate technology to replace SRAM, assuming appropriate dimensioning given an operating threshold voltage. The write current of spin transfer torque (STT)-MRAM is a known limitation; however, this has been recently mitigated by leveraging perpendicular magnetic tunneling junctions. In this article, we present a comprehensive comparison of spin transfer torque-MRAM (STT-MRAM) and SRAM cache set banks. The non-volatility of STT-MRAM allows the definition of new instant on/off policies and leakage current optimizations. Through our experiments, we demonstrate that STT-MRAM is a candidate for the memory hierarchy of embedded systems, due to the higher densities and reduced leakage of MRAM. We demonstrate that adopting STT-MRAM in L1 and L2 caches mitigates the impact of higher write latencies and increased current draw due to the use of MRAM. With the correct system-on-chip (SoC) design, we believe that STT-MRAM is a viable alternative to SRAM, which minimizes leakage current and the total power consumed by the SoC.

Keywords: semiconductors; VLSI; SoC; memory; non-volatile memory (NVM); Magnetic random access memory (MRAM); embedded systems; memory hierarchy

1. Introduction

The deep submicron era creates new constraints, including short channel effects (SCEs), dramatically increased leakage currents, lithography issues, reduced control of thresholds, increased sensitivity to variations in the process and environmental parameters [1]. These obstacles threaten the scaling of complementary metal–oxide–semiconductor (CMOS) devices, following the evolution according to Moore's law. A large number of switching devices are currently being explored. However, one important feature of the new emerging devices and circuits must be compatibility with conventional CMOS [2]. Furthermore, in memory design, we are achieving a so-called "design wall", caused by the technological limitations of shrinking the cell technology of these mainstream memory cells. This landscape motivated the surge of a number of non-volatile memory (NVM) technologies, such as spin transfer torque magnetic random access memory (STT-MRAM), Phase-Change RAM (PCM or PCRAM) and Resistive RAM (RRAM or ReRAM), there is also less discussed NVM alternatives like Thermally Assisted Switching MRAM (TAS-MRAM) and Ferroelectric RAM (FeRAM, F-RAM or FRAM) among others. The PCRAM, ReRAMand STT-MRAM are considered by the International Technology Roadmap for Semiconductors (ITRS) as the most promising candidates to take over mainstream memory technologies. In Table 1, an overview of the main characteristics of those technologies is given.

Table 1. A comparison of non-volatile memory (NVM) technologies [2–4]. STT-MRAM, spin-transfer torque magnetic RAM; pSTT, perpendicular STT; TAS-MRAM; FeRAM.

Technology	**Minimal cell size** (F^2)	**Endurance (cycles)**	**Read latency** (ns)	**Write latency** (ns)
SRAM	150	ϕ	2	2
STT-MRAM	20	10^{16}	5	5–30
pSTT-MRAM[1]	ϕ	ϕ	3	3
TAS-MRAM [2]	ϕ	10^{12}	30	30
NAND	4	10^4	10^4	10^6
NOR	10	10^5	15	10^3
FeRAM	22	10^{12}	40	65
ReRAM	30	10^5	100	100
PCM	4	10^{12}	12	100

[1] Data obtained from Toshiba; [2] Data provided by Crocus.

MRAM densities (depending on the magnetic tunnel junction (MTJ) geometry) are approximately four to eight times higher then SRAM; however, the access time is on the order of three to ten times longer, depending on the MRAM technology. Some of the most recent research results were presented by

Toshiba [5] concerning perpendicular STT (pSTT); they demonstrated an access time of approximately 4 ns and a bit energy read/write almost equivalent to SRAM.

MRAM memory is based on the magnetic tunneling junction (MTJ). MTJ is the cell memory of MRAM. The information in an MTJ is stored as the magnetic orientation in one of the two ferromagnetic layers. An MTJ is a nanopillar composed of two ferromagnetic (FM) thin films (CoFeB) separated by an oxide barrier (MgO), as shown in Figure 1a). The resistance of MTJ depends on the relative orientation of the magnetization in the two FM layers. In standard applications, the magnetization of one FM layer (reference layer) is commonly pinned, whereas the other (storage) layer is free to have a parallel (P) or anti-parallel (AP) orientation, determining the parallel (R_P) or anti-parallel (R_{AP}) MTJ resistance. The difference between these two resistances defines the tunnel magneto-resistance (TMR) ratio, $\triangle R/R = (R_{AP} - R_P)/R_P$. In recent decades, a great deal of research effort has been invested in trying to improve the TMR ratio of MTJs (from 10% at the beginning to over 600% nowadays) and in making them more attractive for integration with CMOS [6–8]. The structure of an MTJ is presented in Figure 1.

Figure 1. We observe in the image the three generations of magnetic tunnel junction (MTJs): (**A**) FIMS; (**B**) TAS and (**C**) STT. All cases are denoted as in-plane anisotropy [9].

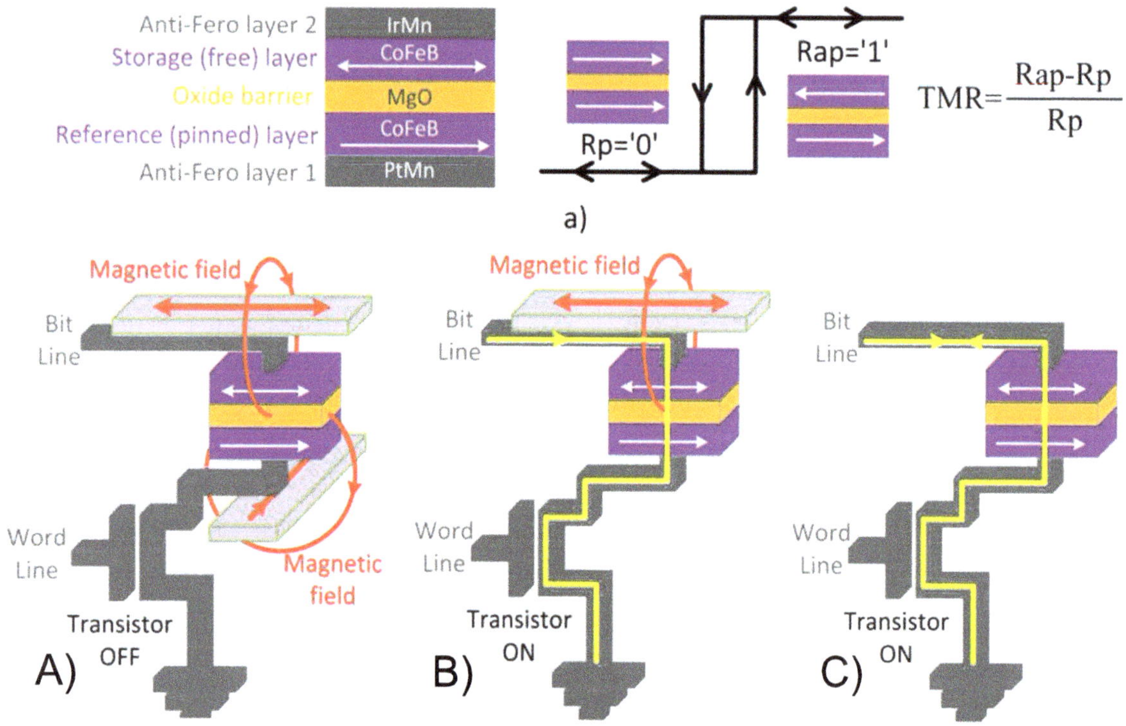

The main objective of this publication is to propose an embedded processor evaluation flow based on STT-MRAM for the memory hierarchy. We strongly believe that STT-MRAM could be a valuable memory technology to embed in the next generation of system-on-chip (SoC) . We evaluate STT-MRAM performance, the energy and silicon area and demonstrate that for L1 and L2 , the STT-MRAM brings interesting features for a set of applications. Section 2 describes the methodology flow proposed. Section 3 gives a series of result comparisons between STT-MRAM and SRAM. Section 4 demonstrates that STT-MRAM is a new technology that should be considered for embedded memory hierarchy.

2. Methodology Evaluation Flow

In order to evaluate the impact of STT-MRAM applied in the memory hierarchy, we propose a full methodology flow, as depicted in Figure 2.

Figure 2. Evaluation flow.

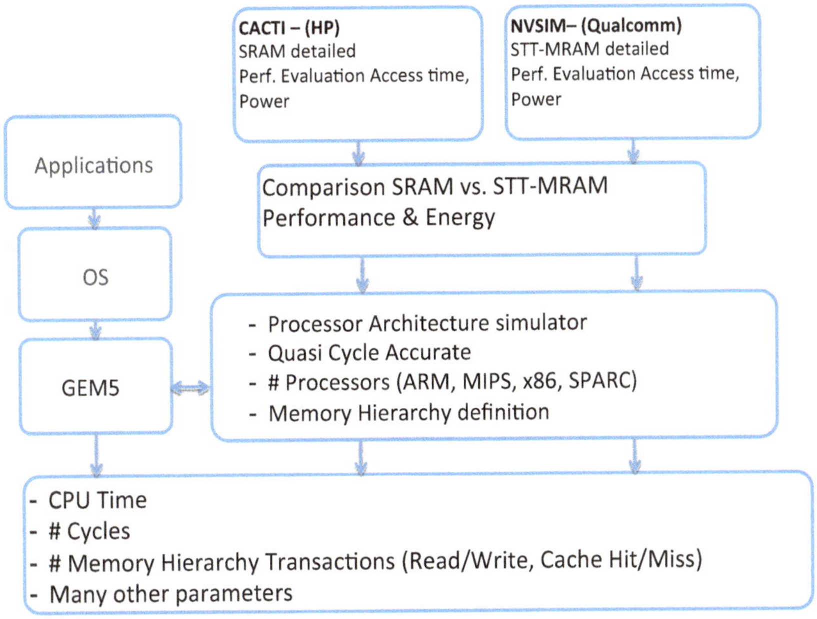

For the flow, we defined a processor architectural simulator, in this case, the GEM5 [10]. The GEM5 simulator currently supports a variety of Instruction Set Architecture (ISAs) like Alpha, ARM, MIPS, PowerPC, SPARC and X86. The simulator modularity allows these different ISAs to plug into the generic CPU models and the memory system, without having to specialize one for the other. In our particular case, we adopted the ARM ISA v7 available in GEM5. Specifically, we used the Application Binary Interface (ABI) compatible with the Cortex-A9/A15; our cross-compiler generates binaries specifically for that target.

Our evaluation methodology provides a flow to determine the overall processor system architecture performance, given the memory hierarchy specifications. Tunable parameters include cache size, L1 and L2 latencies and main memory size. Leveraging an architectural simulator, we are able to extract all memory transactions: numbers of read/write accesses into the L1 and L2, cache hits and misses, among other statistical observations. The use of GEM5 (a quasi-cycle accurate simulator) allows us to evaluate different memory sizing strategies, cache policies and the effects of latency at different levels of the memory hierarchy. Combining the memory hierarchy statistics with the intrinsic parameters, we could infer the power consumption, given that one knows the energy cost for the memory bank operation. This way, it is possible to evaluate the overall performance of a built-in SoC using MRAM or SRAM.

Our objective is to compare the use of SRAM against STT-MRAM in the caches of the embedded processor memory hierarchy. For this reason, for SRAM and STT-MRAM, it is necessary to obtain the features of these memories (latency read/write access time, power consumption) to calibrate the GEM5 simulator. For doing this, we pass through our intrinsic analyses flow, which can be performed using NVSim.

For SRAM performances, we can use the memory simulator, CACTI [11–15] or NVSim [16]. CACTI is an integrated simulator, which, based on the technology node, is able to provide accurate information about cache and memory access time, cycle time, area, leakage and the dynamic power model. CACTI was evaluated comparing its results against SPICE simulations.

Concerning STT-MRAM, the memory simulator is based on a modified environment of CACTI: the NVSim simulator [16]. NVSim is a circuit-level model simulator for NVM memory performance, energy and area estimation, which supports various NVM technologies, including STT-RAM, PCRAM, ReRAM and legacy NAND Flash, as well as the SRAM. NVSim was validated comparing its results with industrial NVM prototypes [16], and it is expected to boost architecture-level NVM-related studies.

In the end, we are able to provide for a given application a clear comparison of performance between STT-MRAM and SRAM for an embedded processor architecture. For this article, specifically, we modeled and simulated both SRAM and MRAM using only NVSim. Therefore, all of our physical parameters are generated using NVSim. According to our methodology flow, you can use any one; for us and this publication, we focused on the NVSim only.

3. STT-MRAM *vs.* SRAM Memory Bank Comparison

In this section, we present our results and analysis comparing the memory intrinsic parameters of STT-MRAM and SRAM. We provide details regarding power, area and an overall performance comparison. A comparison is made between memory banks of equivalent sizes to understand how each technology impacts the performance of the memory hierarchy in a given system. The target technology considered for CMOS and STT-MRAM is a 45 nm node (low-power 45 nm CMOS process (low power performance (LOP))) In Table 2, we provide the technology model features adopted for STT-MRAM and SRAM. Detailed information is available in [17,18].

Table 2. Memory Cell features models, used to simulate the electrical and physical memory for bank at 45 nm node technology at 300 K of temperature operation.

	STT-MRAM	SRAM
Cell Area	$10F^2$	$146F^2$
Aspect Ratio	1.0	1.46
R_p	7100	ϕ
R_{ap}	15,600	ϕ
Read Current	20 µA	ϕ
Write Current	29 µA	ϕ
Write Pulse	10 ns	ϕ
Access CMOS Width	6F	1.3F
NMOS Width	ϕ	2.0F
PMOS Width	ϕ	1.2F

Figures 3–8 depict the relations of power, latency, dynamic power, leakage and area for memory banks built using the low power performance (LOP) 45 nm technology process.

In Table 3, depicted in Figure 3, we observe that increases to the memory size correspond to an area increase, as expected. However, the area of SRAM memory banks increases exponentially, while that of STT-MRAM increases linearly. One reason is the memory density: as we double the memory capacity, we indicate that the STT-MRAM density is 10–15 times higher than SRAM density according to [18] (the explanation is in STT-MRAM; the MTJ at 45 nm has an area of 10 F^2, while the SRAM cell has 146 F^2). Therefore, increases in the amount of memory available for the architecture organization and the impact regarding the density in SRAM are much more critical than with STT-MRAM. They are critical because the slightest increase has an impact on the total die area used, and such an impact using STT-MRAM is not so problematic. To demonstrate this, assume passing cache memory from 2 to 4 MB; in SRAM, we pass from 5.6643 mm^2 to 11.2673 mm^2, while with STT-MRAM, from 0.6183 mm^2 to 0.9958 mm^2 assuming only the surface area into account. This same pattern is observed for all of the parameters, with more or less variation according to the parameter.

This comparison is based on the physical parameters of memory banks only, without taking into account bank interconnections at system-level architectures.

Figure 3. Total area (mm^2).

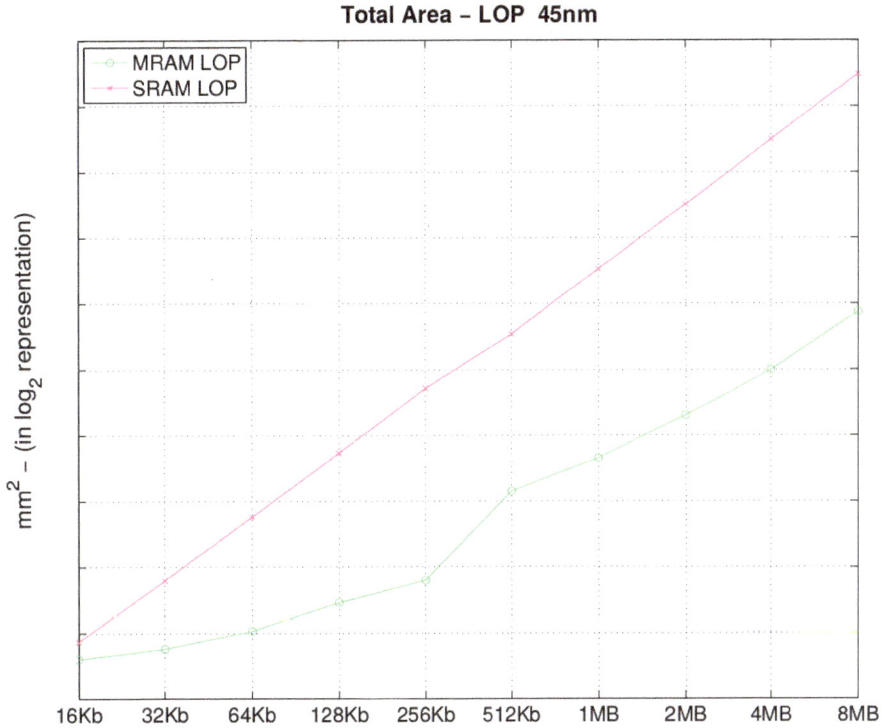

Table 3. Total area (mm^2).

Size	MRAM	SRAM	Size	MRAM	SRAM
16 KB	0.04	0.05	512 KB	0.27	1.44
32 KB	0.05	0.10	1 MB	0.39	2.87
64 KB	0.06	0.21	2 MB	0.61	5.66
128 KB	0.08	0.41	4 MB	0.99	11.26
256 KB	0.10	0.81	8 MB	1.83	22.34

In Figure 4 and Table 4, we observe that the drawback of STT-MRAM memory banks is the write latency, given our model of MTJ. It is a fact that with current STT-MRAM technology, write latency is an issue that jeopardizes the adoption of STT-MRAM into memory hierarchy. In Figure 4, we observe that the write latency gap between STT-MRAM and SRAM is narrowing as we increase the bank size. In section 4, we evaluate the impact of this latency difference, on an architectural simulator executing an application on top of the memory hierarchy configured according to the presented analysis.

Figure 4. Cache total write latency (ns).

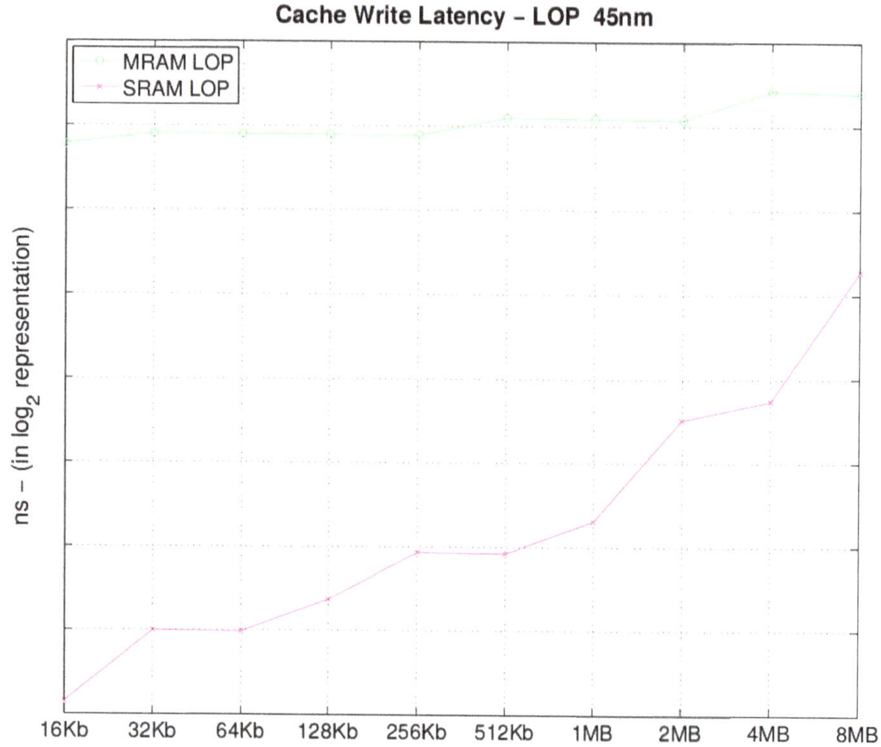

Table 4. Cache total write latency (ns).

Size	MRAM	SRAM	Size	MRAM	SRAM
16 KB	109.44	1.09	512 KB	135.87	3.73
32 KB	118.73	1.97	1 MB	134.73	4.88
64 KB	118.20	1.95	2 MB	133.60	11.24
128 KB	117.72	2.55	4 MB	170.29	13.16
256 KB	117.19	3.76	8 MB	167.69	38.38

In Figure 5 and Table 5, we observe one of the major advantages of MRAM, the leakage power, considering both the TAG and DATA arrays of a cachememory set bank. Leakage power is one of the biggest strengths of STT-MRAM technology, which makes it appealing, even with a higher write latency. The main interest in STT-MRAM is that no leakage is induced by the physical device, the MTJ. Given that information is stored only in a magnetic orientation, leakage is created due to the wiring and metal tracks for decoding and sense amplifiers to read the stored information in the STT-MRAM cells. Furthermore, observing Figure 5, one can see that with a 8 MB MRAM, we have almost the same leakage as an SRAM LOP of 128 KB.

Figure 5. Cache total leakage power (mW).

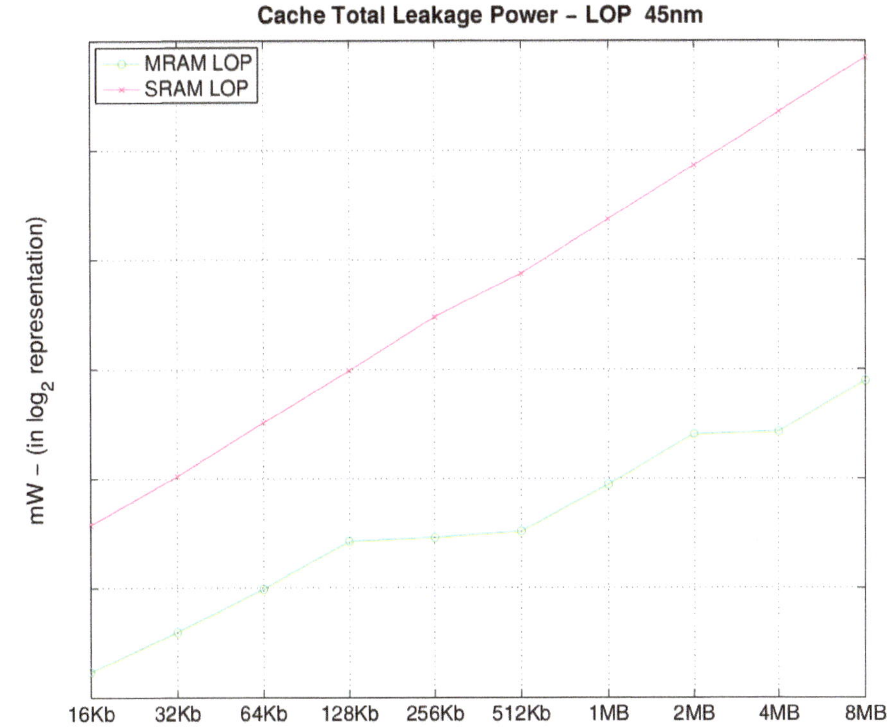

Table 5. Cache total leakage power (mW).

Size	MRAM	SRAM	Size	MRAM	SRAM
16 KB	0.02	0.14	512 KB	0.13	3.42
32 KB	0.03	0.26	1 MB	0.23	6.80
64 KB	0.06	0.51	2 MB	0.44	13.35
128 KB	0.11	0.99	4 MB	0.45	26.57
256 KB	0.12	1.96	8 MB	0.87	52.54

In Table 6 and Figure 6, we present the energy consumption of a hit-access into the cache memory, during the search for the information in the memory bank (disregarding the latency of the Memory Management Unit (MMU) block search). To obtain a hit or even a miss, for an address in the cache, we need to perform a search first in the cache content. Therefore, for the hit, only the read mode is activated, explaining the reason why the cache hit energy is lower for MRAM.

Table 6. Cache hit dynamic energy (nJ).

Size	MRAM	SRAM	Size	MRAM	SRAM
16 KB	0.03	0.006	512 KB	0.15	0.13
32 KB	0.03	0.01	1 MB	0.14	0.24
64 KB	0.03	0.02	2 MB	0.14	0.526
128 KB	0.03	0.03	4 MB	0.16	0.96
256 KB	0.03	0.07	8 MB	0.16	2.02

Figure 6. Cache hit dynamic energy (nJ).

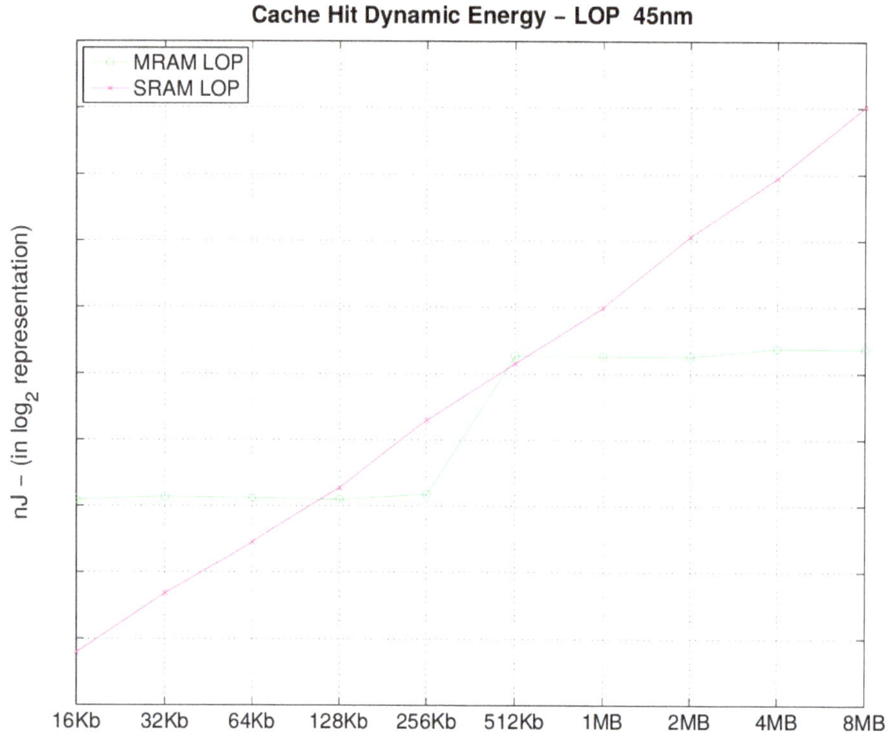

In Table 7 and in Figure 7, we observe that STT-MRAM consumes 91× more dynamic energy for a 16 KB memory bank; when the size of the DATA arrays increases, the difference decreases to 6.9× (in 8 MB) of the write dynamic energy. This is due to the fact that an increase in memory bank size causes a linear increase in dynamic energy for STT-MRAM and an exponential increase when using SRAM. It is important to remark that if we consider the ratio between dynamic energy of 1 MB memory and dynamicenergy of 16 KB memory for STT-MRAM, this ratio is near one. Considering SRAM, this ratio is about six. For the STT-MRAM, the dynamic power consumption is less sensitive to memory size when compared to SRAM. It is easy to identify that the initial energy of STT-MRAM is currently higher than SRAM. However, with the evolution in materials and MTJ manufacturing process, it is expected that the initial dynamic power consumption will decrease to the point where STT-MRAM becomes comparable or even better than SRAM [4,5]. However, for bank sizes larger than 8 MB, the observation is that by increasing the memory bank size, the write dynamic energy of SRAM will become equal to or higher than STT-MRAM at some point. Although the static power consumption is already better than SRAM, the last barriers to STT-MRAM achieving viability of mass employment is the dynamic energy consumption optimization.

Table 7. Cache write dynamic energy (nJ).

Size	MRAM	SRAM	Size	MRAM	SRAM
16 KB	0.14	0.001	512 KB	0.15	0.008
32 KB	0.14	0.002	1 MB	0.15	0.009
64 KB	0.14	0.003	2 MB	0.17	0.01
128 KB	0.14	0.004	4 MB	0.17	0.01
256 KB	0.14	0.005	8 MB	0.20	0.02

Figure 7. Cache write dynamic energy (nJ).

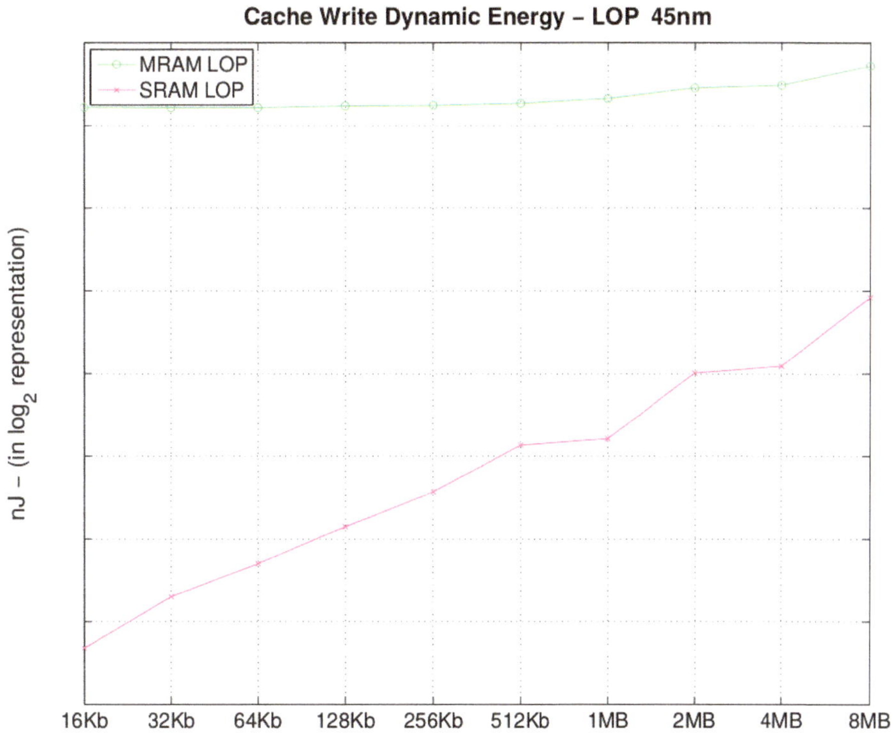

In this section, we presented the power characteristics of a set of memory banks by comparing STT-MRAM *versus* SRAM. The obtained results demonstrate that besides the drawbacks regarding latency for writing, STT-MRAM can drastically reduce the power consumption of a cache set memory bank in an embedded system or a high performance system.

Furthermore, the leakage power of an STT-MRAM memory is worth the drawbacks in performance given the gains with power consumption reduction in most of the cases.

In the next section, we are going to apply these results to a real application, demonstrating the performance of STT-MRAM when applied within the memory hierarchy.

4. STT-MRAM Applied in Cache Hierarchy

To evaluate our proposed methodology flow, we present an application case study based on the X.264 video encoder. The processor used for this case study is a 32-bit Reduced Instruction Set Computing (RISC), dual-issue superscalar, out-of-order, speculating dynamic length pipeline (8–11 stages) processor, which is modeled after the ARMv7 architecture. The processor has a clock frequency of 1.5 GHz, and we are running the Linux Operating System on the processor. The video that we wish to encode has a frame rate of 30 FPS and has a resolution of 1280×720.

We also evaluated the impact of STT-MRAM for L2 cache, and we will compare its characteristics with a similar system employing SRAM. Table 8 summarizes the properties of 45 nm SRAM and STT-MRAM memory banks used as a 2-MB eight-way associative L2 cache. Furthermore, the detailed parameters introduced here were extracted during the intrinsic analysis evaluation discussed in the previous section [17].

Table 8. Details regarding the architecture and the levels of memory hierarchy.

Parameter	Features
Processor	32 bits RISC Processor - 8-11 stages pipeline - 2 instructions per cycle
L1 Caches	64 Kbyte SRAM - 4-way set associative, 2 ns access latency - 32 byte per cache line
L2 Caches	2 Mbyte SRAM - 8-way set associative, ~20 ns access latency - 32 byte per cache line

The performance comparisons among memory cache banks, eight-way associative, 2-MB SRAM and STT-MRAM in the technology node of 45 nm on L2 cache are described in Table 9.

Table 9. Memory banks characteristics.

Field		SRAM	MRAM
Area	Total area	5.6 mm^2	2.2 mm^2
	Data array area	5 mm^2	1.8 mm^2
	Tag array area	0.63 mm^2	0.39 mm^2
Timing	Cache hit latency	18.8 ns	70.1 ns
	Cache miss latency	2.9 ns	66.0 ns
	Cache write latency	10.1 ns	75.1 ns
Power	Hit dynamic energy	1.07 nJ	0.213 nJ
	Miss dynamic energy	1.07 nJ	0.213 nJ
	Write dynamic energy	0.03 nJ	0.22 nJ
	Total leakage power	1326.7 mW	26.5 mW
	Data array leakage Power	1180.6 mW	24.3 mW
	Tag array leakage Power	146.1 mW	2.2 mW
	hit(ns)	18.8	70.8
	response (ns)	10.1	75.1

4.1. L2 Cache Exploration for a High Performance System

Considering X.264, the experimental results described in Table 9 are obtained based on the execution of the X.264 benchmark using the Linux OS on top of the Gem5 calibrated with the memory bank latency for each technology.

Based on Table 10, we observe that the total CPU time increases from 16.2 s to 17.1 s, which is tolerable for the X.264 application. Given the fact that the STT-MRAM hit latency is 2.7× higher than SRAM, yet we only see a 5% increase in execution time, we conclude that an increased latency of a hit in the L2 cache has a slight impact on the total CPU time.

A major benefit for using STT-MRAM is the lower energy consumption compared to SRAM. STT-MRAM is roughly 50× more energy efficient compared to SRAM, and this is primarily attributed to the lower leakage current of STT-MRAM as seeing in Table 11.

Table 10. L2-cache dynamic energy estimation.

	SRAM	STT-MRAM
CPU time (s)	16.2	17.11
Write back total (Write)	5,879,046	5,944,740
Overall access (Read)	21,113,987	22,309,621
Write energy per cache line	22.8 nJ	170.6 nJ
Read energy per cache line	957.7 pJ	150.4 pJ
Total write energy (J)	0.13	1.01
Total read energy (J)	0.020	0.0033

Table 11. Static power consumption.

	SRAM	MRAM
Execution time (s)	16.2	17.1
Static power	1326.7 mW	26.5 mW
Total energy (J)	21.49	0.45315

Current implementations of STT-MRAM consume more dynamic energy than SRAM, noted in Table 10. Although STT-MRAM has this disadvantage, STT-MRAM consumes less energy overall when the dynamic and static energy is calculated for our case study. STT-MRAM write operations consume 7.5× more energy than SRAM and 6× more energy for read operations when compared to SRAM. In the case study that we are presenting, STT-MRAM consumes 1.25× more dynamic energy than SRAM for the given workload.

To put our observation in perspective, we look into similar works that achieved similar results through different techniques, like in [19], for example, a 2-MB L2 SRAM cache could be replaced with an 8-MB L2 STT-MRAM cache, using roughly the same silicon die area. In this particular case, the increase on the cache size was not enough to compensate for the penalty, due to the cache access delay. By employing write buffers and a novel cache access policy, the authors managed to achieve a similar performance while reducing the power consumption on the overall application (comprising all of the memory hierarchy) by almost 74%.

They also presented a hybrid MRAM/SRAM cache organization, having 31 sets implemented in STT-MRAM and one set implemented in SRAM. The write-intensive data is kept in the SRAM part, in order to mitigate the higher write delay. A method for determining which datum is suitable for being placed in the SRAM set is also discussed.

In this section, we have demonstrated the interest in using an L2 STT-MRAM cache for high performance applications. In the next section, we will discuss the usage of L1 STT-MRAM cache.

4.2. L1 Cache Exploration for a Low Performance System

Similar to the evaluation of STT-MRAM in a Level-2 cache, we now present an evaluation of STT-MRAM in a Level-1 cache of a microprocessor targeted for an embedded system. Specifically, the target is low performance, where constraints are different, like size, latency, power consumption, but mostly latencies are a predominant factor given the demand for the performance of a CPU. Our goal is to determine if STT-MRAM is a suitable candidate technology to replace SRAM for an L1 cache given that the two share the same silicon fingerprint.

The baseline configuration is quite simple. It consists of a single processor having a single cache level and a large external memory capacity, which is an assumption that can be considered for many systems. In contrast to our previous work in [20], we now assume that the STT-MRAM density is four-times the SRAM [21]. We are then comparing, for instance, a 4-KB SRAM-based cache with a 16-KB MRAM-based cache. For this set of experiments, we assumed a latency of three clock cycles during each cache access. This means that the processor will stall upon each cache request, waiting for the data to become available. We also assumed a latency of 1,000 cycles for the external memory to make the first word available and 10 cycles for each subsequent word, while doing burst reading [22]. In the same manner, as shown in Figure 8, where a 128-KB SRAM cache is compared with its 512-KB STT-MRAM counterpart, the latter shows comparable performance to the smaller, yet faster, SRAM.

Figure 8. Same silicon area STT-MRAM *versus* SRAM L1 cache execution time comparison: 1-KB SRAM *versus* 4 KB.

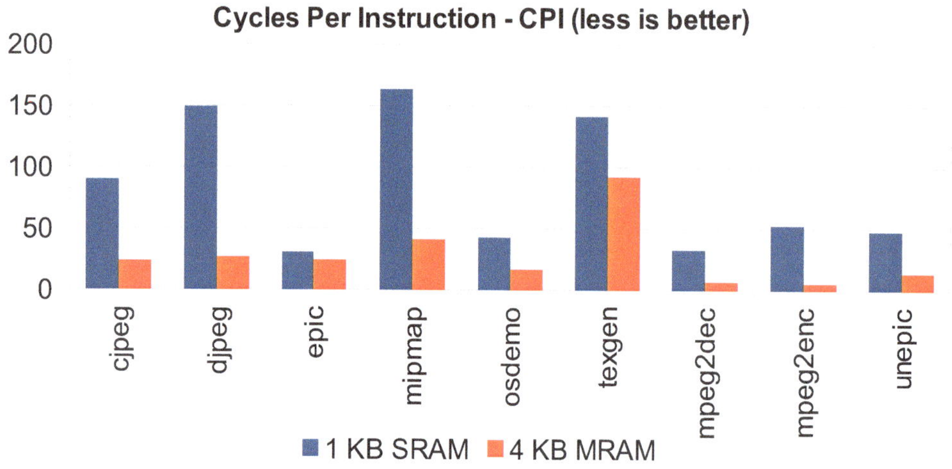

In Figure 9, we set an STT-MRAM L1 cache of 512 KB, and we compare it with an L1 128-KB SRAM cache size. It is shown that, for most benchmarks, they are comparable in terms of performance.

In order to generalize this conclusion, let us then define the CPI penalty as the increase in the CPI caused by replacing an SRAM cache with an STT-MRAM cache using the same silicon area, as follows:

$$CPI_{penalty} = 1 - \frac{CPI_{MRAM}}{CPI_{SRAM}} \tag{1}$$

Based on the Cycles Per Instruction (CPI) penalty ($CPI_{penalty}$), in Figure 10, the best case, the worst case and the average performance over the benchmark set is shown as a function of the cache capacity. Given that our assumptions are valid, STT-MRAM does present a CPI gain rather than a CPI penalty for most cases. Once the cache capacity is large enough to contain the whole benchmark data, the CPI gain turns into a penalty that can no longer be compensated for if no specific technique is employed.

Figure 9. Same STT-MRAM silicon area *versus* SRAM L1 cache execution time comparison: 128-KB SRAM *versus* 512-KB MRAM.

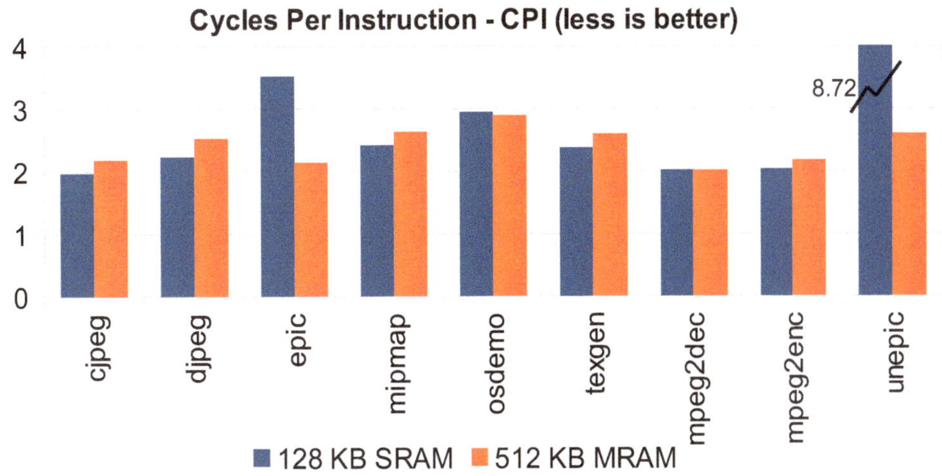

Figure 10. Overview of the CPI penalty: best case, worst case and average of the Mediabench benchmarks' performance.

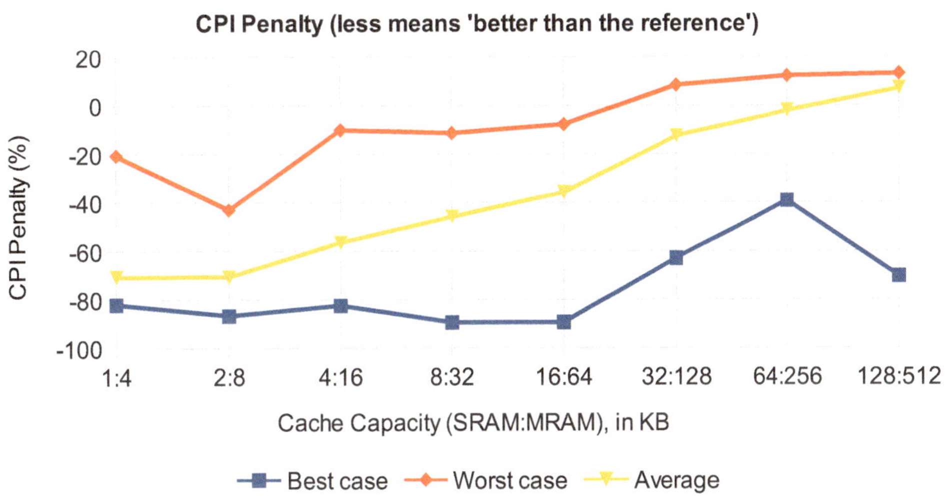

5. Conclusion

In this article, we have demonstrated that STT-MRAM can replace SRAM in the memory hierarchy of a microprocessor. We achieve this conclusion based on our memory hierarchy evaluation methodology flow. Using this flow, we have investigated the performance and power characteristics of STT-MRAM, when used as a replacement for SRAM. Although STT-MRAM has higher latencies, the lower leakage power of STT-MRAM, as compared to SRAM, makes it an attractive candidate replacement technology. Current results indicate that it could be a solution to address the rising power consumption of CMOS circuits. The use of STT-MRAM enables the possibility of new techniques for the implementation of power-saving mechanisms. The non-volatility could be explored to power-off the devices whenever they are idle. Furthermore, non-volatile memory arrays do not need refreshing, reducing the dynamic power and leakage. Independently of our results, the physical properties used in our evaluation of STT-MRAM produce two possible pathways for integrated circuit design. In one, we can conserve the total silicon die area and increase the amount of memory at least four-fold, or we could maintain the same amount of memory, but increase the production yield of the circuit four-fold by switching from SRAM to STT-MRAM. Given our evaluation and the benefits STT-MRAM brings to integrated circuit design, we conclude that STT-MRAM is a strong candidate to replace SRAM in the memory hierarchy of microprocessors.

Acknowledgments

We would like to thank Guillaume Prenat and Gregory Di Pendina, from Spintec at the Commissariat à l'Énergie Atomiquee (CEA) in Grenoble, for their assistance on the physical details regarding the STT-MTJ. ANR, due to the grant of the SPIN and MRAM based Architecture for Reliable and low power Systems (MARS) projects, made this research and all research effort regarding STT-MRAM in France possible. Also, to Adam Manzanares, my colleague for his time reviewing the text, fixing and improving it and suggestions made, that enriched the final text and presentation of this work, Thank you very much Adam.

Author Contributions

Luis Vitorio Cargnini was responsible for the MTJs physics analyses and simulation and STT-MRAM memory banks elaboration. The sections related to architectural evaluation using Gem5, CACTI and NVSim performing the analyses, combining results and compiling the data to put together in this work, also the main text of this publication.

Lionel Torres also worked reviewing and discussing the data from the MTJs regarding the physical aspects and posterior usage to build memory banks, whose data were used into Gem5 for architectural evaluation. He also supervised the work previously done using SuperScalar for the "L1 Cache Exploration for a Low Performance System" section, reviewing this entire work. He is the Spintronics MRAM head researcher at LIRMM.

Raphael Brum was responsible for the section entitled "L1 Cache Exploration for a Low Performance System", this work began originally with his evaluation using SuperSCalar that later was migrated to Gem5.

Sophiane Senni reviewed the text and the Gem5 analyses, he is currently pursuing the same approach applied to TAS-MRAM.

Gilles Sassatelli has a major involvement in the works using Gem5, regarding the aspects of how to use and add the STT-MRAM banks models in the Gem5 to perform the architectural analyses

Conflict of Interest

The authors declare no conflict of interest.

References

1. Rabaey, J. *Low Power Design Essentials*; Series on integrated circuits and systems; Springer: Berkeley, California, USA, 2009.
2. ITRS. *Emerging Research Devices*; Publisher: New York, USA, 2012; pp. 1–83.

3. Kim, W.; Park, S.I.; Zhang, Z.; Yang-Liauw, Y. Forming-Free Nitrogen-Doped AlO X RRAM with Sub-μA Programming Current. In Proceedings of the 2011 Symposium on VLSI Technology (VLSIT); Honolulu, HI, USA, 14–16 June 2011.
4. Yoda, H.; Fujita, S.; Shimomura, N.; Kitagawa, E.; Abe, K.; Nomura, K.; Noguchi, H.; Ito, J. Progress of STT-MRAM Technology and the Effect on Normally-off Computing Systems. In Proceedings of the 2012 IEEE International Electron Devices Meeting (IEDM); San Francisco, CA, USA, 10–13 December 2012; p. 11.
5. Kitagawa, E.; Fujita, S. STT-MRAM cuts power use by 80%. Available online: http://www.eetimes.com/document.asp?doc_id=1280753 (accessed on 13 October 2013).
6. Kawahara, T.; Ito, K.; Takemura, R.; Ohno, H. Spin-transfer torque RAM technology: Review and prospect. *Microelectron. Reliab.* **2012**, *52*, 613–627.
7. Chen, P.; Feng, G.; Shull, R. Use of half metallic Heusler alloys in CoFeB/MgO/Heusler alloy tunnel junctions. *IEEE Trans. Magn.* **2013**, *49*, 4379–4382.
8. Sterwerf, C.; Meinert, M.; Schmalhorst, J.M.; Reiss, G. High TMR ratio in Co_2FeSi and Fe_2CoSi based magnetic tunnel junctions. *IEEE Trans. Magn.* **2013**, *49*, 4386–4389.
9. Singer, P. *IEDM: Nanoelectronics provide a path beyond CMOS—ElectroIQ*; Extension Media, Solid State Technology: San Francisco, CA, USA, 2012.
10. Binkert, N.; Sardashti, S.; Sen, R.; Sewell, K.; Shoaib, M.; Vaish, N.; Hill, M.D.; Wood, D.A.; Beckmann, B.; Black, G.; *et al.* The Gem5 simulator. *ACM SIGARCH Comput. Archit. News* **2011**, *39*, 1–7.
11. Wilton, S.J.; Jouppi, N.P. *An Enhanced Access and Cycle Time Model for on-Chip Caches*; Digital Western Research Laboratory (HP): Palo Alto, California, USA, 1993.
12. Muralimanohar, N.; Ahn, J.H.; Jouppi, N.P. Memory Modeling with CACTI. In *Processor and System-on-Chip Simulation*; Leupers, R., Temam, O., Eds.; Springer Science + Business Media LLC: Cambridge, Massachusetts, USA 2010; Vol. III, pp. 229–242.
13. Muralimanohar, N.; Balasubramonian, R.; Jouppi, N.P. CACTI 6.0: A Tool to Model Large Caches. In Proceedings of the International Symposium on Microarchitecture, Chicago, IL, USA, April 21 2009.
14. Reinman, G.; Jouppi, N.P. *CACTI 2.0: An Integrated Cache Timing, Power, and Area Model*; Technical report; Compaq Computer Corporation Western Research Laboratory: Palo Alto, CA, USA, February 2000.
15. Shivakumar, P.; Jouppi, N.P. *CACTI 3.0: An Integrated Cache Timing, Power, and Area Model*; Technical report; Compaq Computer Corporation Western Research Laboratory: Palo Alto, CA, USA, 2001.
16. Dong, X.; Xu, C.; Xie, Y.; Jouppi, N.P. NVSim: A circuit-level performance, energy, and area model for emerging nonvolatile memory. *IEEE Trans. Comput.-Aided Des. Integr. Circuits Syst.* **2012**, *31*, 994–1007.
17. Cargnini, L.V. MRAM Applied to Embedded Processors Architecture and Memory Hierarchy. Ph.d. Thesis, Laboratoire de Informatique, Robotique et Microelectronique de Montpellier (LIRMM) Université de Montpellier (UM2), Montpellier, France, 2013.

18. Kang, S.H. *Embedded STT–MRAM for Mobile Applications: Enabling Advanced Chip Architectures*; Qualcomm Inc., University of California, San Diego, CA, USA, 11–13 April 2010; pp. 1–23.
19. Sun, G.; Dong, X.; Xie, Y.; Li, J.; Chen, Y. A Novel Architecture of the 3D Stacked MRAM L2 Cache for CMPs. In *High Performance Computer Architecture*; HPCA: Raleigh, NC, USA, 14–18 February 2009; pp. 239–249.
20. Zhao, W.; Zhang, Y.; Lakys, Y.; Klein, J.O.; Etiemble, D.; Revelosona, D.; Chappert, C.; Torres, L.; Cargnini, L.; Brum, R.; *et al.* Embedded MRAM for High-Speed Computing. In Proceedings of the IEEE/IFIP 19th International Conference on VLSI and System-on-Chip (VLSI-SoC), Hong Kong, China, 3–5 October 2011; pp. 37–42.
21. Mackay, K. *TAS, TAS+STT-MRAM and Magnetic Logic Unit*; Technical report; Crocus Technology, e-NVM conference, Gardanne, Provence-Alpes-Côte d'Azur, France, 2011.
22. JEDEC Solid State Technology Association. *Double Data Rate (DDR) SDRAM Standard*; JESD79F; JEDEC Committee JC-42.3; JEDEC: Arlington, VA, USA, February 2008.

13

A Survey of Neural Front End Amplifiers and Their Requirements toward Practical Neural Interfaces

Eric Bharucha, Hassan Sepehrian and Benoit Gosselin *

Biomedical Microsystems Laboratory, Electrical and Computer Engineering Department, Université Laval, 1065 avenue de la Médecine, Québec (Québec) G1V 0A6, Canada;
E-Mails: eric.bharucha.1@ulaval.ca (E.B.); Hassan.sepehrian.1@ulaval.ca (H.S.)

* Author to whom correspondence should be addressed; E-Mail: benoit.gosselin@gel.ulaval.ca

External Editor: Naveen Verma

Abstract: When designing an analog front-end for neural interfacing, it is hard to evaluate the interplay of priority features that one must upkeep. Given the competing nature of design requirements for such systems a good understanding of these trade-offs is necessary. Low power, chip size, noise control, gain, temporal resolution and safety are the salient ones. There is a need to expose theses critical features for high performance neural amplifiers as the density and performance needs of these systems increases. This review revisits the basic science behind the engineering problem of extracting neural signal from living tissue. A summary of architectures and topologies is then presented and illustrated through a rich set of examples based on the literature. A survey of existing systems is presented for comparison based on prevailing performance metrics.

Keywords: neural recording; analog front-end; neural amplifier; low-noise; review

---

1. Introduction

Interfacing electronics with the brain is one of the greatest technical challenges of our era. During the past decade, specialized amplifier circuitry to retrieve weak neural bio-potentials from extracellular microelectrodes has improved dramatically [1]. Such progress enabled to significantly accelerate advances in the field of neural engineering in neural prosthesis, because neural amplifiers are key building blocks of active microelectronic interfaces. Neural interfaces are becoming increasingly important for enhancing our understanding of brain function, and for developing potentially therapeutic and prosthetic applications. Neural interfaces generally aim to utilize the largest number of channels for the study of neurons from specific brain microcircuits. The important design parameters, such as power consumption, noise performance, CMRR and size of neural amplifiers have all improved; and, despite the opposing design options and tradeoffs, many have successfully improved several of these parameters simultaneously. This is in part thanks to innovative design approaches and novel circuit topologies. The objective of this paper is to present a comprehensive review of current designs trends and strategies in neural front-end amplifier circuitry. In the end we glimpse into a comparison of multi-channel systems with a compilation of significant implementations to date.

2. Physiology and Action Potentials, a Brief Review

Action potentials are the basic signal waveform mediating information transfer in neural cells. These action potentials occur because the neural cell membrane reverts its polarization or depolarizes. The occurrence of this event can be understood by the movement of ions across the cell membrane. While at rest, several mechanisms, active and passive, promote the transport of potassium K^+ into the cell; while sodium Na^+ ions are transferred to the exterior. By active and passive, it is implied that they either consume or not the energy storage molecule ATP to mediate their action. Despite the fact that K^+ and Na^+ ions are positively charged; the inside of the cell is more negative than positive. The main cathions involved possess different leakage properties across the cell membrane and other elements, such as negatively charged proteins impose a negatively charged resting potential. Thus, the membrane potential is non-zero and in mammalian nervous systems, the outside of the cell is more positive than the interior, which at rests is approximately around −70 mV. When ionic channels become activated either electrically, biochemically or though other mechanisms, the depolarization that occurs is generally the result of voltage sensitive Na^+ sodium channels opening first. The passage of sodium causes an influx of positive charges into the cell and thus raising the potential there rapidly. The resting potential is restored only when, a few moments later the K^+ channels open up sending potassium ions across the cell membrane and into the external surrounding fluid to re-balance things out. This ionic transit occurs sequentially because the voltage dependent channels for potassium open more slowly than their sodium counterparts. This sequence of events transpires in a timeframe in thc order of the millisecond. A bio-molecular structure, the Na^+/K^+ ATPase pump then restores the resting equilibrium; this molecular ion pump is powered by ATP, the main energy molecule in animal cells [2]. The ion transport model can be understood physically with the electrical circuit model displayed below in Figure 1. In this figure, ionic leakage and active transport can be seen as a set of battery cells in series with variable resistors. The membrane's lipid bilayer can be correctly predicted to act as a capacitor.

The resting potential that ensues is the result of ionic gradients across the cell membrane and can be computed by derivation from the Nernst equation. The famous result is known as the constant field equation first coined by Goldman in 1943. The take home message from an engineering perspective is that nerve signals mainly travel as binary electrical pulses of roughly 0.1 V in amplitude in the 1-ms range and signal amplitude is pulse code modulated. This last point means that intensity is translated as a proportional increase in firing frequency. The resting membrane potential in real cells is estimated as follow:

$$V_m = \frac{RT}{F} ln\left(\frac{p_K[K^+]_o + p_{Na}[Na^+]_o + p_{Cl}[Cl^-]_i}{p_K[K^+]_i + p_{Na}[Na^+]_i + p_{Cl}[Cl^-]_o}\right) \quad (1)$$

in which K^+, Na^+, and Cl^- are the major contributors to the membrane potential. Note that the unit of V_m is the Volt. However, for practical reasons the membrane potential is typically reported in millivolts (mV). If the channels for a given ion (Na^+, K^+, or Cl^-) are closed, then the corresponding relative permeability values can be set to zero. For example, if all Na^+ channels are closed, $p_{Na} = 0$. R is the universal gas constant (8.314 $J{\cdot}K^{-1}{\cdot}mol^{-1}$). T is the temperature in Kelvin (K = °C + 273.15). F is the Faraday's constant (96485 $C{\cdot}mol^{-1}$). p_K is the membrane permeability for K^+. Normally, permeability values are reported as relative permeabilities with p_K having the relative reference value of one (because in most cells at rest p_K is larger than p_{Na} and p_{Cl}). For a typical neuron at rest, $p_K{:}p_{Na}{:}p_{Cl}$ = 1:0.05:0.45. Because permeability is reported in relative terms, permeability values are unitless. p_{Na} is the relative membrane permeability for Na^+. p_{Cl} is the relative membrane permeability for Cl^-. $[K^+]_o$, $[Na^+]_o$ and $[Cl^-]_o$ are the concentrations of K^+, Na^+ and Cl^- on the outside of the cell or extracellular fluid; while $[K^+]_i$, $[Na^+]_i$, $[Cl^+]_i$ are the same ionic concentrations but in the intracellular fluid. During depolarization permeabilities and concentrations are dynamically modulated resulting in the observed waveforms.

Figure 1. The electrical circuit model of the ion transport model.

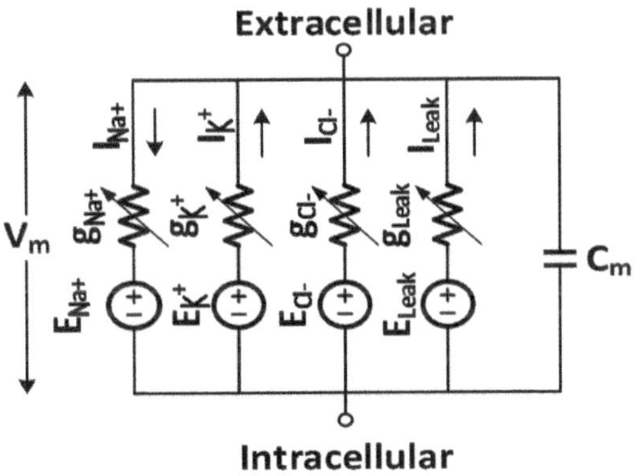

3. Neural Amplifiers Specifications

To get a sneak peak at these action potentials, an amplifying mechanism is required. A neural amplifier must filter neural waveforms to remove DC offsets and augment the resulting signal's amplitude. To provide high signal quality this amplifier must generate sufficient gain, filter appropriate bandwidth, possess high signal-to-noise ratio (SNR) with excellent linearity, and have high common mode and power supply rejection ratios (CMRR and PSRR). The background noise present at the electrode tissue interface is usually in the range of about 10 μV_{rms} or less. This background noise sets the stage for practical neural signal recording. The majority of neural amplifiers that successfully extract *in vivo* action potentials have an input referred noise value below 3–7 μV_{rms} [1,3,4]. Without this controlled noise level, the signal is drowned in a sea of noise. Locally, at the neural cell membrane level, the action potential is in the 100-mV range; but at a relatively short distance away from the cell surface, this potential falls away rapidly. This is why a gain in the 40-dB range is the least amount one can get away with [4]. Action potentials can be argued to have useful frequency content between 100 Hz to about 8 kHz and signal amplitude that can go into single digit microvolt range [1]. It is evident from stimulation studies that several factors are significant; signal characteristics owe some dependency on physiological factors but are also modulated by electrode geometry [5], target neural tissue composition [2] and electrode impedance [6–8]. The presence of external noise as well as internal noise sources, common mode and power supply noise must also be kept in check. Examples of external noise sources are as the electrical activity from muscle contractions, which mostly affects nerve signal recordings and intensified brain activity which can modulate low-frequency baseline potential levels; environmental noise from radio, cellular and electrical equipment in general can also affect performance. In standard MOSFETs, which are prevailing in modern microelectronics, internal noise sources are mainly thermal noise and flicker noise resulting from intrinsic semiconductor properties. Thermal noise can be controlled by modulating transconductance or the W/L ratio. Flicker noise on the other hand, can mainly be improved by increasing gate capacitance and the easiest way to do that is to increase the area of the input pair transistors. In so doing, however, one must use large input transistors. This yet again pushes the design into weak inversion regions [9]. Fortuitously, the weak inversion point is also the optimal current efficiency area of the device. Maintaining low power consumption and low noise thus dictates common design trends. Finally, it should be mentioned that for practical purposes, one must keep chip size within reasonable bounds. Achieving all those specifications simultaneously is a tall order and a challenge that was recognized early on by experimenters and engineers alike.

When it comes to competing engineering options, difficulty arises when trying to compare the performance of different amplifier topologies, architectures and design tradeoffs. One figure of merit that has endured the test of time is the Noise Efficiency Factor (NEF) [10], despite limitations of this comparison metric, and perhaps in lieu of a better alternative, it is still widely used today. It will be used for comparison in later sections.

$$NEF = V_{rms,in}\sqrt{\frac{2 \cdot I_{total}}{4\,\pi\,kT \cdot BW}} \tag{2}$$

4. System Level Specifications

The previous section sets the stage in terms of performance features for a neural amplifier. However, there is a design gap between the single neural amplifier and requirements for a practical system. Indeed additional design constraints are befitting; for starters, if one wishes to observe even a small parcel of living brain function, multi-electrode systems are essential. In multi-electrode systems, one amplifier is often needed for each channel. Four architectures are possible: either a fixed architecture in which one electrode is connected to one amplifier; a semi-static architecture where the number of electrodes is greater than the number of amplifiers; a third architecture where electrode/amplifier assignment can be configured occasionally to account changes in performance; and finally, a real-time dynamic system in which assignment and topology may be reassigned in real-time.

Regardless of architectural choice the requirement for multiple channel observation dictates low power operation, reduced size, and scalability for integration into implantable devices. Yet, the diminutive magnitude of *in vivo* signals requires amplifiers to have substantial gain at low noise and all this demands power. The power conundrum is aggravated by a third limitation; heat dissipation. It is widely accepted that the maximum practical power consumption for a neural implant is in great part set by the maximum heating of its components. Several groups have estimated this limit [11,12], and some normative texts even indirectly dictate it [13]. It has also been shown that a heat flux of 80 mW/cm^2 can be set as a safe limit to uphold [14], in that article, a 4–5 °C elevation was given as an empirical limit based on observed tissue response pathology. In the medical device industry; the acceptable temperature rise limit is usually regulated at ≤2 °C. That limit is typically the maximum tolerated for all commercial implants [13]. Others go even further setting the safety limit at 1 °C or lower [15]. The thermal ceiling imposes a threshold for power dissipation in small device enclosures. Wireless power transmission is also an issue because of inductive heating losses in circuit structures. Furthermore, with inductive charging, the practical charge rate is also limited for commercial medical devices. The reason for this is that electromagnetic field strength exposures are constrained for safety. Within a chip scale device, a practical estimate would typically result in a few mA of available current supply. For a 1.8 V supply for example, this so called limit can be estimated to be 3–5 mA/h or 9 mW/h [11,16]. For useful brain machine interfaces as envisioned by many; a very large number of electrodes will be required. It is expected that systems with thousands of electrodes will emerge within the next two decades [17,18]. Let's suppose as an example such an implantable device with 10 thousand electrodes powered from a state-of-the-art medical grade rechargeable battery with a current capacity of 200 mA/h at 3.7 V. Battery chemistries remain relatively stable and slow changing in medical devices as safety regulations are extremely stringent. If the battery needs to be fully recharged every three to four days; then this leaves us with a power budget about 9 mW, the previously demonstrated limit. With control circuitry, ADC, signal processing, telemetry and heat losses there may be at best half the available power left for neural sensing amplifiers. Theoretically, this would mean a maximum per channel of approximately 450 nW. This is quite a challenge even without considering integration obstacles. Although the 10 thousand electrode count may seem ambitious, similar large count electrode arrays have already been reported and interfaced [19] albeit at a lesser functionality and multiplexed channel conditioning circuitry. Unlike the critical parameters just illustrated for brain machines; this last example was devised for the purpose of cultivated cells in a controlled environment. Regardless of the technology; the goal of these systems is to develop efficient neural interfaces for peering into action potentials mediated brain activity.

5. Circuit Topologies

From the aforementioned specifications, designs can be implemented using various circuit architectures. The most common circuit topologies for implementing low-noise operational transconductance amplifiers (OTAs) are the two stage Op-Amp (Miller OTA, see Figure 2 below) and the current mirror OTA; also there is the folded cascode that allows larger swing margins or reduced supply margins. This last architecture is becoming increasingly popular as the push for sub 1 V supplies becomes prevalent. Other notable architectures include the telescopic cascode amplifier as reported by [20–22]. Source degenerate active loads lead to significantly diminished currents in the folded branch compared to the differential pair [23]. Finally, one last structure that deserves mention is the differential self-biased OTA as reported by [24]. A full treatment comparing the noise performance of these architectures can be found in [25].

Figure 2. Low-noise OTA. (**a**) Miller OTA; (**b**) current mirror OTA; (**c**) folded cascade; (**d**) telescopic and (**e**) differential self-biased OTAs.

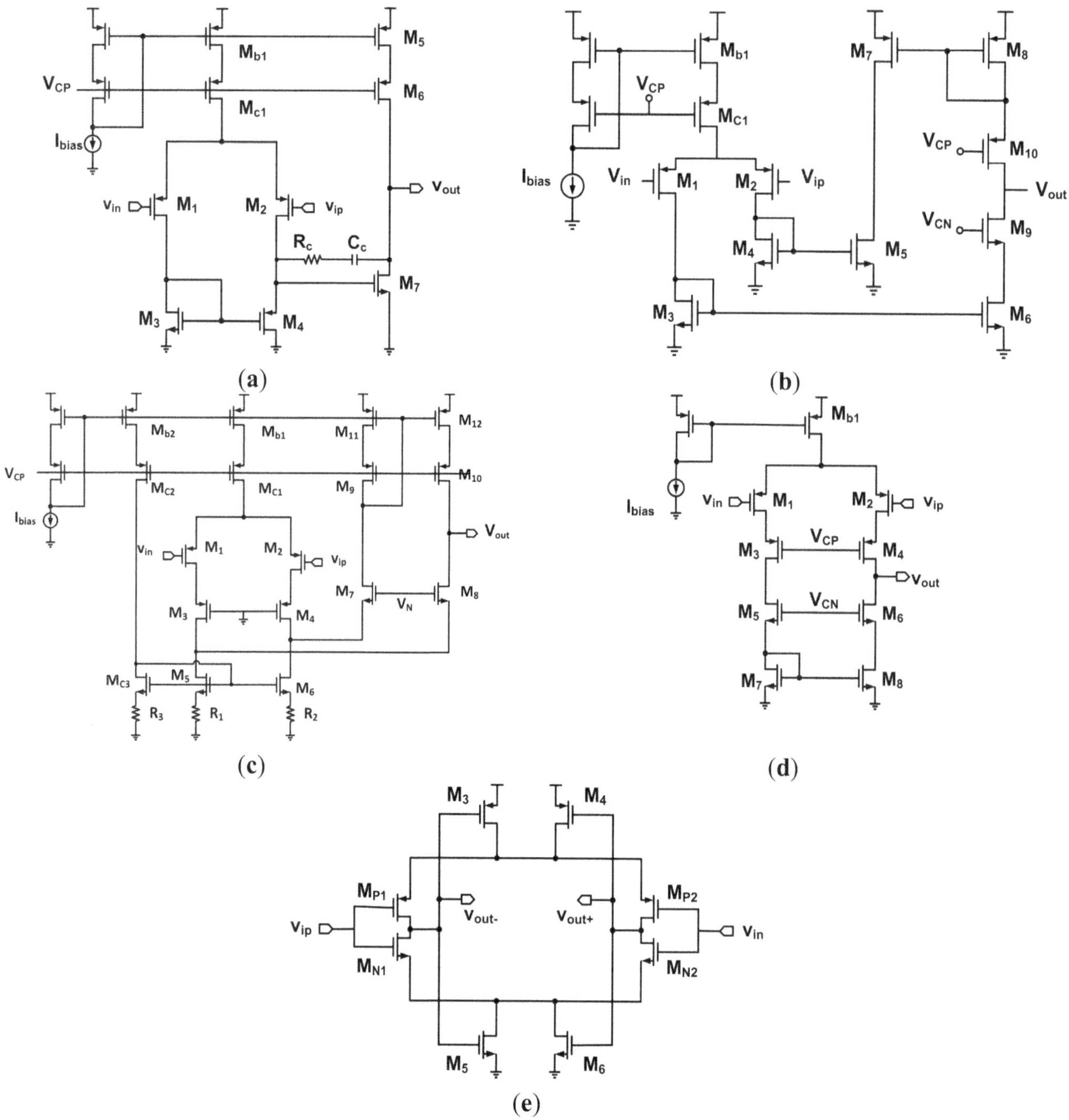

The previously mentioned topologies can be implemented in differential form. In [21], several differential architectures including closed-loop fully differential, telescopic-cascode amplifier;

complementary-input open-loop amplifier and complementary-input closed-loop amplifier architectures are presented and tested. The last one was the most successful achieving a gain of 40.5 dB an NEF of 4.5 and a power consumption of 12.5 μW. The chip area was also impressive being below 0.05 mm^2. The table below summarizes the main characteristics achieved by theses comparative designs. One should note here that the second architecture had a debilitating PSRR. Figure 3 also shows the circuit of the most significant implementation.

Table 1. Comparison of the different structures reported in [21].

Architecture	NEF	Power consumption (μw)	Mid band gain (dB)	Area (mm^2)
Telescopic-cascode Closed-loop Amplifier	4.5	12.5	40.5	0.047
Complementary-input Open loop Amplifier	1.9	0.8	36	0.046
Complementary-input Closed-loop Amplifier	2.9	12.1	40	0.072

The current feedback instrumentation amplifier possesses key features, such as high impedance and CMRR [26]. However, common problems with this architecture include offset in the mV range. Chopping effectively modulates noise and offset to a pre-selected chopping frequency resulting in a white noise base band. The downside as in most switching architecture is a ripple at the output. Different topologies exist for canceling the ripple [27–32] but they require high chopping frequencies. [33] shows that this can be avoided by using a ripple reduction loop. This type of feedback loop reduces unwanted ripple in the low-frequency chopped output filter by effectively introducing a notch filter in the pass band. Although chopper stabilization results in lower noise; the design is ill suited for large scale integration where power consumption and circuit complexity push the design towards simpler architectures when optimization is key.

Figure 3. The efficient complementary input closed-loop fully differential design as reported by [21].

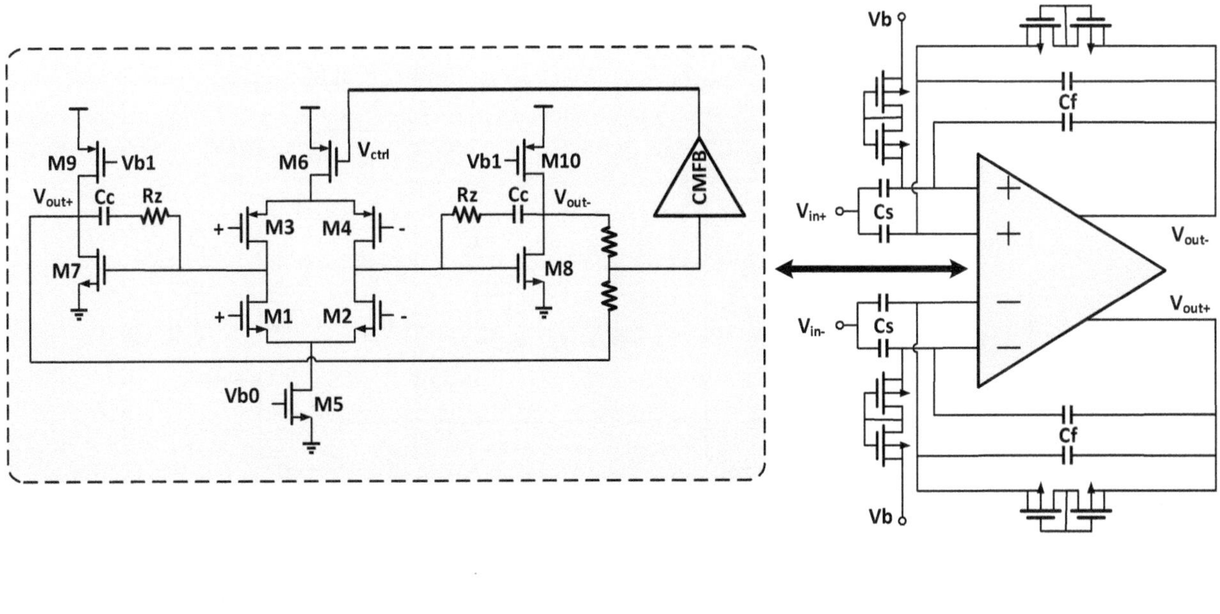

Recently, a NEF of 1.64 was achieved by using the technique of current reuse as described by [34] and [35] (see Figure 4). In this scheme, current is reused between parallel stages to save power consumption and silicon area. Such a scheme consists of multi-staked differential input pairs (Figure 6a) and a current recombination block (Figure 4b), which separates the output currents assigned to the input signals, and an output stage (Figure 4b), which uses recombined currents to generate an output voltage that corresponds to a given input signal. One current source is employed to bias each stack of several differential input stages, in which tail currents are coming from previous stage. The schematic of a two stacked input stage current reuse amplifier is presented in Figure 4a. The second stage (Stage 2 in Figure 4a) is divided into two differential pairs, the input transistors of which are half the size of the input transistors of the first stage. Both stacked input-stages (Stage 1 and Stage 2 in Figure 4) have same g_m. As the total current that flows in second stage is same as the current that flows in the first stage (I_{bias}), the overall g_m in the second stage is equal to the g_m in the first stage. Each output current (small signal currents i_1, i_2, i_3, i_4) is an independent and linear combination of several output currents derived from the corresponding input voltages. The input pairs in the second stacked stage are parallel because they have same inputs (V_{in2+}, V_{in2-}), but they are independent since their tail currents ($I_{bias}/2$) are independent. For example, to generate V_{Out1}, V_{Out1+} and V_{Out1}, the corresponding output currents are re-constructed by summing currents i_1, i_2 and currents i_3, i_4 respectively. V_{Out2} is generated by summing i_1, i_3 currents to construct V_{Out2+}, and currents i_2, i_4 to construct V_{Out2-} (See the schematic of the Output stage in Figure 6b). However, when employed in a multi-channel system, there is some concern that the resulting crosstalk would likely cause issue with increasing channel count. Current reuse as a noise and power reduction method was also discussed in [36].

Figure 4. Two-Stage orthogonal current reuse scheme by [35]. (**a**) The input block consists of two stacked input differential pairs; and (**b**) the output stage includes recombination current mirrors.

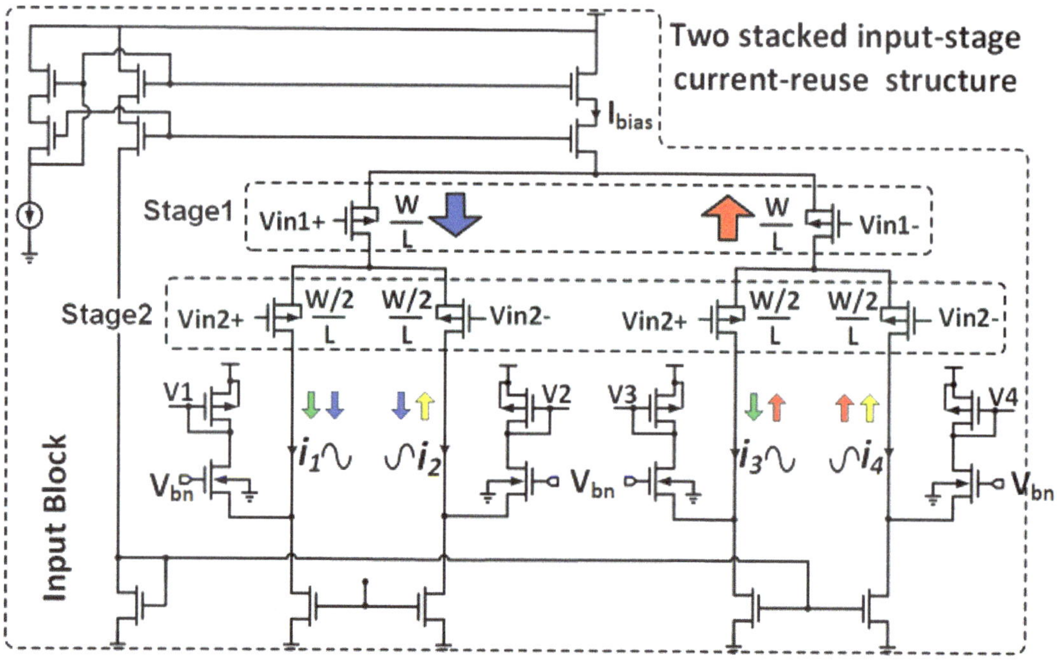

(a)

Figure 4. *Cont.*

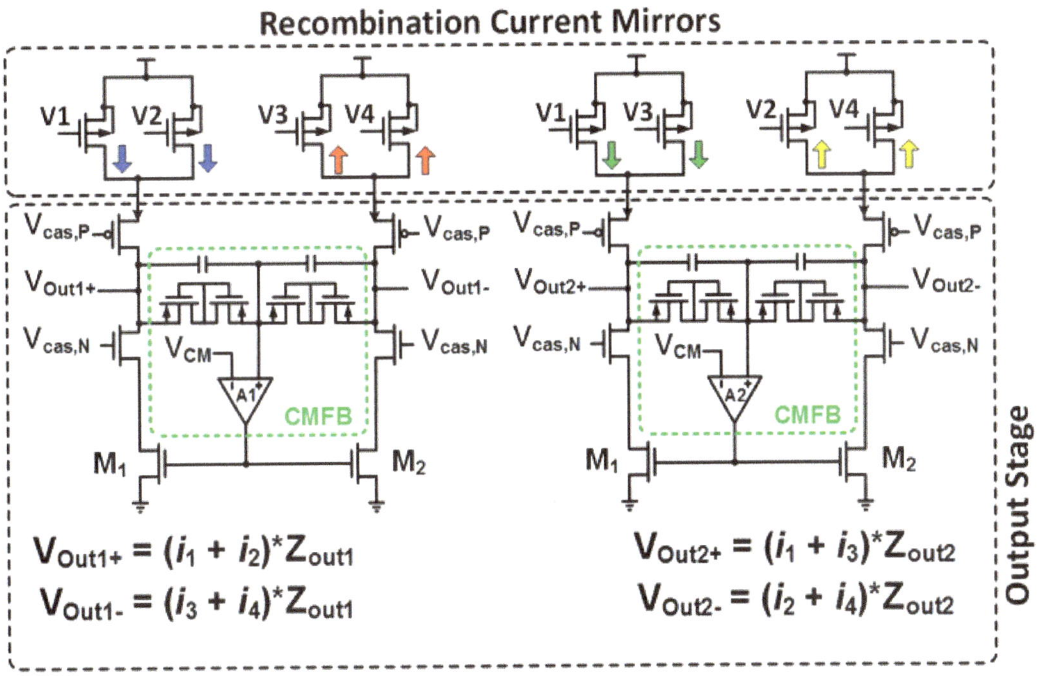

(**b**)

6. System Architectures

6.1. Open-Loop Amplifiers

The above circuit topologies can be used to implement different neural front-end amplifier architectures. Taking a top down approach, a comparison of amplifier topologies is appropriate. The first is the open-loop topology; this structure offers high gains at lowered noise levels it is often implemented adopting some sort of instrumentation *type* layout. Generally open-loop circuits are used for either extremely low power or extremely low noise, but have rarely been reported to achieve both objectives simultaneously. Some good design examples can be found in [37–39] refer to Figure 5 below. It is interesting to note that manually tuned tube versions were used by early electro-physiologists. Open-loop amplifiers are nowadays seldom used in implantable devices. This can be explained by their inherent power and size disadvantages when designing for portable systems. They must often be tuned or screened for performance due to a lack of reproducibility between devices [21]. The problem is mainly due to the gain and CMRR variability that results from imperfect lithographic tolerance between circuit structures. For microelectronics, successful open-loop designs typically use mature, lower resolution, processes as the relative process tolerances are diminished. This approach yields better performance repeatability between channels and devices at the cost of size increase and higher power consumption. Without the lithographic considerations mentioned above, significant discrepancy are observed between simulated and fabricated devices especially for CMRR; the empirical difference in value reaching up to 40 dB from simulations. Modern microelectronic applications are mainly limited to cuff electrode recording where the mechanical layout and nerve implant site require particularly low input noise performance that can be achieved with this architecture. State-of-the-art for tripolar recording systems using this technology is reported in [40]. This amplifier consumes a relatively elevated 310 μA at 3 V but exhibits a close to theoretical limit of 0.68 μV_{rms}. [41] also describes a micropower fully differential instrumentation amplifier capable of monitoring neurochemical modalities by its sensitivity also reaching a low input referred noise of 2 μV_{rms}. The dynamic range of this current mode amplifier spans an impressive six decades from picoamps to microamps and is thus well suited for scientific investigation as it is also wirelessly interfaced. [39] presented an open-loop sub-microwatt amplifier that made use of pseudo-resistors. Their design achieved 36 dB of gain with a bandwidth of 0.3 to 4.7 kHz and had a respectable input referred noise of 3.5 V_{rms} and NEF of 1.8.

Figure 5. Open-loop schematic reported in [38].

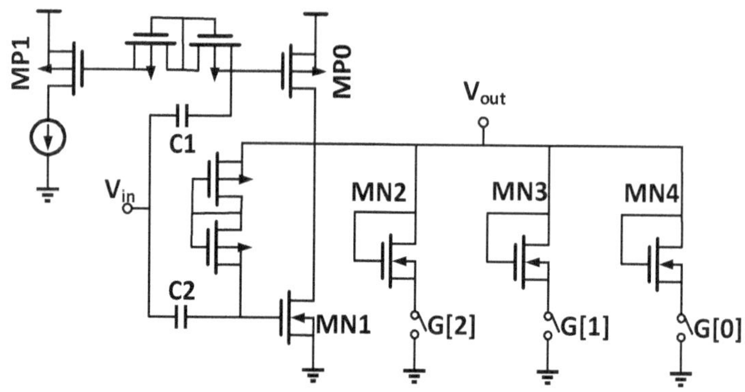

Figure 6. Classic neural amplifier topology with capacitive feedback.

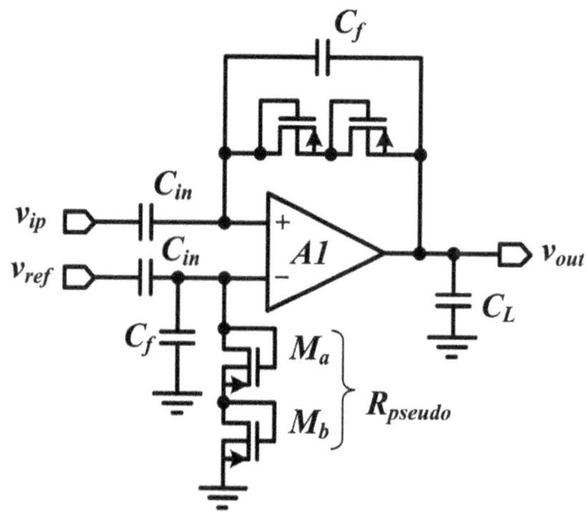

6.2. Capacitive Feedback Topology

In the last decade, engineers have widely borrowed a model that early on, achieved many of the desirable features of neural amplifier. Thus know as capacitive feedback techniques has been the method of choice for many neural amplifier designs, as first reported by [12]. Figure 6 below show the basic topology. It is obvious to see that it is derived from the ubiquitous feedback topology, first introduced by H.S. Black in 1927. Here an impedance element constrains gain by a proportional bandwidth reduction. Increased power consumption is also the cost of "controlling" signal output via the feedback loop and providing a controllable linear response. Typically a small chip scale capacitor is use in conjunction with a very large MOSFET based resistor to create the required low pass response for neural amplifiers in scalable fashion.

The capacitive feedback topology has been optimized for very low-power operation through a dedicated circuit design methodology. The original design achieved a gain of 39.5 dB while posting a power consumption 80 μW and a respectable NEF of four. Modifications and extensive optimization for low-noise resulted through the use of source degenerate current mirrors and high current scaling ratios between input differential pair and folded branch transistors of a folded cascode design, as reported by [23]. The successful results were only possible due to the careful matching of the degenerate mirror. This resulted in a current consumption of 7.56 μW power consumption and an NEF of 2.67 at a gain over 40 dB.

Others have optimized the capacitive feedback topology; [42] reports an ultra-low-power 77-nW bioamplifier, with an NEF of 1.32 and tunable bandwidth, however this circuit also has an input referred noise of 14.3 μV_{rms}. The circuit is a classic capacitive feedback single stage LNA. Similarly, [43] described a 220-nW neural amplifier, also with tunable bandwidth. However, Kim added a buffer, adopting a two-stage architecture with a similarly limiting input noise of 14.5 μV_{rms}. The high input noise and low reported gains (around 30 dB) severely limit the usefulness of these amplifiers. It is important to note here that NEF as a means for comparing amplifiers is practical; but the previous design illustrate its limitation. As seen, a very low NEF can result in bad performance if the input referred noise isn't constrained within the bounds of ambient noise.

More recently [3,11] reported a power consumption of 7.92 μW and an input referred noise of 3.5 μV_{rms}. Also described is an OTA sharing architecture that effectively reduces both chip area and NEF, which was reported to be 3.35. The sharing architecture, in essence, commonly replaces C_f, from Figure 6 above and the MOS pseudo-resistance composed of M_a and M_b in the reference branch input for all channels (see Figure 6) of a classic capacitive feedback design.

6.3. Active Feedback Topology

The third significant topology discussed is the active feedback topology. In this topology a segment of the signal spectrum is filtered and used as a feedback canceling element for suppressing that signal component. This is a very efficient way of extracting low-frequency components from the signal of interest and makes for very stable systems. This method is commonly used in other biomedical engineering applications such as EEG. Figure 7 displays the salient features of this topology. Another advantage of this architecture is the reduction in size that it allows by reducing the capacitor size compared to the capacitive feedback topology [44]. The slowly drifting DC offset is also eliminated in this way. Muller also followed suit on this design by modifying the active feedback with a digital to an analog converter achieving further reductions in size [45] however most of the improvement there shifts analog hardware to a digital platform. Wei's work [46] upholds the arguments of efficiency for this architecture by achieving a 6 μW, an NEF of 3.1, a bandwidth of 8.9 kHz and a mid-band gain of 46 dB occupying a mere 0.022 mm^2 [47] also showed an interesting and novel local feedback at the output technique for reducing DC output. Yet another approach is to design an amplifier in current mode active feedback as demonstrated by [48] and [49]. [48] Achieved a 13 μW power consumption with a 55 dB gain with a 10 kHz bandwidth and a final size of 0.76 mm^2. One interesting feature is that no large value resistor or capacitor needs to be used to achieve a high-frequency cutoff of 0.3 Hz.

Figure 7. Active feedback LNA topology.

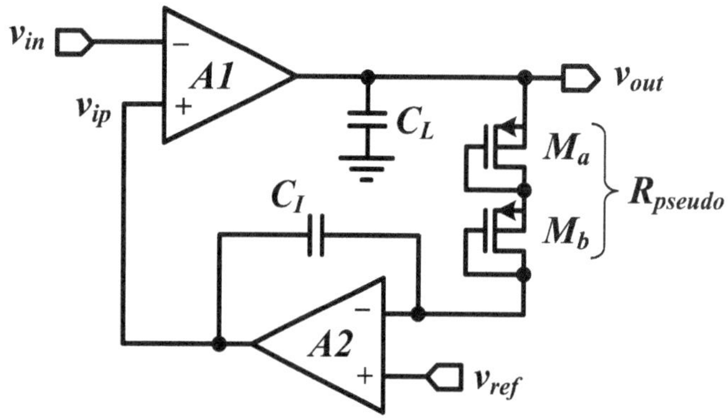

A comparative study of a capacitive feedback, a Miller integrator feedback and a capacitive amplifier feedback networks showed that for a given input noise the capacitive feedback topology edged out other designs in terms of NEF, area and power [4]. In this work; area, NEF and power consumption were constrained and plotted against noise in an iterative process until they reached preset gain and frequency specifications. Although the study is interesting, this topic is likely to be mooted in future, as the study was theoretical and other critical design parameters like CMRR and DC rejection performance were not analyzed or observed. Furthermore, it was demonstrated above that NEF can be a pitfall, especially when used as a target specification due to existing power and noise antagonism.

6.4. Pseudo-Resistors

In many of the previous sections we briefly touched upon the use of pseudo-resistors. The large input impedances required for measuring biopotentials as well as the requirement to amplify signals in a range starting from the sub-Hz range requires large resistors. However such resistors are impractical because of the large size that would result. One clever solution to this problem is the use of pseudo-resistors. Such devices provide a means of generating large impedance using biased MOSFET channels. Pseudo-resistor design techniques are extensively discussed in [50] they make use of cross-coupled pseudo-resistors to tune band pass circuits and use fixed versions for the input block. Other design examples can also be found in [51–53]. Pseudo-resistors offer high impedance with limited space use, a desirable feature, additionally they offer tuning possibilities. [54] demonstrated full bandwidth adjustments using pseudo-resistors. Linearity of pseudo-resistors is also an issue. [51,55] addressed the linearity issue by using a cross-coupled circuit with a referenced bias and a sine wave that only turns on one of the pseudo-resistor MOSFETs at a time thereby eliminating voltage variations and the non-linear resistive effects that result from this.

6.5. Adaptive Topology

One emerging method for reduction power consumption in amplifier circuits is to have variable performance features or tunable devices. This allows the modulation overall power duty cycling depending on functional needs. This can take the shape of topology changes and broad range tuning [56,57] uses similar techniques to modulate slew rate and phase margin depending on circuit use. [28] reports an asynchronously adjustable data output rate. [58] describes duty cycling and time division techniques for improving front-end overall power efficiency; using such techniques a LNA with power consumption of 2 μW was achieved for a 32 KHz bandwidth and an excellent NEF of 1.3. A sophisticated power scheduling scheme is often used in medical devices as exemplified in [59] were a duty cycling is applied to the higher power consumption circuits. In this example sensor conditioning amplifiers and data telemetry have a modulated power scheme. One caveat is that power cycling applied front end circuits require fast settling times such that the sensor signal can be sampled in the shortest time following a power on of the circuit. This reduced settling time requirement usually translates into higher current consumption. A fine balance must therefore be struck between the selected designs tradeoffs, as the combination of benefits does not follow a linear trend when it comes to power economy.

7. Integrated Systems Based on Neural Amplifiers

Merging subsets of the above techniques has resulted in an impressive collection of complete neural acquisition systems. Many such implementations have been reported over the years. In chronological order of reporting for high performance multi-channel systems a list of relevant papers is reported in Table 2 below. Such systems are not so easily compared as they have different aims and inherently different design optimization schemes. Selected systems had to have a bandwidth that permitted neural recording at least up to 4 KHz. For current relevance; our survey was limited to the previous seven years. The main architecture and topology details are listed for reference. Also listed are the NEF, the Gain, the supply voltage and current, the CMRR and the area. Some data were estimated from available information. At the time of writing, the highest numbers of amplifier channels are in the low hundreds as reported by [20] and [60].

8. Discussion

Better designs begin with a better understanding of the problem at hand. This is the reason why we started with an overview of physiological mechanisms involved in neural firing. Three main points are worth noting. At the cell surface nerve signals typically have an amplitude of about 100 mV, their period is of the order of the millisecond and signals are often pulse code modulated. It was then noted that a noise performance in the range of ≤5 μV_{rms} was a key factor in achieving a quality neural signal. This last requirement comes with the caveat that despite the increased need for power, heat must be limited such that the devices maintains a ΔT within 2 °C. Adding these constraints, power and chip size are limited. For power, we argued that a 450 nW/ch is a singularity point after which practical large-scale systems would become possible.

From these specifications, classic circuit topologies were reviewed including the Miller OTA, the current mirror OTA, the folded cascode, telescopic and self-biased as well as their derivatives. Finally we delved into current feedback, current reuse and chopping architecture as means for improved CMRR, power consumption and noise control.

In the following section our attention is focused on system level architectures, the principal ones described are: open-loop, capacitive feedback, active feedback and adaptive or variable circuits. Open-loop circuits provide the best noise performance but at the cost of higher power consumption and a lack of inter device repeatability. Capacitive feedback is the most popular topology at present because of the good compromises between competing design objectives. The main features for active feedback are spectral separation and stability. Later, we report on the necessary trend of power scaling, and power management that results in optimized circuit topology.

Finally significant complete systems from the past seven years were listed in a comprehensive table (Table 2) lending to easy comparison of surveyed designs. It can be remarked that increased channel count devices typically are still an order of magnitude too high in power consumption when compared with the requirements we aimed for initially (450 nW) see [20,60–62]. In state-of-the-art devices, the channel count is also an order or two of magnitude below what we would needed for complex neural circuit observation. Typically, noise and gain are marginally within the aforementioned requirements of 40 dB and about 5 μV_{rms} noise; but it's a fine balance between the two [62]. Diminishing size is the metric that has scaled the best in the past years starting with great leaps in 2009 by [60] and [63,64].

With the coming bounty of channels and signals, comes the imperative of signal processing. In that line of thought, the aim is to decipher the underpinnings of neural circuitry and not observe the action potential in a traditional scientific approach. Rather, a pragmatic engineering perspective is to extract only the required firing information before transmission. In that effort, many have labored on the ever increasing need for such spike sorting circuitry. Some reference on the topic have included: [61,63–66].

Although great strides have been made in the past decade, a manufacturable, high-performance, low noise sub-microwatt neural amplifier building block remains an elusive quest. The competing demands for size, power, speed, noise control and integration illustrated still amount to a significant challenge even with today's tools.

9. Conclusions

Much of the past has focused on this microscopic aspect of nerve cells function resulting in a wealth of information regarding action potential mechanisms. We can now understand and accurately simulate cellular interaction quite well. Humans have also observed the macroscopic effect of neural circuitry through behavior analysis and brain physiology as far back as the earlier part of the last century. Current trends provide great optimism that solution for a practical neural front end is within grasp. When beginning the design of a front end amplifier one must ask: Is the design objective to observe details of the action potential waveform or detect the presence of this action potential? Answering this question has significant impacts on design decisions. The former approach has guided neuroscience of the past 70 years while the later seeks to exploit this new information. However to obtain the "in between" part, we need to focus on the significant portions of the signal processing chain and thus the required electronics. We presented not only specifications for achieving efficient neural amplifiers but also the intricate requirements to achieve practical brain interface systems. It would be impractical to design a multi-million transistor digital circuit by modeling every transistor simultaneously. Based on that premise, sensing all signal at high resolution simultaneously is not likely the correct approach to decipher the mechanisms of the brain.

Table 2. Comparison of previously reported multi-channel neural front-end amplifier and systems. Some of the acronyms and contractions used are defined as follows: N/A is used to indicate that information is not available, missing or cannot be computed from published values. CMRR is the "Common Mode Rejection Ratio"; HP stands for "high pass" in the filter sense; LNA stands for "Low Noise Amplifier"; NEF is the "Noise Efficiency Factor" as defined in [10]; OTA stands for "Operational Transconductance Amplifier"; BW stands for "Band Width".

Author	Year of publication	Circuit topology/Architecture	Number of channels	Gain (dB)	NEF	Frequency range (Hz)	Noise (μV_{rms})	Supply voltage (V)	Supply current (μA)	CMRR (dB)	Power consumption per Ch. (Front-end only) or * entire chip (μW)	Area per Ch./Entire chip (mm^2)
Harrison [67]	2007	Two-stage with added g_m-C HP filter/Capacitive feedback	100	60	10.47	300–5 k	5.1	N/A	12.8	N/A	13500 *	N/A/5.9 × 4.7
Chae [61]	2008	Two-stage/Capacitive feedback differential	128	40	8	0.1–20 k	4.9	1.65	40.0	90	6000*	N/A/8.8 × 7.2
Aziz [60]	2009	Two-stage/Capacitive feedback + transconductance LNA	256	48–68	4.6	0.01–5 k	7	3	5.0	N/A	15	0.04/3.5 × 4.5
Gosselin [68]	2009	Active feedback/Multiple stage	16	70	4.9	100–9.2 k	5.4	1.8	4.8	45	8.6	0.05/2.304
Mollezadeh [41]	2009	Two stage/Capacitive feedback	16	39.6	2.9*	0.2–8.2 k	1.94	3.3	8.0	70	26.4	0.107/N/A
Sodagar [64]	2009	Two stage/Capacitive feedback	64	59.5	21.3	24m–9.1 k	8	1.8	41.7	N/A	75	0.072/N/A
Liew [69]	2009	Two-stage/Capacitive feedback	16	45.7–60.5	2.16	0.23–7.8 k	4.43	1	3.8	58	3.77	N/A
Miranda [70]	2010	N/A	32	N/A	N/A	1–4.5 k	5	N/A	N/A	N/A	142000 *	N/A/24000
Perlin [71]	2010	Two-stage/Capacitive feedback	64	60	11.42	<10–9.1 k	4.8	1.5	33.3	N/A	50	0.098/N/A
Shahrokhi [20]	2010	Fully differential/Telescopic	128	33	5.55	10–5 k	6.08	3	2.8	N/A	8.4	0.02/N/A

Table 2. *Cont.*

Author	Year of publication	Circuit topology/Architecture	Number of channels	Gain (dB)	NEF	Frequency range (Hz)	Noise (μV_{rms})	Supply voltage (V)	Supply current (μA)	CMRR (dB)	Power consumption per Ch. (Front-end only) or * entire chip (μW)	Area per Ch./Entire chip (mm^2)
Greenwald [72]	2011	Two-stage with Common mode feedback/Differential	16	40	4.61	N/A–8.2 k	3	3.3	12.5	N/A	41.25	N/A/(3.4 × 2.5)
Aceros [73]	2011	Two stage/Capacitive feedback	32	45.6	115.66	0.1–7.8 k	8.5	3	14.5	N/A	43.8	N/A
Al-Ashmouny [17]	2011	Two stage/Capacitive feedback	16	52.4–79.8	2.9	0.1–17 k	6.76	0.9	3.7	60	3.3	0.07/N/A
Jo-Yu [74]	2011	Two stage/Capacitive feedback	4	51.9	2.79	2.38–12.9 k	4.7	−0.9/+0.9	2.9	N/A	5.22	N/A
Lopez [75]	2011	Folded Cascode OTA multiple stages/Capacitive Feedback	16	40–60	12.12	2.6–6.2 k	2.9	3.3	70.0	>63	231	N/A/25.2
Majidzadeh [11]	2011	Shared reference structure/Capacitive feedback	N/A	39.4	3.35	10–7.2 k	3.5	1.8	4.4	70.1	7.92	0.065/N/A
Rouse [76]	2011	N/A	96	variable	N/A	5 k	N/A	1.7–2.2	2.5	>80	4.25–5.5	N/A
Szuts [77]	2011	N/A	64	65	N/A	10–4.5 k	4	3	N/A	N/A	N/A	N/A
Wattanapanitch [78]	2011	Three-stage—Source degenerate active loads/Capacitive feedback	32	49–66	4.4–5.9	350–11.6 k	5.4–11.2	1.8	3–11.11	62	5.4–20	0.03/N/A
Lo [79]	2011	Two stage/Capacitive feedback	64	47–59	3	0.5–12 k	3.8	1.2	5.0	N/A	6	N/A/(3 × 4)

Table 2. *Cont.*

Author	Year of publication	Circuit topology/Architecture	Number of channels	Gain (dB)	NEF	Frequency range (Hz)	Noise (μV_{rms})	Supply voltage (V)	Supply current (μA)	CMRR (dB)	Power consumption per Ch. (Front-end only) or * entire chip (μW)	Area per Ch./Entire chip (mm^2)
Zoladz [80]	2011	Two stage—folded cascode/Capacitive feedback	64	60	21.28	0.1–12 k	3.7	1.65	15.2	48	25	N/A/(5 × 5)
Yin [81]	2012	Two-stage/Capacitive feedback	100	46	3.3	0.1–7.8 k	2.83	3	20	60	60	N/A/(5.2 × 4.9)
Gao [82]	2012	Fully differential/Capacitive feedback, with switch-cap filtering	96	40–56	6.62	1–10 k	2.2	1.2	56.7	N/A	68	0.26/(5 × 5)
Dong [83]	2013	Two-stage/Capacitive Feedback—fully differential	100	52	1.57	1–10 k	3.2	0.45	162.2	73	73	N/A/25
Johnson [34]	2013	Orthogonal Current-Reuse/ Capacitive feedback	4	40	1.64	19.9 k BW	3.7	1.5	2.6	78	3.9	0.125/N/A
Kmon [56]	2013	Two stage—folded cascode/Capacitive feedback	8	48/60	4.6	0.3–9 k	5	1.8	6.1	48	11	0.065/N/A
Zou [55]	2013	Low Noise OTA/Capacitive feedback multi-stage	100	-	1.9	0.001–5.1 k	4	1.8	6.4	60+	11.6	N/A/28.2
Yin, [84]	2013	Two-stage/Capacitive feedback	100	46	3.3	0.1–7.8 k	2.83	3	20	60	60	N/A/(5.2 × 4.9)
Lopez [62]	2014	Two-stage/Capacitive feedback	55/455	29.5/72	3.08	0.2–6 k	3.2	1.8	3.9	60	7.02	0.19/N/A
Sepehrian [35]	2014	Orthogonal Current-Reuse/ Capacitive feedback	4	45.2–59.7	4.37	10.02 k BW	3.28	1.8	2.27	76	4.1	0.035/N/A

Acknowledgments

This work was supported, in part, by the Natural Sciences and Engineering Research Council of Canada, the Fonds de recherche du Québec—Nature et technologies and by the Microsystems Strategic Alliance of Quebec.

Author Contributions

In this work, Éric Bharucha conducted an in-depth survey of the literature, and he wrote the initial draft of the paper. He was responsible for comparing the presented circuit topologies and discussing their advantages and disadvantages, and for drawing figures. Hassan Sepehrian and Benoit Gosselin contributed to the writing of the paper, they drew figures and they proofread the paper.

Conflicts of Interest

The authors declare no conflict of interest.

References

1. Jochum, T.; Denison, T.; Wolf, P. Integrated circuit amplifiers for multi-electrode intracortical recording. *J. Neural Eng.* **2009**, *6*, doi:10.1088/1741-2560/6/1/012001.
2. Shepherd, G.M. *Neurobiology*, 2nd ed.; Oxford University Press: New York, NY, USA, 1988.
3. Majidzadeh, V. A Micropower Neural Recording Amplifier with Improved Noise Efficiency Factor. In Proceedings of the European Conference on Circuit Theory and Design, Antalya, Turkey, 23–27 August 2009; pp. 319–322.
4. Ruiz-Amaya, J. A Review of Low-Noise Amplifiers for Neural Applications. In Proceedings of the 2nd Circuits and Systems for Medical and Environmental Applications Workshop (CASME'10), Merida, Mexico, 13–15 December 2010.
5. Krasteva, V.; Papazov, S. Estimation of current density distribution under electrodes for external defibrillation. *BioMed. Eng. OnLine* **2002**, *1*, doi:10.1186/1475-925X-1-7.
6. Ackermann, D.M.; Bhadra, N.; Foldes, E.L.; Wang, X.F.; Kilgore, K.L. Effect of nerve cuff electrode geometry on onset response firing in high-frequency nerve conduction Block. *IEEE Trans. Biomed. Circuits Syst.* **2010**, *18*, 658–665.
7. Ackermann, D.M.; Foldes, E.L.; Bhadra, N.; Kilgore, K.L. Electrode Design for High Frequency Block: Effect of Bipolar Separation on Block Thresholds and the Onset Response. In Proceedings of the Annual International Conference of the IEEE Engineering in Medicine and Biology Society (EMBC'09), Minneapolis, MN, USA, 3–6 September 2009; pp. 654–657.
8. Agnew, W.F.; McCreery, D.B. *Neural Prostheses: Fundamental Studies*; Prentice Hall: Englewood Cliffs, NJ, USA, 1990.
9. Baker, R.J. *CMOS: Circuit Design, Layout, and Simulation*; Wiley: Hoboken, NJ, USA, 2011.
10. Steyaert, M.S.J.; Sansen, W.M.C. A micropower low-noise monolithic instrumentation amplifier for medical purposes. *IEEE J. Solid State Circuits* **1987**, *22*, 1163–1168.
11. Majidzadeh, V.; Schmid, A.; Leblebici, Y. Energy efficient low-noise neural recording amplifier with enhanced noise efficiency factor. *IEEE Trans. Biomed. Circuits Syst.* **2011**, *5*, 262–271.

12. Harrison, R.R.; Charles, C. A low-power low-noise CMOS amplifier for neural recording applications. *IEEE J. Solid Stare Circuits* **2003**, *38*, 958–965.
13. British Standards Institution. *Active Implantable Medical Devices. General Requirements For Safety, Marking and Information to Be Provided by the Manufacturer*; British Adopted European Standard, B. EN45502–1:1998; BSI: London, UK, 1998.
14. Seese, T.M.; Harasaki, H.; Saidel, G.M.; Davies, C.R. Characterization of tissue morphology, angiogenesis, and temperature in the adaptive response of muscle tissue to chronic heating. *Lab. Investig.* **1998**, *78*, 1553.
15. IEEE. *IEEE Standard for Safety Levels With Respect to Human Exposure to Radio Frequency Electromagnetic Fields, 3 kHz to 300 GHz*; IEEE Std C95.1–2005 (Revision of IEEE Std C95.1–1991); IEEE: Piscataway, NJ, USA, 2006; pp. 1–238.
16. Silay, K.M.; Dehollain, C.; Declercq, M. Numerical Analysis of Temperature Elevation in the Head Due to Power Dissipation in a Cortical Implant. In Proceedings of the 30th Annual International Conference of the IEEE Engineering in Medicine and Biology Society (EMBC'08), Vancouver, BC, USA, 20–25 August 2008; pp. 951–956.
17. Al-Ashmouny, K.; Sun-Il, C.; Euisik, Y. A 4 mW/Ch Analog Front-End Module with Moderate Inversion and Power-Scalable Sampling Operation for 3-D Neural Microsystems. In Proceedings of the IEEE Biomedical Circuits and Systems Conference (BioCAS'11), San Diego, CA, USA, 10–12 November 2011; pp. 1–4.
18. Olsson, R.H.; Wise, K.D. A three-dimensional neural recording microsystem with implantable data compression circuitry. *IEEE J. Solid State Circuits* **2005**, *40*, 2796–2804.
19. Urs, F. An 11k-Electrode 126-Channel High-Density Microelectrode Electrode to Interact with Electrogenic Cells. In Proceedings of the IEEE International Solid-State Circuits Conference Digest of Technical Papers (ISSCC'07), San Francisco, CA, USA, 11–15 February 2007; pp. 158–593.
20. Shahrokhi, F.; Abdelhalim, K.; Serletis, D.; Carlen, P.L.; Genov, R. The 128-channel fully differential digital integrated neural recording and stimulation interface. *IEEE Trans. Biomed. Circuits Syst.* **2010**, *4*, 149–161.
21. Zhang, F.; Holleman, J.; Otis, B.P. Design of ultra-low power biopotential amplifiers for biosignal acquisition applications. *IEEE Trans. Biomed. Circuits Syst.* **2012**, *6*, 344–355.
22. Saberhosseini, S.S.; Zabihian, A.; Sodagar, A.M. Low-Noise OTA for Neural Amplifying Applications. In Proceedings of the 8th International Caribbean Conference on Devices, Circuits and Systems (ICCDCS'12), Playa del Carmen, Mexico, 14–17 March 2012; pp. 1–4.
23. Wattanapanitch, W.; Fee, M.; Sarpeshkar, R. An energy-efficient micropower neural recording amplifier. *IEEE Trans. Biomed. Circuits Syst.* **2007**, *1*, 136–147.
24. Chae, M.; Kim, J.; Liu, W. Fully-differential self-biased bio-potential amplifier. *Electron. Lett.* **2008**, *44*, 1390–1391.
25. Gosselin, B. Recent advances in neural recording microsystems. *Sensors* **2011**, *11*, 4572.
26. Qinwen, F.; Huijsing, J.H.; Makinwa, K.A. A 21 nV/ $Hz^{-1/2}$ Chopper-Stabilized Multipath Current-Feedback Instrumentation Amplifier with 2 mV Offset. In Proceedings of the IEEE International Solid-State Circuits Conference Digest of Technical Papers (ISSCC'10), San Francisco, CA, USA, 7–11 February 2010; pp. 80–81.

27. Wu, R.; Makinwa, K.A.A.; Huijsing, J.H. A chopper current-feedback instrumentation amplifier with a 1 mHz 1/f noise corner and an AC-coupled ripple reduction loop. *IEEE J. Solid State Circuits* **2009**, *44*, 3232–3243.
28. Wei, T.; Chenxi, H.; Dongsoo, K.; Martini, B.; Culurciello, E. 4-Channel Asynchronous Bio-Potential Recording System. In Proceedings of the 2010 IEEE International Symposium on Circuits and Systems (ISCAS), Paris, France, 30 May–2 June 2010; pp. 953–956.
29. Wu, R.; Huijsing, J.H.; Makinwa, K.A.A. A current-feedback instrumentation amplifier with a gain error reduction loop and 0.06% untrimmed gain error. *IEEE J. Solid State Circuits* **2011**, *46*, 2794–2806.
30. Belloni, M.; Bonizzoni, E.; Fornasari, A.; Maloberti, F. A Micropower chopper-CDS operational amplifier. *IEEE J. Solid Stare Circuits* **2010**, *45*, 2521–2529.
31. Kusuda, Y. A 5.9 nV/Hz$^{-1/2}$ Chopper Operational Amplifier with 0.78 mV Maximum Offset and 28.3 nV/°C Offset Drift. In Proceedings of the IEEE International Solid-State Circuits Conference Digest of Technical Papers (ISSCC'11), San Francisco, CA, USA, 20–24 February 2011; pp. 242–244.
32. Burt, R.; Zhang, J. A micropower chopper-stabilized operational amplifier using a SC notch filter with synchronous integration inside the continuous-time signal path. *IEEE J. Solid Stare Circuits* **2006**, *41*, 2729–2736.
33. Qinwen, F.; Huijsing, J.; Makinwa, K. A Capacitively Coupled Chopper Instrumentation Amplifier with a +/−30 V Common-Mode Range, 160 dB CMRR and 5 mV Offset. In Proceedings of the IEEE International Solid-State Circuits Conference Digest of Technical Papers (ISSCC'12), San Francisco, CA, USA, 19–23 February 2012; pp. 374–376.
34. Johnson, B.; Molnar, A. An orthogonal current-reuse amplifier for multi-channel sensing. *IEEE J. Solid Stare Circuits* **2013**, *48*, 1487–1496.
35. Sepehrian, H.; Mirbozorgi, A.; Gosselin, B. A Low-Power Current-Reuse Analog Front-End for Multi-Channel Neural Signal Recording. In Proceedings of the 12th IEEE International NEWCAS conference (NEWCAS'14), Trois-Rivieres, Quebec, Canada, 22–25 June 2014.
36. Liu, L.; Zou, X.; Goh, W.L.; Je, M. Comparative Study and Analysis of Noise Reduction Techniques for Front-End Amplifiers. In Proceedings of the 13th International Symposium on Integrated Circuits (ISIC'11), Singapore, Singapore, 12–14 December 2011; pp. 555–558.
37. Baru Fassio, M.D. Implantable Signal Amplifying Circuit for Electroneurographic Recording. U.S. Patent 6996435, 2006.
38. Rieger, R.; Schuettler, M.; Pal, D.; Clarke, C.; Langlois, P.; Taylor, J.; Donaldson, N. Very low-noise ENG amplifier system using CMOS technology. *IEEE Trans. Neural Syst. Rehabil. Eng.* **2006**, *14*, 427–437.
39. Holleman, J.; Otis, B. A Sub-Microwatt Low-Noise Amplifier for Neural Recording. In Proceedings of the Annual International Conference of the IEEE Engineering in Medicine and Biology Society (EMBC'07), Lyon, France, 22–26 August 2007; p. 3930.
40. Demosthenous, A.; Pachnis, I.; Jiang, D.; Donaldson, N. An integrated amplifier with passive neutralization of myoelectric interference from neural recording tripoles. *IEEE Sens. J.* **2013**, *13*, 3236–3248.

41. Mollazadeh, M.; Murari, K.; Cauwenberghs, G.; Thakor, N.V. Wireless micropower instrumentation for multimodal acquisition of electrical and chemical neural activity. *IEEE Trans. Biomed. Circuits Syst.* **2009**, *3*, 388–397.
42. Kazerouni, I.A.; Dehrizi, H.G.; Isfahani, S.M.M.; Zhuo, Z.; Baghaei-Nejad, M.; Li-Rong, Z. A 77 nW Bioamplifier with a Tunable Bandwidth for Neural Recording Systems. In Proceedings of the IEEE Asia Pacific Conference on Circuits and Systems (APCCAS'10), Kuala Lumpur, Malaysia, 6–9 December 2010; pp. 36–39.
43. Kim, J.; Chae, M.S.; Liu, W. A 220 nW Neural Amplifier for Multi-Channel Neural Recording Systems. In Proceedings of the IEEE International Symposium on Circuits and Systems (ISCAS'09), Taipei, Taiwan, 24–27 May 2009; pp. 1257–1260.
44. Gosselin, B.; Sawan, M.; Chapman, C.A. A low-power integrated bioamplifier with active low-frequency suppression. *IEEE Trans. Biomed. Circuits Syst.* **2007**, *1*, 184–192.
45. Muller, R.; Gambini, S.; Rabaey, J.M. A 0.013 mm(2), 5 mu W, DC-coupled neural signal acquisition IC with 0.5 v supply. *IEEE J. Solid Stare Circuits* **2012**, *47*, 232–243.
46. Wei, Z.; Li, H.; Youguang, Z. A Low-Noise Integrated Bioamplifier with Active DC Offset Suppression. In Proceedings of the IEEE Biomedical Circuits and Systems Conference (BioCAS'09), Beijing, China, 26–28 November 2009; pp. 5–8.
47. Castro, P.; Silveira, F. High CMRR Power Efficient Neural Recording Amplifier Architecture. In Proceedings of the IEEE International Symposium on Circuits and Systems (ISCAS'11), Rio de Janeiro, Brazil, 15–18 May 2011; pp. 1700–1703.
48. Chung-Yu, W.; Wei-Ming, C.; Liang-Ting, K. A CMOS power-efficient low-noise current-mode front-end amplifier for neural signal recording. *IEEE Trans. Biomed. Circuits Syst.* **2013**, *7*, 107–114.
49. Ferrari, G.; Gozzini, F.; Molari, A.; Sampietro, M. Transimpedance amplifier for high sensitivity current measurements on nanodevices. *IEEE J. Solid State Circuits* **2009**, *44*, 1609–1616.
50. Rezaee-Dehsorkh, H.; Ravanshad, N.; Lotfi, R.; Mafinezhad, K. A Linear Tunable Amplifier for Implantable Neural Recording Applications. In Proceedings of the IEEE 54th International Midwest Symposium on Circuits and Systems (MWSCAS'11), Seoul, South Korea, 7–10 August 2011; pp. 1–4.
51. Zou, X.D.; Xu, X.Y.; Yao, L.B.; Lian, Y. A 1-V 450-nW fully integrated programmable biomedical sensor interface chip. *IEEE J. Solid Stare Circuits* **2009**, *44*, 1067–1077.
52. Chaturvedi, V.; Amrutur, B. A Low-Noise Low-Power Noise-Adaptive Neural Amplifier in 0.13 um CMOS Technology. In Proceedings of the 24th International Conference on VLSI Design (VLSI Design'11), Chennai, India, 2–7 January 2011; pp. 328–333.
53. Yao, K.-W. Design of A Neural Recording Amplifier with Tunable Pseudo Resistors. In Proceedings of the IEEE International SOC Conference, Taipei, Taiwan, 26–28 September 2011; pp. 376–379.
54. Yin, M.; Ghovanloo, M. A Low-Noise Preamplifier with Adjustable Gain and Bandwidth for Biopotential Recording Applications. In Proceedings of the IEEE International Symposium on Circuits and Systems (ISCAS'07), New Orleans, LA, USA, 27–30 May 2007; pp. 321–324.
55. Zou, X.; Liu, L.; Cheong, J.H.; Yao, L.; Li, P.; Cheng, M.-Y.; Goh, W.L.; Rajkumar, R.; Dawe, G.S.; Cheng, K.-W.; *et al.* A 100-channel 1-mW implantable neural recording IC. *IEEE Trans. Circuits Syst. I Regul. Pap.* **2013**, *60*, 2584–2596.

56. Kmon, P.; Grybos, P. Energy efficient low-noise multichannel neural amplifier in submicron CMOS process. *IEEE Trans. Circuits Syst. I Regul. Pap.* **2013**, *60*, 1764–1775.
57. Al-Ashmouny, K.; Sun-Il, C.; Euisik, Y. A 8.6 mW 3-bit Programmable Gain Amplifier for Multiplexed-Input Neural Recording Systems. In Proceedings of the Annual International Conference of the IEEE Engineering in Medicine and Biology Society (EMBC'11), Boston, MA, USA, 30 August–3 September 2011; pp. 2945–2948.
58. Gosselin, B. Approaches for the Efficient Extraction and Processing of Neural Signals in Implantable Neural Interfacing Microsystems. In Proceedings of the 33th Annual International Conference of the IEEE Engineering in Medicine and Biology Society (EMBC'11), Boston, MA, USA, 30 August–3 September 2011; pp. 5855–5859.
59. Park, H.; Kiani, M.; Hyung-Min, L.; Jeonghee, K.; Block, J.; Gosselin, B.; Ghovanloo, M. A wireless magnetoresistive sensing system for an intraoral tongue-computer interface. *IEEE Trans. Biomed. Circuits Syst.* **2012**, *6*, 571–585.
60. Aziz, J.N.Y.; Abdelhalim, K.; Shulyzki, R.; Genov, R.; Bardakjian, B.L.; Derchansky, M.; Serletis, D.; Carlen, P.L. 256-channel neural recording and delta compression microsystem with 3D electrodes. *IEEE J. Solid Stare Circuits* **2009**, *44*, 995–1005.
61. Chae, M.; Liu, W.; Zhi, Y.; Chen, T.; Kim, J.; Sivaprakasam, M.; Yuce, M. A 128-Channel 6 mW Wireless Neural Recording IC with On-the-Fly Spike Sorting and UWB Tansmitter. In Proceedings of the IEEE International Solid-State Circuits Conference (ISSCC'08), San Francisco, CA, USA, 3–7 February 2008; pp. 146–603.
62. Lopez, C.M.; Andrei, A.; Mitra, S.; Welkenhuysen, M.; Eberle, W.; Bartic, C.; Puers, R.; Yazicioglu, R.F.; Gielen, G. An Implantable 455-Active-Electrode 52-Channel CMOS Neural Probe. In Proceedings of the IEEE International Solid-State Circuits Conference Digest of Technical Papers (ISSCC'13), San Francisco, CA, USA, 17–21 February 2013; pp. 288–289.
63. Gosselin, B.; Sawan, M. A low-power integrated neural interface with digital spike detection and extraction. *Analog Integr. Circuits Signal Process.* **2009**, *64*, 3–11.
64. Sodagar, A.M.; Perlin, G.E.; Ying, Y.; Najafi, K.; Wise, K.D. An implantable 64-channel wireless microsystem for single-unit neural recording. *IEEE J. Solid State Circuits* **2009**, *44*, 2591–2604.
65. Gosselin, B.; Sawan, M. Adaptive Detection of Action Potentials Using Ultra Low-Power CMOS Circuits. In Proceedings of the IEEE Biomedical Circuits and Systems Conference (BioCAS'08), Baltimore, MD, USA, 20–22 November 2008; pp. 209–212.
66. Gosselin, B.; Sawan, M.; Kerherve, E. Linear-phase delay filters for ultra-low-power signal processing in neural recording implants. *IEEE Trans. Biomed. Circuits Syst.* **2010**, *4*, 171–180.
67. Harrison, R.R.; Watkins, P.T.; Kier, R.J.; Lovejoy, R.O.; Black, D.J.; Greger, B.; Solzbacher, F. A low-power integrated circuit for a wireless 100-electrode neural recording system. *IEEE J. Solid State Circuits* **2007**, *42*, 123–133.
68. Gosselin, B.; Ayoub, A.E.; Roy, J.F.; Sawan, M.; Lepore, F.; Chaudhuri, A.; Guitton, D. A mixed-signal multichip neural recording interface with bandwidth reduction. *IEEE Trans. Biomed. Circuits Syst.* **2009**, *3*, 129–141.
69. Liew, W.-S.; Zou, X.; Yao, L.; Lian, Y. A 1-V 60 mW 16-Channel Interface Chip for Implantable Neural Recording. In Proceedings of the IEEE Custom Integrated Circuits Conference (CICC'09), San Jose, CA, USA, 13–16 September 2009; pp. 507–510.

70. Miranda, H.; Gilja, V.; Chestek, C.A.; Shenoy, K.V.; Meng, T.H. HermesD: A high-rate long-range wireless transmission system for simultaneous multichannel neural recording applications. *IEEE Trans. Biomed. Circuits Syst.* **2010**, *4*, 181–191.
71. Perlin, G.E.; Wise, K.D. An ultra compact integrated front end for wireless neural recording microsystems. *J. Microelectromechanical Syst.* **2010**, *19*, 1409–1421.
72. Greenwald, E.; Mollazadeh, M.; Hu, C.; Tang, W.; Culurciello, E.; Thakor, N. A VLSI neural monitoring system with ultra-wideband telemetry for awake behaving subjects. *IEEE Trans. Biomed. Circuits Syst.* **2011**, *5*, 112–119.
73. Aceros, J.; Ming, Y.; Borton, D.A.; Patterson, W.R.; Nurmikko, A.V. A 32-Channel Fully Implantable Wireless Neurosensor for Simultaneous Recording from Two Cortical Regions. In Proceedings of the Annual International Conference of the IEEE Engineering in Medicine and Biology Society (EMBC'11), Boston, MA, USA, 30 August–3 September 2011; pp. 2300–2306.
74. Jo-Yu, W.; Kea-Tiong, T. A Band-Tunable, Multichannel Amplifier for Neural Recording with AP/LFP Separation and Dual-Threshold Adaptive AP Detector. In Proceedings of the Annual International Conference of the IEEE Engineering in Medicine and Biology Society (EMBC'11), Boston, MA, USA, 30 August–3 September 2011; pp. 1847–1850.
75. Lopez, C.M.; Braeken, D.; Bartic, C.; Puers, R.; Gielen, G.; Eberle, W. A 16-Channel Low-Noise Programmable System for the Recording of Neural Signals. In Proceedings of the IEEE International Symposium on Circuits and Systems (ISCAS'11), Rio de Janeiro, Brazil, 15–18 May 2011; pp. 1451–1454.
76. Rouse, A.G.; Stanslaski, S.R.; Cong, P.; Jensen, R.M.; Afshar, P.; Ullestad, D.; Gupta, R.; Molnar, G.F.; Moran, D.W.; Denison, T.J. A chronic generalized bi-directional brain-machine interface. *J. Neural Eng.* **2011**, *8*, 036018.
77. Szuts, T.A.; Fadeyev, V.; Kachiguine, S.; Sher, A.; Grivich, M.V.; Agrochao, M.; Hottowy, P.; Dabrowski, W.; Lubenov, E.V.; Siapas, A.G.; *et al.* A wireless multi-channel neural amplifier for freely moving animals. *Nat. Neurosci.* **2011**, *14*, 263–269.
78. Wattanapanitch, W.; Sarpeshkar, R. A low-power 32-channel digitally programmable neural recording integrated circuit. *IEEE Trans. Biomed. Circuits Syst.* **2011**, *5*, 592–602.
79. Lo, Y.-K.; Liu, W.; Chen, K.; Tsai, M.-H.; Hsueh, F.-L. A 64-Channel Neuron Recording System. In Proceedings of the Annual International Conference of the IEEE Engineering in Medicine and Biology Society (EMBC'11), Boston, MA, USA, 30 August–3 September 2011; pp. 2862–2865.
80. Zoladz, M.; Kmon, P.; Grybos, P.; Szczygiel, R.; Kleczek, R.; Otfinowski, P.; Rauza, J. Design and Measurements of Low Power Multichannel Chip for Recording and Stimulation of Neural Activity. In Proceedings of the Annual International Conference of the IEEE Engineering in Medicine and Biology Society (EMBC'12), San Diego, CA, USA, 28 August–1 September 2012; pp. 4470–4474.
81. Ming, Y.; Borton, D.A.; Aceros, J.; Patterson, W.R.; Nurmikko, A.V. A 100-Channel Hermetically Sealed Implantable Device for Wireless Neurosensing Applications. In Proceedings of the IEEE International Symposium on Circuits and Systems (ISCAS'12), Seoul, Korea, 20–23 May 2012; pp. 2629–2632.

82. Gao, H.; Walker, R.M.; Nuyujukian, P.; Makinwa, K.A.A.; Shenoy, K.V.; Murmann, B.; Meng, T.H. HermesE: A 96-channel full data rate direct neural interface in 0.13 μm CMOS. *IEEE J. Solid Stare Circuits* **2012**, *47*, 1043–1055.
83. Dong, H.; Yuanjin, Z.; Rajkumar, R.; Dawe, G.; Minkyu, J. A 0.45 V 100-Channel Neural-Recording IC with Sub-mW/Channel Consumption in 0.18 mm CMOS. In Proceedings of the IEEE International Solid-State Circuits Conference Digest of Technical Papers (ISSCC'13), San Francisco, CA, USA, 17–21 February 2013; pp. 290–291.
84. Yin, M.; Borton, D.A.; Aceros, J.; Patterson, W.R.; Nurmikko, A.V. A 100-channel hermetically sealed implantable device for chronic wireless neurosensing applications. *IEEE Trans. Biomed. Circuits Syst.* **2013**, *7*, 115–128.

14

An Efficient Radio Access Control Mechanism for Wireless Network-On-Chip Architectures

Maurizio Palesi [1,*], Mario Collotta [1], Andrea Mineo [2] and Vincenzo Catania [2]

[1] Faculty of Engineering, University of Enna, Kore, Cittadella Universitaria, Enna 94100, Italy; E-Mails: Maurizio.Palesi@unikore.it (M.P); Mario.Collotta@unikore.it (M.C.)

[2] Department of Electrical, Electronics and Computer Engineering, University of Catania, Viale A. Doria, 6, Catania 95125, Italy; E-Mails: andrea.mineo@dieei.unict.it (A.M.); vincenzo.catania@dieei.unict.it (V.C.)

* Author to whom correspondence should be addressed; E-Mail: maurizio.palesi@unikore.it

Academic Editor: Alexander Fish

Abstract: Modern systems-on-chip (SoCs) today contain hundreds of cores, and this number is predicted to reach the thousands by the year 2020. As the number of communicating elements increases, there is a need for an efficient, scalable and reliable communication infrastructure. As technology geometries shrink to the deep submicron regime, however, the communication delay and power consumption of global interconnections become the major bottleneck. The network-on-chip (NoC) design paradigm, based on a modular packet-switched mechanism, can address many of the on-chip communication issues, such as the performance limitations of long interconnects and integration of large number of cores on a chip. Recently, new communication technologies based on the NoC concept have emerged with the aim of improving the scalability limitations of conventional NoC-based architectures. Among them, wireless NoCs (WiNoCs) use the radio medium for reducing the performance and energy penalties of long-range and multi-hop communications. As the radio medium can be accessed by a single transmitter at a time, a radio access control mechanism (RACM) is needed. In this paper, we present a novel RACM, which allows one to improve both the performance and energy figures of the WiNoC. Experiments, carried out on both synthetic and real traffic scenarios, have shown the effectiveness of the proposed RACM. On average, a 30% reduction in communication delay and a 25% energy savings have been observed when the proposed RACM is applied to a known WiNoC architecture.

Keywords: network-on-chip; on-chip wireless communication; access control mechanism; performance evaluation; power analysis; energy efficiency

---

1. Introduction

The evermore increasing number of cores integrated into a multiprocessor system-on-chip (MPSoC) has exacerbated the role played by the on-chip communication system [1]. Nowadays, the network-on-chip (NoC) design paradigm is considered as the most viable communication infrastructure for dealing with the performance, energy and reliability issues of many-core system architectures [2]. An NoC is composed by IP cores and switches connecting them by communication channels.

An important performance limitation in traditional NoCs arises from planar metal interconnect-based multi-hop communications, wherein the data transfer between two distant cores causes high latency and power consumption. In fact, beyond a certain length, wireless links are more energy efficient than conventional metal wires [3]. Hence, the performance improvements by using long-range wireless links will be more than that using wired links [3]. For this reason, wireless NoC (WiNoC) architectures, which use a wireless backbone upon the traditional wire-based NoC, have been recently proposed [4,5]. WiNoCs can be seen as a natural evolution of conventional NoCs to deal with the scalability issues that characterize the next generation of many-core architectures. It is foreseen that such architectures will open new opportunities in the direction of meeting critical performance and energy requirements and in the context of real-time and multimedia applications, which usually have to be supported by specific hardware resources and *ad hoc* management schemes.

A WiNoC is an extension of a conventional electric NoC in which, in addition to the conventional routers, there are a number of radio hubs that use the radio medium for communicating with each other. Such radio hubs are formed by a router augmented with a wireless interface. Thus, long-range communication is realized by a mixed path that involves the use of both electric links and radio communications. For instance, if a node A has to communicate with a node B, which would involve many hops in a traditional NoC, in a WiNoC, such communication can be realized as follows: (1) node A sends the packet to the closest radio hub, $RH1$, through the electric NoC; (2) $RH1$ sends the packet to the radio hub, $RH2$, closest to node B, by means of wireless communication; and (3) $RH2$ sends the packet to node B through the electric NoC. In this way, the number of hops needed for the communication from node A to node B can be reduced with a consequent reduction in terms of communication delay and energy consumption.

In current WiNoC implementations, due to feasibility and cost-related issues, the availability of a single radio channel is often considered [6]. That is, there can be a single transmitting radio hub at a time. For this reason, an access control mechanism for determining the radio hub authorized to use the radio media is needed.

The radio access control mechanism (RACM) used in the WiNoC context has a critical impact on communication performance and power metrics. Further, its implementation must not affect the cost and power figures of the communication fabric. For this reason, in current WiNoC implementations, a simple token ring-based architecture is used for implementing the RACM [6], as shown in Figure 1. A token circulates through the radio hubs and enables the current radio hub that holds the token to use the radio medium in transmission mode for a certain amount of clock cycles. The amount of cycles assigned to each radio hub (hold cycles) is defined at design time based on the communication bandwidth requirements of the applications mapped onto the WiNoC. The ring architecture is favored for its low wiring overhead, which is important, as the radio hubs are usually placed far away from each other over the chip.

In this paper, we propose a new RACM for WiNoC architectures that allows one to improve the communication performance and power without losing the advantages of the low wiring overhead of the ring architecture. The basic idea behind the proposed RACM is to dynamically tune the hold time assigned to each radio hub by guaranteeing that the maximum token round time is below a given threshold. Experiments, carried out on both synthetic and real traffic scenarios, prove the effectiveness of the proposed RACM. As compared to the conventional RACM, the use of the proposed RACM allows one to reduce, on average, the communication delay by 30% and to save 25% of the energy.

2. Related Work

Access control is the process of mediating every requests to resources and data maintained by a system and determining whether the request should be granted or denied. Several access control mechanisms for arbitrating the use of the radio medium have been proposed in the literature in different contexts, including on-chip, off-chip or macro-world wireless networks.

In the context of radio access control mechanisms, they can be divided in two main classes, namely contention-based and TDMA (time division multiple access)-based protocols [7–9]. The contention-based protocols allow nodes independent access to the shared wireless medium [7,9]. In the TDMA-based protocols, the system time is divided into time slots. Each of the nodes has its own time slot assigned and may access the shared medium only in this time slot. This allows avoiding collisions, idle listening and scheduling the sleep of the transceiver, without additional overhead.

The radio access control mechanism (RACM) typically used in WiNoC architectures is based on TDMA [6,10–12] in which time slots are regulated by means of a token, which circulates among the transmitting stations [13]. Another peculiarity of WiNoC architectures is the number of available radio channels. Current, real WiNoC implementations are limited to the use of a single radio channel. In fact, implementing multiple radio channels, in current technologies, would be infeasible due to the area overhead of the antennas (multiple channels requires multiple antennas). However, when, in the future, multiple radio channels in the same chip will be available, the proposed RACM can be still applied. In fact, multiple channels results in multiple rings in which a different instance of the RACM can be used in each of them. In the current literature, several studies have analyzed how performance and energy metrics would scale with the number of radio channels. For instance, in [14], a multi-channel WiNoC system based on UWB transceivers is proposed. The authors in [6] have presented a mm metal zigzag antenna for designing a mm-wave WiNoC architecture and then extended this single channel mWNoC design to a multichannel mWNoC. The authors in [15] have proposed a WiNoC system by using sub-THz antennas deposited in a polyimide layer in a 32-nm CMOS process. Such an infrastructure provides distinct channels in the frequency range of 100–500 GHz. Carbon nanotube (CNT) antennas have been proposed to achieve the terahertz frequency range [16]. The CNT antennas can provide much higher communication data rates due to their small size, resulting in ultra-low area overhead. Using such technology, non-overlapping wireless channels, each with a 10-Gb/s bandwidth and 0.33 pJ/bit of energy, can be achieved [12].

3. Radio Access Control Mechanism

The radio medium can be accessed by a single radio hub at a time. Thus, a radio access control mechanism (RACM) must be used for avoiding multiple radio hubs having access to the radio medium simultaneously. A token-based approach is usually used for implementing the RACM. Basically, a token is passed through every radio hub, enabling the radio hub owning the token to use the radio medium for a maximum number of clock cycles. Such a token-based mechanism can be implemented by means of a ring network topology connecting the radio hubs, as shown in Figure 1.

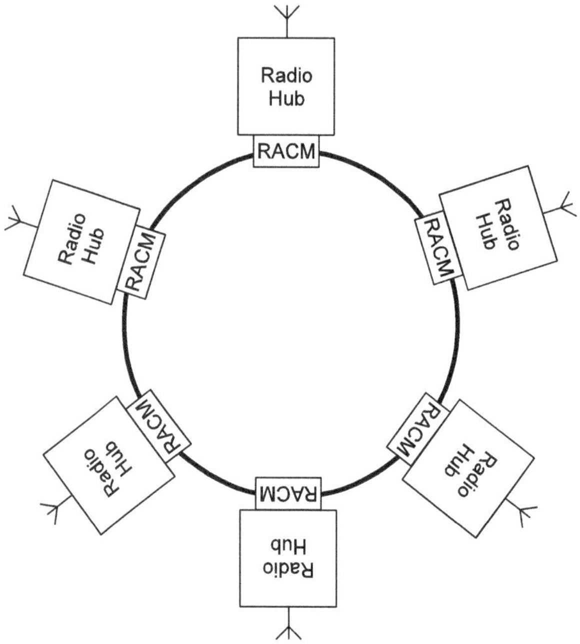

Figure 1. Conventional ring-based topology for implementing the radio access control mechanism in a wireless network-on-chip (WiNoC).

Although different implementations of the RACM are possible, we focus on distributed architectures implemented by a ring network topology. The main motivation of this choice is related to their suitability for being used in the on-chip context, in which silicon area (*i.e.*, cost) and energy consumption are inalienable design metrics. In fact, a ring network topology requires the minimum number of short point-to-point links among the radio hubs, as compared to other topologies or a centralized control structure, which would bring to the wiring congestion issues and complications during placement and the routing phases of the design flow.

3.1. Conventional RACM

The conventional RACM used in WiNoCs [13] can be implemented by means of a simple finite state machine, whose behavior, for a generic node of the ring, is summarized by Algorithm 1. The generic node of the ring has three inputs and two outputs. The radio hub uses input $request$ to request the access to the radio channel. If the ring node owns the token, the output $grant$ is asserted, and the radio hub is allowed to use the radio medium in transmission mode. The $grant$ output is kept high, as long as the $hold$ input is high, but for a maximum number of cycles that does not exceed a certain threshold, referred to as the maximum hold cycles, MHC. When the $hold$ input is unset or MHC cycles are elapsed, the token is passed to the next node.

Algorithm 1 Conventional radio access control mechanism (RACM) implemented by a node of the ring.

Require: $request$, $hold$, $token\_in$
Ensure: $grant$, $token\_out$

```
if request and owns_token then
    grant ← true
else
    grant ← false
end if
if owns_token then
    if hold then
        hold_count ← hold_count + 1
        if hold_count > MHC then
            token_out ← token_in
            owns_token ← false
        end if
    else
        token_out ← token_in
        owns_token ← false
    end if
else if token_in then
    owns_token ← true
    hold_count ← 0
end if
```

Thus, if we indicate with MHC_i the maximum hold count for the i-th node of the ring, the generic j-th radio hub is guaranteed to have granted the access to the radio medium for MHC_j cycles every $\sum_{i \neq j} MHC_i$ cycles in the worst case. The main drawback of this approach is that the generic i-th radio hub is limited to using the radio medium for MHC_i cycles at most, even if the rest of the radio hubs do not consume their assigned budget of clock cycles MHC_j ($j \neq i$). Such a situation can strongly affect the performance of the network, as will be shown in the experimental section. Please note that the $hold\_count$ and $owns\_token$ registers of each radio hub are initialized during the reset period. Specifically, the $hold\_count$ register is initialized with zero, whereas $owns\_token$ is initialized with true for the $radio\_hub$ with ID 0 and with false for the other radio hubs.

Another conventional RACM does not use the hold counter and releases the token only when all of the flits of the packet have been transmitted. We will refer to this conventional RACM as conventional RACM without holding.

In the next subsection, we will present a novel RACM that dynamically tunes the MHC_i based on the actual communication profile.

3.2. Proposed RACM

In the conventional RACM, the maximum number of cycles assigned to the generic i-th radio hub for using the radio medium when it owns the token, MHC_i, is statically determined and remains the same, irrespective of the traffic conditions. In practical situations, however, there are radio hubs that do not use their MHC cycles entirely, whereas there are others that entirely use their MHC cycles. Such unbalanced utilization can be seen as an opportunity for performance improvement. Thus, the basic idea of the proposed RACM is redistributing the unused clock cycles among the radio hubs that have entirely used their MHC. More precisely, let us indicate with H_i the number of cycles that the radio hub i has held the token in the current round. Thus, the number of unused cycles by radio hub i in the current round is $S_i = MHC_i - H_i$. In the proposed RACM, the total number of unused cycles, $S = \sum_i S_i$, is redistributed among the radio hubs in proportion to the utilization of their MHC in the previous round.

The pseudocode implementing the proposed RACM is summarized by Algorithm 2. In this case, the token holds a set of information that is updated and used by every RACM module in the ring. In particular, $token.S$ stores the total number of unused cycles in the previous round. $token.SC$ is used for accumulating the unused cycles while the token circulates through the RACM modules. $token.U[i]$ stores the number of cycles for which the radio hub i held the token in the previous round (*i.e.*, the utilization of the radio medium). Finally, $token.MU$ stores the maximum utilization observed at the previous round. As can be observed, the S, MU and SC fields of the token are initialized during the reset period and when the token completes a round, *i.e.*, when the token returns to the RACM module with id zero (Lines 3–5). Before passing the token to the next RACM module (Lines 17 and 23), the U and SC fields of the token are updated with the current utilization and with the total number of unused cycles up to the current radio hub, respectively. Differently from Algorithm 1, here, the current radio hub can hold the token for a maximum number of cycles, which depends on its utilization and on the total number of unused cycles in the previous round (Line 16). Please note that, like for the conventional RACM, the generic j-th radio hub is guaranteed to have granted the access to the radio medium every $\sum_{i \neq j} MHC_i$ cycles in the worst case.

Algorithm 2 Proposed RACM implemented by a node of the ring.

Require: $request, hold, token\_in$

Ensure: $grant, token\_out$

if $owns\_token$ **then**
  if $id = 0$ **then**
    $token.S \leftarrow token.SC$
    $token.MU \leftarrow \max\{token.U\}$
    $token.SC \leftarrow 0$
  end if
  if $request$ **then**
    $grant \leftarrow true$
  else
    $grant \leftarrow false$
  end if
end if
if $owns\_token$ **then**
  if $hold$ **then**
    $hold\_count \leftarrow hold\_count + 1$
    if $hold\_count > MHC + \frac{token.U[id]}{token.MU} \times token.S$ **then**
      $token.U[id] \leftarrow hold\_count$
      $token.SC \leftarrow token.SC + MHC - hold\_count$
      $token\_out \leftarrow token$
      $owns\_token \leftarrow false$
    end if
  else
    $token.U[id] \leftarrow hold\_count$
    $token.SC \leftarrow token.SC + MHC - hold\_count$
    $token\_out \leftarrow token$
    $owns\_token \leftarrow false$
  end if
else if $token\_in$ **then**
  $token \leftarrow token\_in$
  $owns\_token \leftarrow true$
  $hold\_count \leftarrow 0$
end if

3.3. Synthesis of the RACM Module

Figure 2 shows the interface of the module implementing the proposed RACM. It is connected to the radio hub with the conventional three signals, namely, *request*, *hold* and *grant*. The *request* signal is asserted from the radio hub when it wants to access the radio medium. As soon as it gets the grant, it asserts the *hold* signal, until it has something to transmit and the *grant* signal is set. The RACMs are connected to each other by means of signals S, SC, MU and U, which correspond to the fields of the token, as discussed in Section 3.2. Specifically, S encodes the total number of unused cycles in the previous round. SC encodes the number of unused cycles in the current round up to the current RACM module. MU encodes the maximum utilization (in terms of cycles) observed at the previous round. Finally, U encodes the number of cycles for which each radio hub has held the token in the previous round. Both the conventional and proposed RACM modules have been designed in VHDL, synthesized and evaluated with Synopsys Design Compiler and mapped on a 28-nm CMOS standard cell library from TSMC. An operating clock frequency of 1 GHz has been considered. For the proposed RACM, the S, SC, MU and U signals have been considered eight bits wide, which results in a limit of 256 clock cycles for the maximum hold count (MHC in Algorithm 2).

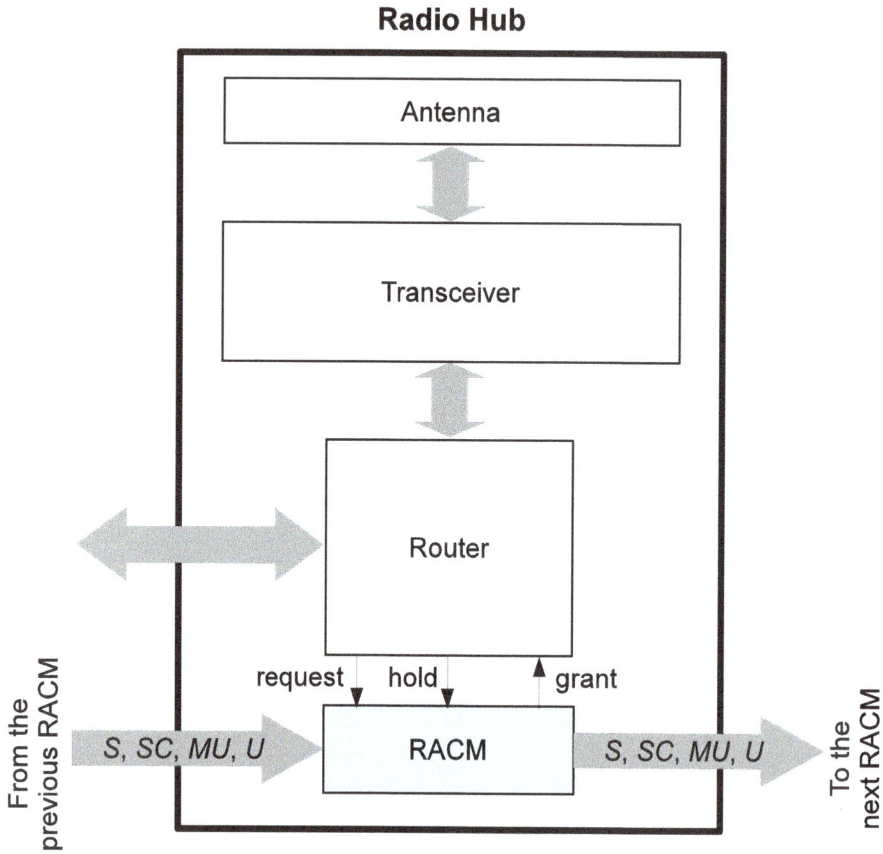

Figure 2. Interface of the proposed RACM module.

Figure 3 shows the area and power breakdown of the radio hub implementing the conventional (with and without hold) and the proposed RACM. As can be observed, the contribution of the RACM module, both in terms of silicon area and power, is negligible compared to the transceiver and the router. In fact, the RACM module accounts for less than 1% of the total radio hub area and power, and its contribution increases to about 3% when the proposed mechanism is used. However, as will be discussed in the

Experiments Section, even if the proposed RACM increases the total power of the radio hub, the total communication energy decreases due to the improvement of the performance metrics.

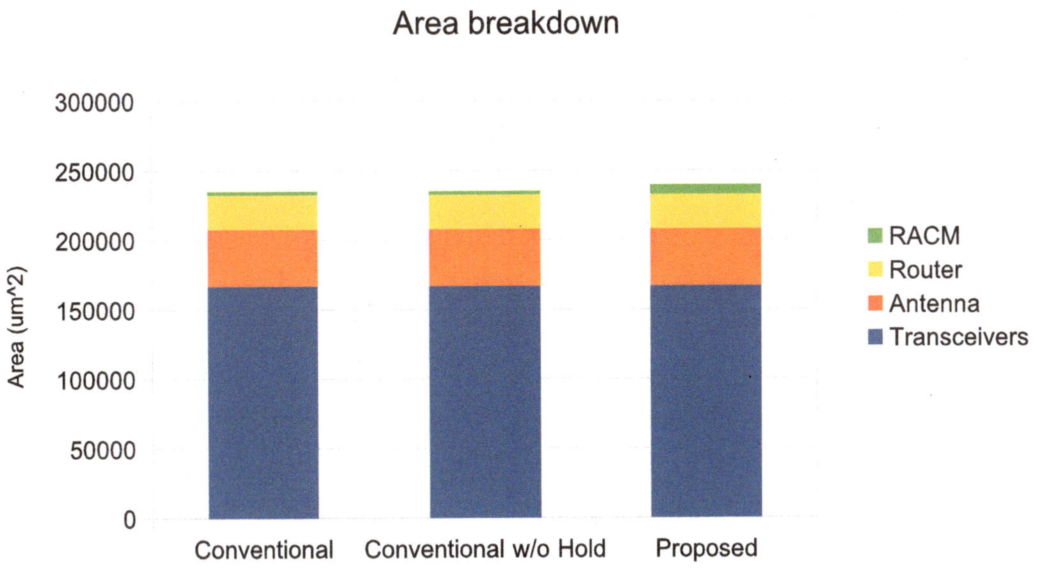

(**a**)

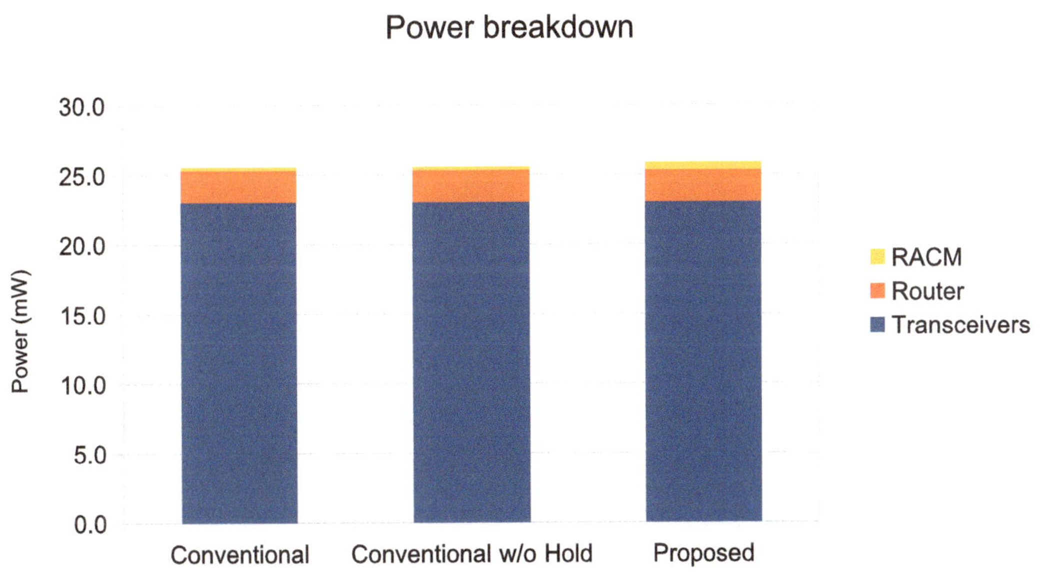

(**b**)

Figure 3. Area (**a**) and power (**b**) breakdown of the radio hub.

4. Experiments

In this section, we compare the proposed RACM with the conventional RACM. We implemented the proposed mechanism in an extended version of Noxim [17], which supports wireless communications and iWise64 architecture [18], assuming a single radio channel. Noxim is a cycle-accurate NoC simulator that allows one to estimate power and performance figures. The basic elements that form the WiNoC, namely routers, radio hubs, network interfaces and links, have been modeled in VHDL, synthesized and mapped on a 28-nm CMOS standard cell library from TSMC. The Synopsys Design Compiler has been used for extracting power figures for different configurations and different utilization factors. The energy dissipated by wireline links has been obtained through HSPICE simulations taking into consideration the length and layout of the wireline links.

In the analysis, a WiNoC architecture implemented into a $20 \text{ mm} \times 20 \text{ mm}$ silicon die has been considered. A zigzag antenna has been accurately modeled and characterized with Ansoft HFSS [19] (High Frequency Structural Simulator). HFSS is a leading commercial finite element method (FEM) field solver that simulates 3D structures and produces S-parameters and radiation patterns. We considered a high resistivity $\rho = 5 \text{ K}\Omega \cdot \text{cm}$ SOI with a substrate thickness of $350 \ \mu\text{m}$ and $30 \ \mu\text{m}$ for the oxide (SiO_2). The antennas are situated at an elevation of $2 \ \mu\text{m}$ from the substrate, compatible with the guidelines reported in [20] for reducing the interference with others metal structures ([20] demonstrates that the interference due to other metallic structures is negligible by following such rules). The zigzag antenna has a thickness of $2 \ \mu\text{m}$ and an axial length of $2 \times 340 \ \mu\text{m}$ for operating at 60 GHz. From HFSS simulation, we obtain the scattering parameters used for computing the Friis formula and then for calculating the attenuation introduced by the wireless medium. The detailed description of the transceiver circuit is out of the scope of this paper. However, the transceiver was designed in [6].

The aforementioned data have been used for back-annotating the NoC simulator in such a way to obtain power and energy figures. The conventional and the proposed RACMs are compared under both synthetic traffic scenarios and using communication traces extracted from the execution of SPLASH-2 benchmarks [21].

Figure 4 shows the average communication delay *versus* packet injection rate (pir) for a 64-node WiNoC architecture under a uniform traffic scenario and a random packet size ranging from 4 to 16 flits. The average packet delay is the average of the delays experienced by the packets during the simulation. The packet delay is computed by measuring the difference of the clock cycle in which the packet is generated and injected into the local buffer of the source router and the clock cycle in which the tail flit of the packet is received from the destination node. Let d_i be the delay experienced by the i-th packet and N the total number of injected packets. Then, the average packet delay is $\sum_{i=1}^{N} d_i/N$. We consider five different MHC values (1, 2, 4, 8 and 16 clock cycles). As can be observed, for a given pir value, when the proposed RACM is used, the average communication delay decreases. In this experiment, the packets' sizes are randomly generated between two and 16 flits. The average delay improves as MHC increases, and no relevant improvements are observed when MHC is greater than eight clock cycles. This is due to the fact that, on average, eight clock cycles are enough for a radio hub to drain a packet. Thus, as MHC decreases, the probability of a radio hub interrupting the transmission of the flits of a long packet increases. This is because the radio hub has to pass the token as soon as it consumes all of its assigned clock cycles (*i.e.*, MHC clock cycles in the case of the conventional RACM and a variable number of cycles, $\geq MHC$, in the case of the proposed RACM). Then, the radio hub has to wait again for the token to finalize the transmission of the packet. Now, recalling that the communication delay is the average number of clock cycles elapsed from the time in which the header flit of the packet is injected into the network to the time in which the tail flit of the packet is received by the destination core, interrupting the transmission of a packet has a strong impact on the communication delay. The possibility of dynamically and selectively tuning the MHC of any radio hub based on the current traffic condition provided by the proposed RACM allows one to dramatically improve the overall performance of the network.

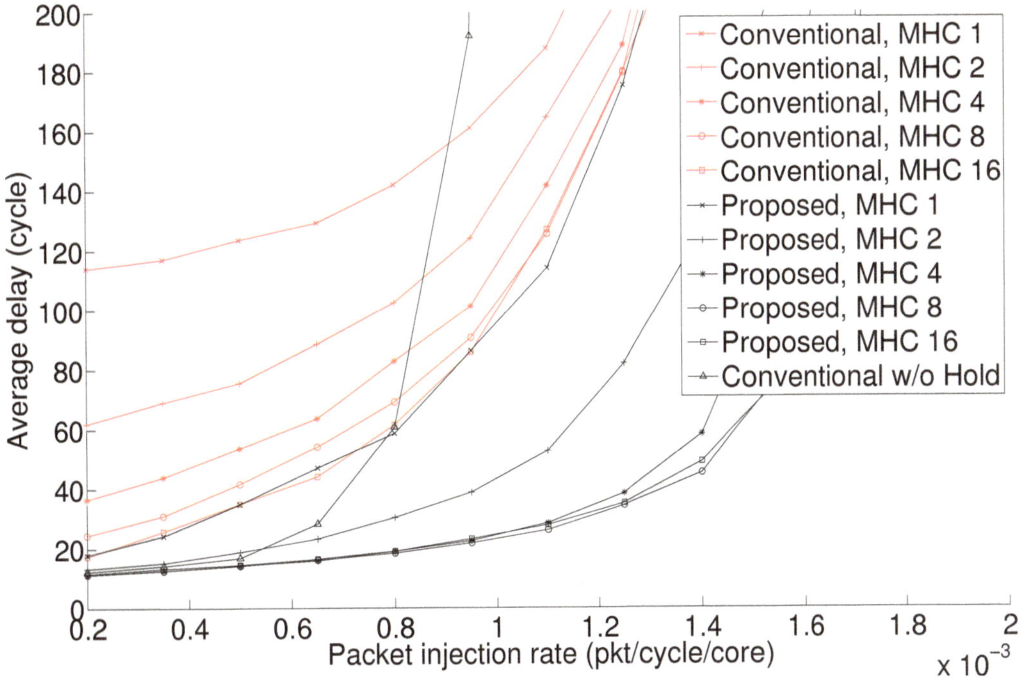

Figure 4. Average communication delay *versus* the packet injection rate (pir) under a uniform traffic scenario.

Figure 4 also shows the performance results when the conventional RACM without holding is used. As can be observed, for low packet injection rates, it outperforms the conventional RACM. However, as soon as the pir increases, the communication latency quickly increases, and the saturation point is reached before compared to the case in which the other RACMs are used. Such a behavior can be explained as follows. In a low-traffic condition, the fact that a radio hub keeps the token reserved for the entire duration of the transmission of the packet (irrespective of the packet size) does not affect the token waiting time of the other radio hubs, since the contention for the use of the radio medium is low. As soon as the injection load increases, the number of requests of the radio hubs for accessing the radio medium increases, and the token waiting time increases, as the radio hub that owns the token is not forced to release it.

Figure 5 summarizes the performance improvement obtained when the proposed RACM is used. Two performance metrics are considered as follows: the percentage increase of the saturation pir and the percentage decrease of the average communication delay measured at a pir value corresponding to 50% of the bisection bandwidth for that specific traffic scenario. The saturation pir is defined as the maximum pir value above which the variation of the throughput deviates from its linear behavior. For all of the experiments, we consider four-flit input buffers for routers and eight-flit input buffers for routers with the wireless interface (*i.e.*, the radio hubs). For the sake of clarity, we consider the case with $MHC = 8$. As can be observed, as compared to the conventional RACM, the proposed RACM allows one to obtain relevant performance improvement. For the considered traffic scenarios, on average, the saturation pir increased by 34% and the communication delay decreased by 29%. As expected, even higher performance improvements are observed when the proposed RACM is compared with a conventional RACM without holding. On average, the saturation pir increased by 44% and the communication delay decreased by 76%.

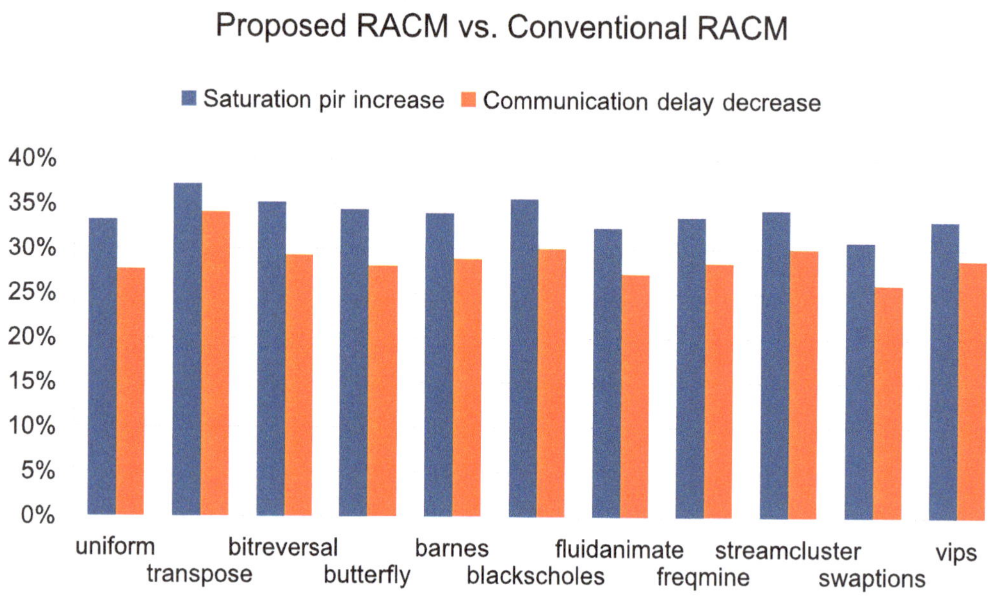

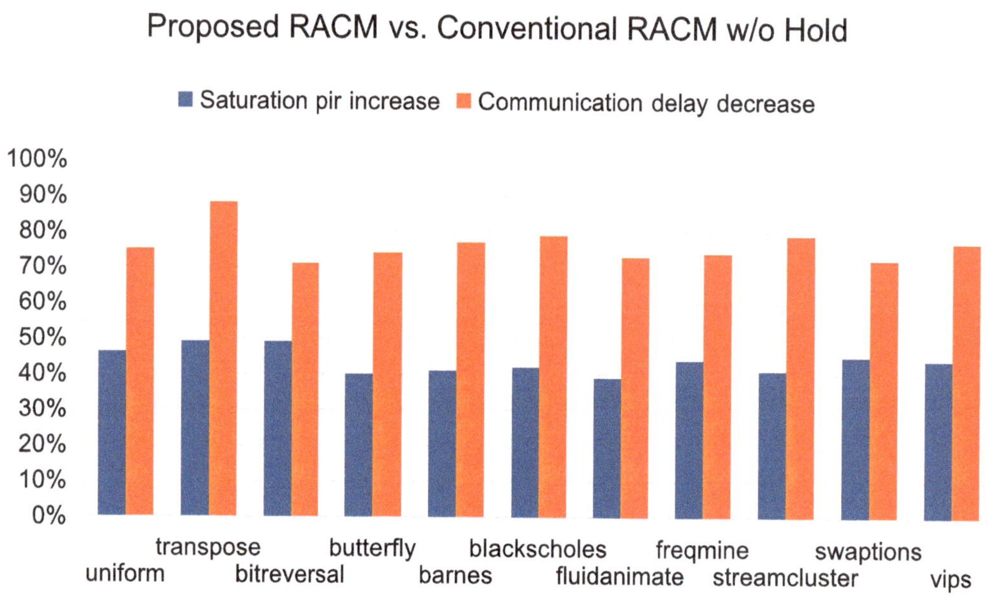

Figure 5. Percentage increase of the saturation pir and percentage decrease of the average communication delay when the proposed RACM is used.

The effectiveness of the proposed RACM can also be assessed by observing the grant probability, that is the probability that a generic radio hub requests the access to the radio medium and the request is granted, or it has been already granted the access to it, and it can hold the access for additional cycles. Figure 6 shows the grant probability for different packet injection rates under uniform traffic. The grant probability represents the fraction of times a radio hub requires the access to the radio medium and the access is granted. As can be observed, for a given pir, when the proposed RACM is used, the probability of a radio hub being granted its request to access the radio medium or holding the radio medium for additional cycles is higher than that observed when the conventional RACM is used. The same figure shows the grant probability when the conventional RACM without holding is used. In this case, by grant probability it is meant the probability that a generic radio hub requests the access to the radio medium and the request is granted. As already observed in Figure 4 for a low injection rate, the conventional RACM without holding performs well, and this is also reflected by its grant probability. However, its performance quickly degrades as soon as the injection rate increases. In fact, since the radio hub that holds the token is not forced to release the token until it has not exhausted the transmission of the packet, the other radio hubs ready for transmission will experience a high waiting time. Based on this, the grant probability rapidly decreases as the pir increases.

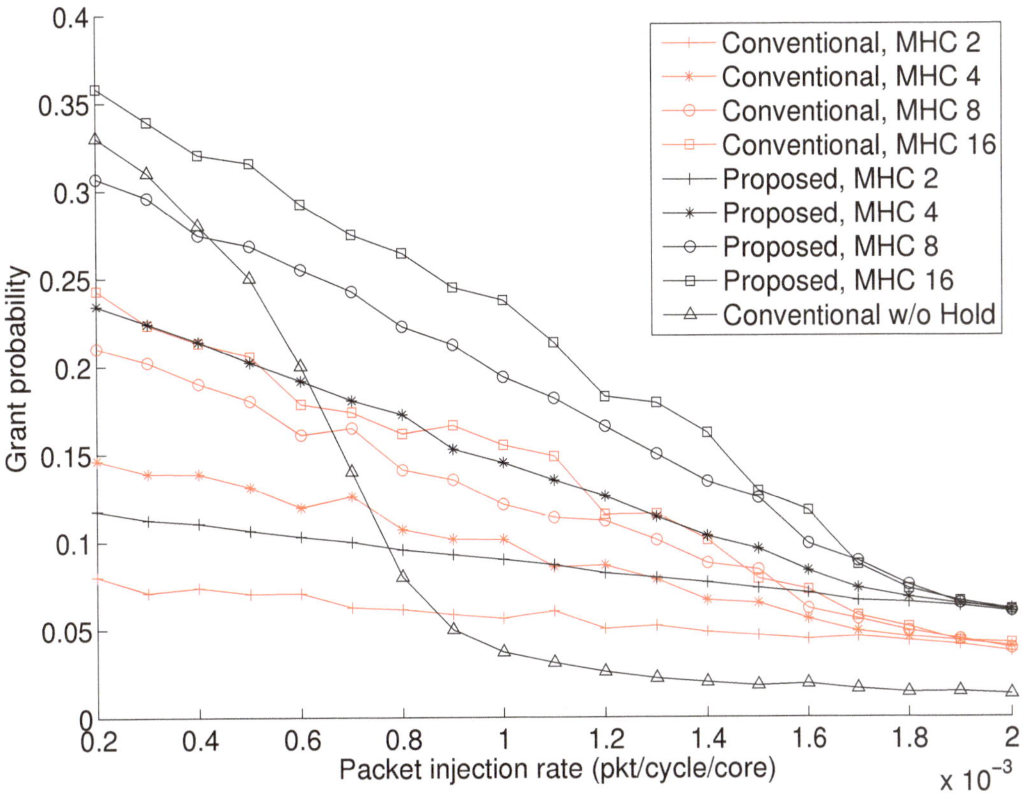

Figure 6. Grant probability under uniform traffic.

As discussed in Section 3.3, the proposed RACM introduces an overhead, both in terms of area and energy, as the implementation of the mechanism requires additional logic, which is not present in the conventional RACM. However, although the average power dissipated by the proposed RACM module is higher than that of the conventional RACM, the reduction of the average communication delay (as shown in Figure 5) results in a reduction of the execution time with a consequent reduction of the total energy consumption. Figure 7 shows the communication energy savings when the proposed RACM is used. Energy figures have been computed without taking into account the energy contribution of the cores, *i.e.*, only communication energy is considered. As can be observed, on average, 25% of the energy savings is obtained when the proposed RACM is used, and this goes up to 32% when compared to the conventional RACM without holding. Both the proposed RACM and the conventional RACM have been applied considering a radio hub implementing the transceiver proposed in [22] and also used in [18]. For this, we estimated a power consumption of 7 mW to 23 mW for the minimum and maximum transmitting power (depending on the physical distance between the source and the destination node), respectively. These correspond to energy per bit ranging from 0.42 to 1.4 pJ/bit.

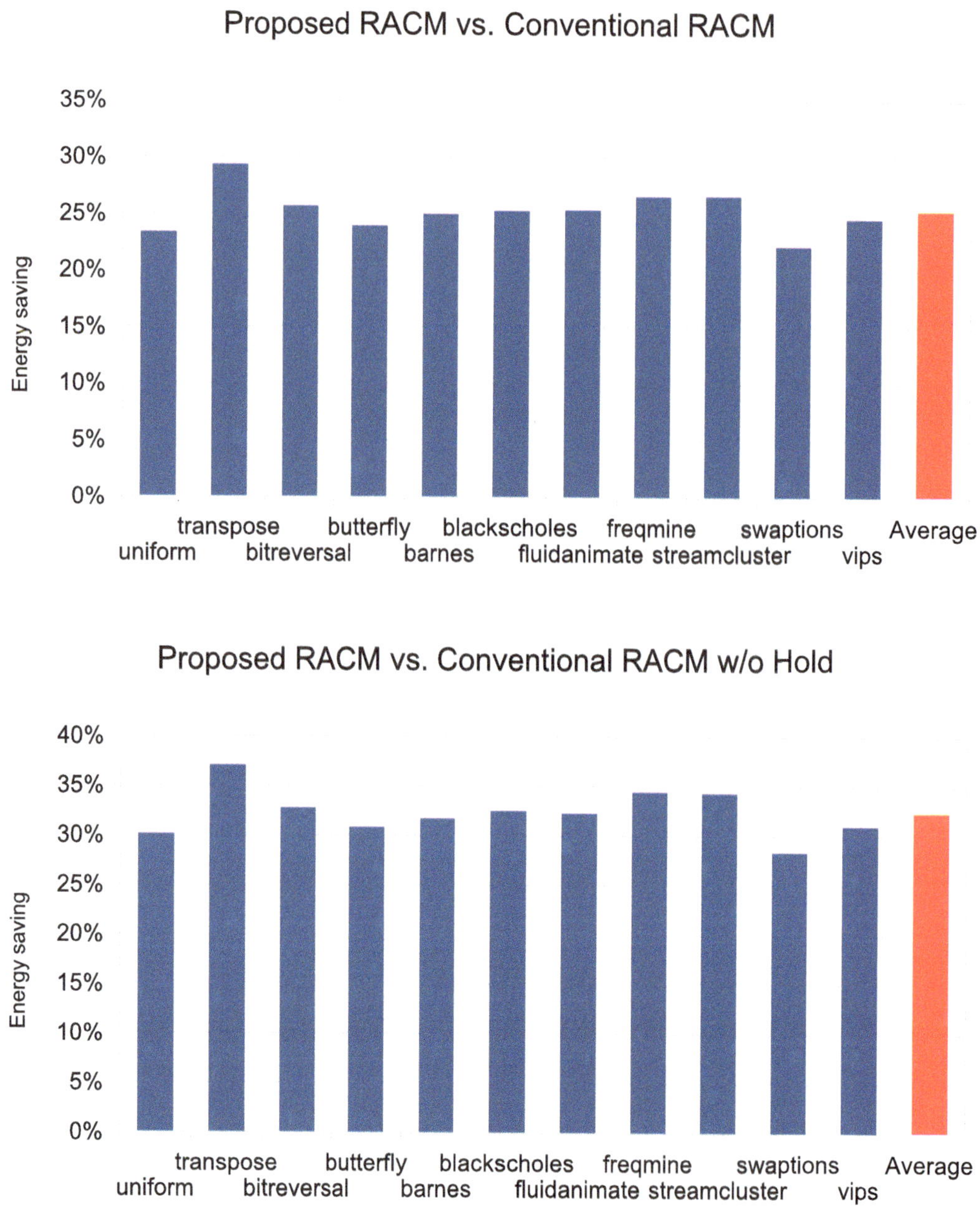

Figure 7. Communication energy savings when the proposed RACM is used.

Figure 8 shows the measured energy per bit under different traffic scenarios in the two cases in which the traditional RACM is used and when it is replaced by the proposed RACM. As can be observed, the use of the proposed RACM results in significant energy per bit savings for all of the benchmarks considered in the experiments.

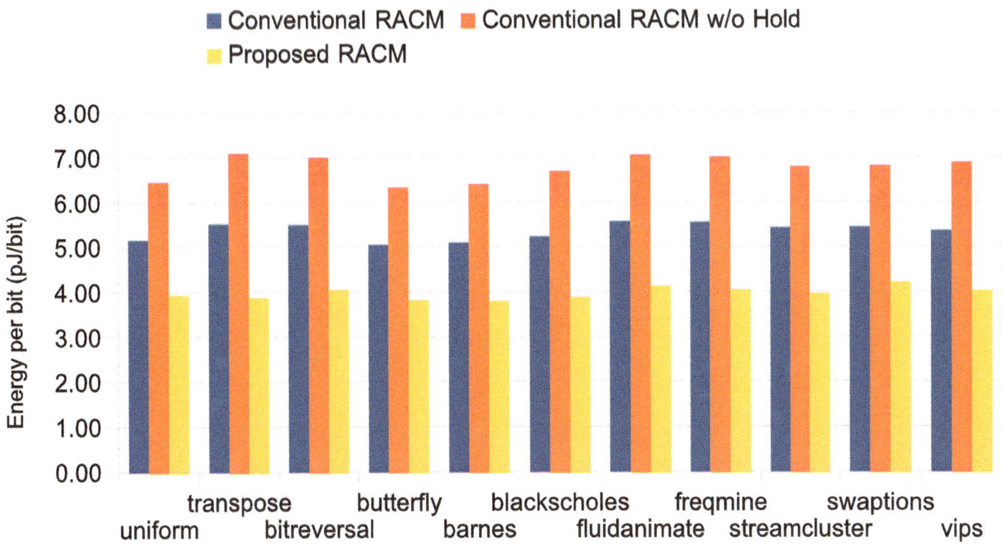

Figure 8. Measured energy per bit when the traditional RACM and the proposed RACM are used.

Finally, Figure 9 shows the energy savings for different network loads. We considered the synthetic traffic scenarios (uniform, transpose, bit-reversal and butterfly), and for each of them, we measured the energy consumption as the injected load increases. As can be observed, for a low injection load, the conventional RACM is more energy efficient than the proposed RACM (negative values of energy savings). This is due to the fact that for low loads, the access to the shared radio medium does not represent a bottleneck for communication latency, as the level of contention for it is low. Thus, since the energy overhead of the proposed RACM is not compensated by the performance improvement (which is limited at low loads), the percentage of energy savings is negative. Things change drastically as soon as the traffic load increases. In fact, since the proposed RACM shifts the saturation load, the performance improvement becomes dominant with a consequent energy savings. Such an effect is even more evident when the proposed RACM is compared with the conventional RACM without holding. As can be observed, as soon as the load increases, the energy savings increases much more quickly than the previous case, due to the performance limitation exhibited by the conventional RACM without holding.

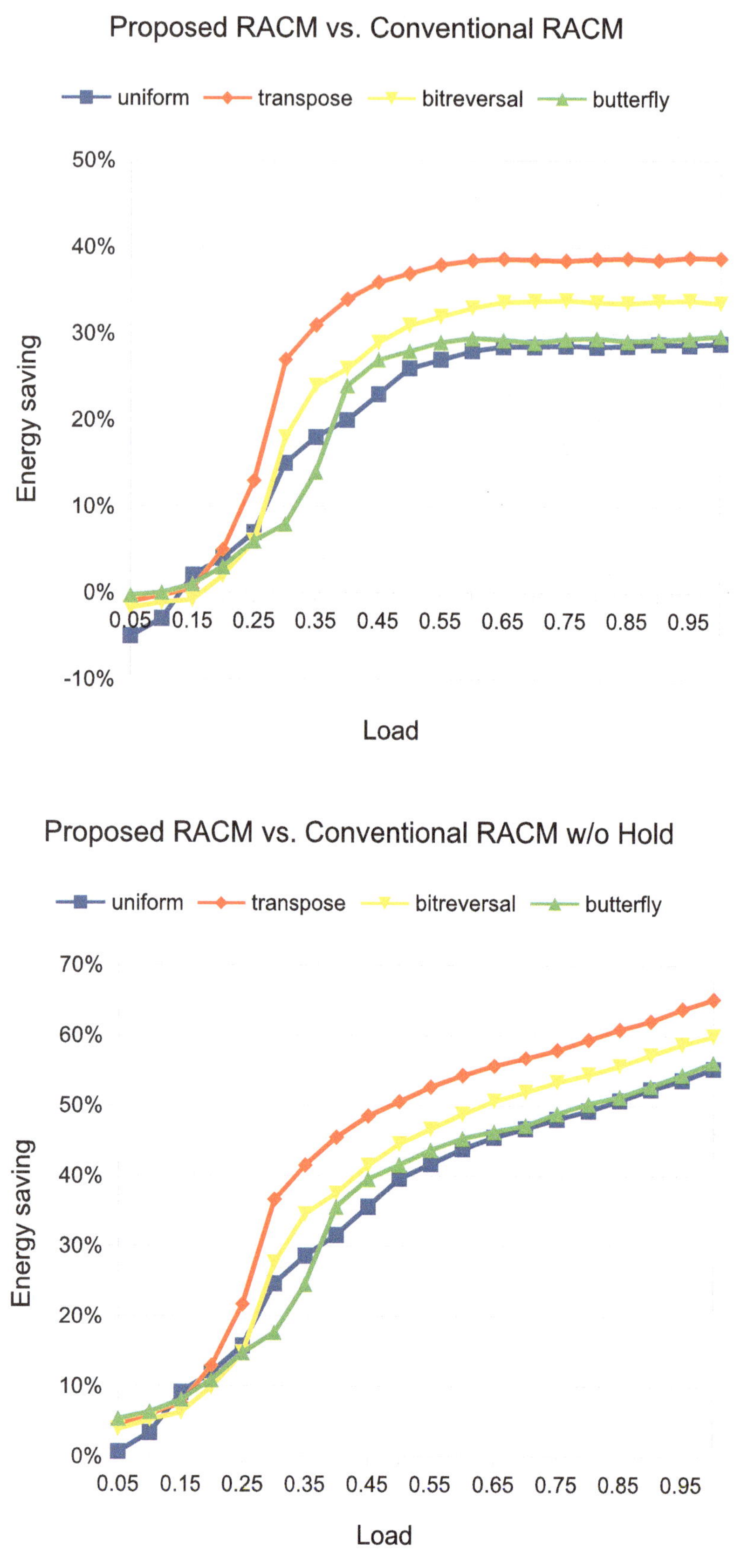

Figure 9. Energy *versus* injected load.

5. Conclusions

In a WiNoC architecture, the routers augmented with a wireless interface, namely radio hubs, can transmit over a single radio channel (or over a limited set of radio channels, typically ranging from one to four) [6,14,18]. Thus, for each radio channel, a radio access control mechanism (RACM) is used for arbitrating the access to the radio medium. Generally, the RACM is implemented by means of a token-based mechanism. Basically, a token circulates through the radio hubs and enables the radio hub owning the token to use the radio medium for a certain number of clock cycles. In this paper, we have presented a novel RACM that allows one to improve the utilization of the radio medium compared to the conventional RACM implemented in current WiNoCs proposed in the literature. The basic idea behind the proposed RACM is based on distributing the unused clock cycles by the radio hubs in the previous round among all of the radio hubs proportionally to how they utilized the radio medium in the previous round. Experimental results, carried out on both synthetic and real traffic scenarios, have shown the effectiveness of the proposed RACM, both in terms of performance and energy consumption. On average, a 30% reduction in communication delay and a 25% energy savings has been observed under both synthetic and real traffic scenarios.

Author Contributions

Maurizio Palesi was responsible for authoring this paper, the design and test of the proposed mechanism, and the experimental analysis. Mario Collotta, Andrea Mineo and Vincenzo Catania, review the proposed mechanism and edit this paper.

Conflicts of Interest

The authors declare no conflict of interest.

References

1. Manferdelli, J.L.; Govindaraju, N.L.; Crall, C. Challenges and opportunities in many-core computing. *Proc. IEEE* **2008**, *96*, 808–815.
2. Ivanov, A.; Micheli, G.D. The network-on-chip paradigm in practice and research. *IEEE Des. Test Comput.* **2005**, *22*, 399–403.
3. Chang, K.; Deb, S.; Ganguly, A.; Yu, X.; Sah, S.P.; Pande, P.P.; Belzer, B.; Heo, D. Performance evaluation and design trade-offs for wireless network-on-chip architectures. *ACM J. Emerg. Technol. Comput. Syst.* **2012**, *8*, 1–23.
4. Zhao, D.; Wang, Y. Sd-mac: Design and synthesis of a hardware-efficient collision-free qos-aware mac protocol for wireless network-on-chip. *IEEE Trans. Comput.* **2008**, *57*, 1230–1245.
5. Deb, S.; Ganguly, A.; Pande, P.; Belzer, B.; Heo, D. Wireless noc as interconnection backbone for multicore chips: Promises and challenges. *IEEE J. Emerg. Sel. Top. Circuits Syst.* **2012**, *2*, 228–239.

6. Deb, S.; Chang, K.; Yu, X.; Sah, S.P.; Cosic, M.; Ganguly, A.; Pande, P.P.; Belzer, B.; Heo, D. Design of an energy-efficient CMOS-compatible NoC architecture with millimeter-wave wireless interconnects. *IEEE Trans. Comput.* **2013**, *62*, 2382–2396.
7. Nguyen, C.K.; Kumar, A. An energy-aware medium-access-control protocol with frequent sleeps for wireless sensor networks. In Proceedings of the IEEE Symposium on Computers and Communications, Murcia, Spain, 27–30 June 2005; pp. 386–391.
8. Chiras, T.; Paterakis, M.; Koutsakis, P. Improved medium access control for wireless sensor networks-a study on the smac protocol. In Proceedings of the IEEE Workshop on Local and Metropolitan Area Networks, Crete, 18, September, 2005; pp. 1–5.
9. Liu, A.; Yu, H.; Li, L. An energy-efficiency and collision-free mac protocol for wireless sensor networks. In Proceedings of the IEEE Vehicular Technology Conference, Stockholm, Sweden, 30 May–1 June 2005; pp. 1317–1322.
10. Carloni, L.; Pande, P.; Xie, Y. Networks-on-chip in emerging interconnect paradigms: Advantages and challenges. In Proceedings of the ACM/IEEE International Symposium on Networks-on-Chip, San Diego, CA, USA, 10–13 May 2009; pp. 93–102.
11. Wang, C.; Hu, W.-H.; Bagherzadeh, N. A wireless network-on-chip design for multicore platforms. In Proceedings of the 19th Euromicro International Conference on Parallel, Distributed and Network-Based Processing (PDP), Ayia Napa, Cyprus, Greece, 9–11 February 2011; pp. 409–416.
12. Ganguly, A.; Chang, K.; Deb, S.; Pande, P.; Belzer, B.; Teuscher, C. Scalable hybrid wireless network-on-chip architectures for multicore systems. *IEEE Trans. Comput.* **2011**, *60*, 1485–1502.
13. Kumar, A.; Peh, L.-S.; Jha, N.K. Token flow control. In Proceedings of the International Symposium on Microarchitecture, Lake Como, Italy, 8–12 November 2008, pp. 342–353.
14. Zhao, D.; Wang, Y.; Li, J.; Kikkawa, T. Design of multi-channel wireless NoC to improve on-chip communication capacity. In Proceedings of the IEEE/ACM International Symposium on Networks on Chip (NoCS), Pittsburgh, PA, 1–4 May 2011; pp. 177–184.
15. Lee, S.-B.; Tam, S.-W.; Pefkianakis, I.; Lu, S.; Chang, M.F.; Guo, C.; Reinman, G.; Peng, C.; Naik, M.; Zhang, L.; *et al.* A scalable micro wireless interconnect structure for CMPs. In Proceedings of the 15th Annual International Conference on Mobile Computing and Networking, MobiCom ’09, New York, NY, USA, 20–25 September 2009; pp. 217–228.
16. Kempa, K.; Rybczynski, J.; Huang, Z.; Gregorczyk, K.; Vidan, A.; Kimball, B.; Carlson, J.; Benham, G.; Wang, Y.; Herczynski, A.; *et al.* Carbon nanotubes as optical antennae. *Adv. Mater.* **2007**, *19*, 421–426, .
17. Fazzino, F.; Palesi, M.; Patti, D. Noxim: Network-on-Chip simulator. Available online: http://noxim.sourceforge.net (accessed on 20 March 2015).
18. DiTomaso, D.; Kodi, A.; Kaya, S.; Matolak, D. iWise: Inter-router wireless scalable express channels for network-on-chips (NoCs) architecture. In Proceedings of the 2011 IEEE 19th Annual Symposium on High Performance Interconnects (HOTI), Santa Clara, CA, USA, 24–26 August 2011; pp. 11–18.
19. Ansoft HFSS. Available online: http://www.ansys.com/ (accessed on 20 March 2015).

20. Seok, E.; Kenneth, K. Design rules for improving predictability of on-chip antenna characteristics in the presence of other metal structures. In Proceedings of the IEEE 2005 International Interconnect Technology Conference, Burlingame, CA, USA, 6–8 June 2005; pp. 120–122.
21. Woo, S.C.; Ohara, M.; Torrie, E.; Singh, J.P.; Gupta, A. The SPLASH-2 programs: Characterization and methodological considerations. In Proceedings of the International Symposium on Computer Architecture, Santa Margherita Ligure, Italy, 22–24 June 1995; pp. 24–36.
22. Daly, D.; Chandrakasan, A. An energy-efficient ook transceiver for wireless sensor networks. *IEEE J. Solid-State Circuits* **2007**, *42*, 1003–1011.

Permissions

All chapters in this book were first published in JLPEA, by MDPI; hereby published with permission under the Creative Commons Attribution License or equivalent. Every chapter published in this book has been scrutinized by our experts. Their significance has been extensively debated. The topics covered herein carry significant findings which will fuel the growth of the discipline. They may even be implemented as practical applications or may be referred to as a beginning point for another development.

The contributors of this book come from diverse backgrounds, making this book a truly international effort. This book will bring forth new frontiers with its revolutionizing research information and detailed analysis of the nascent developments around the world.

We would like to thank all the contributing authors for lending their expertise to make the book truly unique. They have played a crucial role in the development of this book. Without their invaluable contributions this book wouldn't have been possible. They have made vital efforts to compile up to date information on the varied aspects of this subject to make this book a valuable addition to the collection of many professionals and students.

This book was conceptualized with the vision of imparting up-to-date information and advanced data in this field. To ensure the same, a matchless editorial board was set up. Every individual on the board went through rigorous rounds of assessment to prove their worth. After which they invested a large part of their time researching and compiling the most relevant data for our readers.

The editorial board has been involved in producing this book since its inception. They have spent rigorous hours researching and exploring the diverse topics which have resulted in the successful publishing of this book. They have passed on their knowledge of decades through this book. To expedite this challenging task, the publisher supported the team at every step. A small team of assistant editors was also appointed to further simplify the editing procedure and attain best results for the readers.

Apart from the editorial board, the designing team has also invested a significant amount of their time in understanding the subject and creating the most relevant covers. They scrutinized every image to scout for the most suitable representation of the subject and create an appropriate cover for the book.

The publishing team has been an ardent support to the editorial, designing and production team. Their endless efforts to recruit the best for this project, has resulted in the accomplishment of this book. They are a veteran in the field of academics and their pool of knowledge is as vast as their experience in printing. Their expertise and guidance has proved useful at every step. Their uncompromising quality standards have made this book an exceptional effort. Their encouragement from time to time has been an inspiration for everyone.

The publisher and the editorial board hope that this book will prove to be a valuable piece of knowledge for researchers, students, practitioners and scholars across the globe.

List of Contributors

Peter Beshay
The Charles L. Brown Department of Electrical and Computer Engineering, University of Virginia, Charlottesville, VA 22904, USA

Joseph F. Ryan
Intel Corporation, Hillsboro, OR 97124, USA

Benton H. Calhoun
The Charles L. Brown Department of Electrical and Computer Engineering, University of Virginia, Charlottesville, VA 22904, USA

Adrian Rothenbuhler
Department of Electrical and Computer Engineering, Boise State University, 1910 University Drive, Boise, ID 83725, USA

Thanh Tran
Department of Electrical and Computer Engineering, Boise State University, 1910 University Drive, Boise, ID 83725, USA

Elisa H. Barney Smith
Department of Electrical and Computer Engineering, Boise State University, 1910 University Drive, Boise, ID 83725, USA

Vishal Saxena
Department of Electrical and Computer Engineering, Boise State University, 1910 University Drive, Boise, ID 83725, USA

Kristy A. Campbell
Department of Electrical and Computer Engineering, Boise State University, 1910 University Drive, Boise, ID 83725, USA

Jianying Zhao
Tianjin University, School of Electronic Information Engineering, 92 Weijin Rd, Nankai Dist., Tianjin, China

Yogesh Ramadass
Texas Instruments, Inc., 12500 TI Blvd., Dallas, TX 75243, USA

Jeffrey Lang
Massachusetts Institute of Technology, EECS Dept, 77 Mass Ave, Cambridge, MA 02139, USA

Jianguo Ma
Tianjin University, School of Electronic Information Engineering, 92 Weijin Rd, Nankai Dist., Tianjin, China

Dennis Buss
Texas Instruments, Inc., 12500 TI Blvd., Dallas, TX 75243, USA
Massachusetts Institute of Technology, EECS Dept, 77 Mass Ave, Cambridge, MA 02139, USA

Kaushik Mazumdar
Charles. L. Brown Department of Electrical and Computer Engineering, University of Virginia, Charlottesville, VA 22904, USA

Mircea R. Stan
Charles. L. Brown Department of Electrical and Computer Engineering, University of Virginia, Charlottesville, VA 22904, USA

Michael Linder

Department of Computer Science and Computer Engineering, University of Arkansas; Fayetteville, AR 72701, USA

Jia Di
Department of Computer Science and Computer Engineering, University of Arkansas; Fayetteville, AR 72701, USA

Scott C. Smith
Department of Electrical and Computer Engineering, North Dakota State University; Fargo, ND 58108, USA

Marc Bocquet
Institut Matériaux Microélectronique Nanosciences de Provence(IM2NP), Aix-Marseille Université, Centre national de la recherche scientifique (CNRS), UMR 7334, 13284 Marseille, France

Hassen Aziza
Institut Matériaux Microélectronique Nanosciences de Provence(IM2NP), Aix-Marseille Université, Centre national de la recherche scientifique (CNRS), UMR 7334, 13284 Marseille, France

Weisheng Zhao
Institut d Electronique Fondamentale(IEF), University of Paris-Sud, Centre national de la recherché scientifique (CNRS), UMR 8622, F91405 Orsay, France

Yue Zhang
Institut d Electronique Fondamentale(IEF), University of Paris-Sud, Centre national de la recherché scientifique (CNRS), UMR 8622, F91405 Orsay, France

Santhosh Onkaraiah
Institut Matériaux Microélectronique Nanosciences de Provence(IM2NP), Aix-Marseille Université, Centre national de la recherche scientifique (CNRS), UMR 7334, 13284 Marseille, France
Commissariat Í Energie Atomique - Laboratoire d́ Electronique et de Technologie de lInformation (CEA-Léti), 38054 Grenoble, France

Christophe Muller
Institut Matériaux Microélectronique Nanosciences de Provence(IM2NP), Aix-Marseille Université, Centre national de la recherche scientifique (CNRS), UMR 7334, 13284 Marseille, France

Marina Reyboz
Commissariat Í Energie Atomique - Laboratoire d́Electronique et de Technologie de lInformation (CEA-Léti), 38054 Grenoble, France

Damien Deleruyelle
Institut Matériaux Microélectronique Nanosciences de Provence(IM2NP), Aix-Marseille Université, Centre national de la recherche scientifique (CNRS), UMR 7334, 13284 Marseille, France

Fabien Clermidy
Commissariat l Energie Atomique - Laboratoire d Electronique et de Technologie de lInformation (CEA-Léti), 38054 Grenoble, France

Jean-Michel Portal

Institut Matériaux Microélectronique Nanosciences de Provence(IM2NP), Aix-Marseille Université, Centre national de la recherche scientifique (CNRS), UMR 7334, 13284 Marseille, France

Andrey Mokhov
School of Electrical and Electronic Engineering, Newcastle University, Newcastle upon Tyne NE1 7RU, UK

Maxim Rykunov
School of Electrical and Electronic Engineering, Newcastle University, Newcastle upon Tyne NE1 7RU, UK

Danil Sokolov and
School of Electrical and Electronic Engineering, Newcastle University, Newcastle upon Tyne NE1 7RU, UK

Alex Yakovlev
School of Electrical and Electronic Engineering, Newcastle University, Newcastle upon Tyne NE1 7RU, UK

Jian (Denny) Lin
Department of Management Information Systems, University of Houston—Clear Lake, Houston, TX 77058, USA

Albert M. K. Cheng
Department of Computer Science, University of Houston, Houston, TX 77204, USA

Wei Song
Department of Computer Science, University of Houston, Houston, TX 77204, USA

Alexander Shapiro
Department of Electrical Engineering, University of Rochester, Rochester, NY 14627, USA

Eby G. Friedman
Department of Electrical Engineering, University of Rochester, Rochester, NY 14627, USA

Yongxun Liu
National Institute of Advanced Industrial Science and Technology (AIST), 1-1-1 Umezono, Tsukuba, Ibaraki 305-8568, Japan

Toshihide Nabatame
National Institute for Materials Science (NIMS), 1-1 Namiki, Tsukuba, Ibaraki 305-0044, Japan

Takashi Matsukawa
National Institute of Advanced Industrial Science and Technology (AIST), 1-1-1 Umezono, Tsukuba, Ibaraki 305-8568, Japan

Kazuhiko Endo
National Institute of Advanced Industrial Science and Technology (AIST), 1-1-1 Umezono, Tsukuba, Ibaraki 305-8568, Japan

Shinichi O'uchi
National Institute of Advanced Industrial Science and Technology (AIST), 1-1-1 Umezono, Tsukuba, Ibaraki 305-8568, Japan

Junichi Tsukada
National Institute of Advanced Industrial Science and Technology (AIST), 1-1-1 Umezono, Tsukuba, Ibaraki 305-8568, Japan

Hiromi Yamauchi
National Institute of Advanced Industrial Science and Technology (AIST), 1-1-1 Umezono, Tsukuba, Ibaraki 305-8568, Japan

Yuki Ishikawa
National Institute of Advanced Industrial Science and Technology (AIST), 1-1-1 Umezono, Tsukuba, Ibaraki 305-8568, Japan

Wataru Mizubayashi
National Institute of Advanced Industrial Science and Technology (AIST), 1-1-1 Umezono, Tsukuba, Ibaraki 305-8568, Japan

Yukinori Morita
National Institute of Advanced Industrial Science and Technology (AIST), 1-1-1 Umezono, Tsukuba, Ibaraki 305-8568, Japan

Shinji Migita
National Institute of Advanced Industrial Science and Technology (AIST), 1-1-1 Umezono, Tsukuba, Ibaraki 305-8568, Japan

Hiroyuki Ota
National Institute of Advanced Industrial Science and Technology (AIST), 1-1-1 Umezono, Tsukuba, Ibaraki 305-8568, Japan

Toyohiro Chikyow
National Institute for Materials Science (NIMS), 1-1 Namiki, Tsukuba, Ibaraki 305-0044, Japan

Meishoku Masahara
National Institute of Advanced Industrial Science and Technology (AIST), 1-1-1 Umezono, Tsukuba, Ibaraki 305-8568, Japan

Masakazu Hioki
Electroinformatics Group, Nanoelectronics Research Institute, National Institute of Advanced
Industrial Science and Technology (AIST), 1-1-1 Umezono, Tsukuba, Ibaraki 305-8568, Japan

Chao Ma
Electroinformatics Group, Nanoelectronics Research Institute, National Institute of Advanced
Industrial Science and Technology (AIST), 1-1-1 Umezono, Tsukuba, Ibaraki 305-8568, Japan

Takashi Kawanami
Department of Information and Computer Science, College of Engineering, Kanazawa Institute of Technology, 7-1 Ohgigaoka, Nonoichi, Ishikawa 921-8501, Japan

Yasuhiro Ogasahara
Electroinformatics Group, Nanoelectronics Research Institute, National Institute of Advanced
Industrial Science and Technology (AIST), 1-1-1 Umezono, Tsukuba, Ibaraki 305-8568, Japan

Tadashi Nakagawa
Electroinformatics Group, Nanoelectronics Research Institute, National Institute of Advanced
Industrial Science and Technology (AIST), 1-1-1 Umezono, Tsukuba, Ibaraki 305-8568, Japan

Toshihiro Sekigawa
Electroinformatics Group, Nanoelectronics Research Institute, National Institute of Advanced
Industrial Science and Technology (AIST), 1-1-1 Umezono, Tsukuba, Ibaraki 305-8568, Japan

Toshiyuki Tsutsumi
Computer Science Course, Fundamental Science and Technology, Graduate School of Science andTechnology, Meiji University, 1-1-1 Higashi-Mita, Tama-ku, Kawasaki-shi, anagawa 214-8571, Japan

Hanpei Koike
Electroinformatics Group, Nanoelectronics Research Institute, National Institute of Advanced
Industrial Science and Technology (AIST), 1-1-1 Umezono, Tsukuba, Ibaraki 305-8568, Japan

Luís Vitório Cargnini
HGST Inc., 3403 Yerba Buena Road, San Jose, CA 95135, USA
LIRMM - UMR CNRS 5506 - University of Montpellier 2, 161 Rue Ada, Montpellier 34095, France

Lionel Torres
LIRMM - UMR CNRS 5506 - University of Montpellier 2, 161 Rue Ada, Montpellier 34095, Franc

Raphael Martins Brum
LIRMM - UMR CNRS 5506 - University of Montpellier 2, 161 Rue Ada, Montpellier 34095, France

Sophiane Senni
LIRMM - UMR CNRS 5506 - University of Montpellier 2, 161 Rue Ada, Montpellier 34095, France

Gilles Sassatelli
HGST Inc., 3403 Yerba Buena Road, San Jose, CA 95135, USA

Eric Bharucha
Biomedical Microsystems Laboratory, Electrical and Computer Engineering Department,
Université Laval, 1065 avenue de la Médecine, Québec (Québec) G1V 0A6, Canada

Hassan Sepehrian
Biomedical Microsystems Laboratory, Electrical and Computer Engineering Department,
Université Laval, 1065 avenue de la Médecine, Québec (Québec) G1V 0A6, Canada

Benoit Gosselin
Biomedical Microsystems Laboratory, Electrical and Computer Engineering Department,
Université Laval, 1065 avenue de la Médecine, Québec (Québec) G1V 0A6, Canada

Maurizio Palesi
Faculty of Engineering, University of Enna, Kore, Cittadella Universitaria, Enna 94100, Italy;

Mario Collotta
Faculty of Engineering, University of Enna, Kore, Cittadella Universitaria, Enna 94100, Italy;

Andrea Mineo
Department of Electrical, Electronics and Computer Engineering, University of Catania, Viale A. Doria, 6, Catania 95125, Italy

Vincenzo Catania
Department of Electrical, Electronics and Computer Engineering, University of Catania, Viale A. Doria, 6, Catania 95125, Italy